Issues in Resource Management and Development in Kenya

ISSUES IN RESOURCE MANAGEMENT AND DEVELOPMENT IN KENYA

Essays in Memory of
Professor Simeon H. Ominde

Edited by
R. A. Obudho and J. B. Ojwang

EAST AFRICAN EDUCATIONAL PUBLISHERS
Nairobi • Kampala • Dar es Salaam

Published by
East African Educational Publishers Ltd.
Brick Court, Mpaka Road/Woodvale Grove, Westlands
P.O. Box 45314, Nairobi

East African Educational Publishers Ltd.
P.O. Box 11542, Kampala

Ujuzi Educational Publishers Ltd.
P.O. Box 31647, Kijito-Nyama, Dar es Salaam

© R. A. Obudho and J. B. Ojwang 2000
All rights reserved

First published 2000

ISBN 9966-25-007-7

Printed and bound in Kenya by Fotoform Ltd.
Muthithi Hse., Muthithi Road, P.O. Box 14681, Nairobi

Contents

Foreword
Preface
Acknowledgements
List of Tables and Figures
About the Editors and Contributors

PART I: PROLOGUE

1. Professor Simeon Hongo Ominde's Contribution to
   Geographical Sciences - *R.S. Odingo* — 3
2. The Worlds of Ojiji: Luo Women, 1900-1990 - *E.S. Atieno Odhiambo* — 8
3. Scholarship in National Development - *J.B. Ojwang* — 17

PART II: SUSTAINABLE USE OF NATURAL RESOURCES

4. Trends in Resource Development in Mountainous
   Environments - *Peter H. Omara-Ojungu* — 29
5. Environmental Stress and Conflicts in Africa: A Case Study of African
   International Drainage Basins - *C.O. Okidi* — 39
6. The Conjunctive Use of Ground and Surface Water Resources
   in Tsetse Infested Areas - *George O. Khroda* — 62

PART III: POPULATION AND HUMAN RESOURCES

7. A System Analysis Approach to Medical Geographic Studies: A Case
   of the Human Trypanosomiasis Disease in Kenya - *Justus I. Mwanje* — 81
8. Nuptial Determinants of Fertility: A Case Study of
   Western Kenya - *E.O. Ayiemba* — 99
9. Mortality Differentials and their Implications for
   Socio-economic Planning in Kenya - *Z. Muganzi* — 113
10. Use of Models in Studying Infant and Child
    Mortality - *J.A.M. Otieno, S.A. Odhiambo, and F.O. Ouma* — 129
11. Anthropological Techniques for Demographic Field
    Studies - *A.B.C. Ocholla-Ayayo* — 160

12. The Health of Migrant Workers in Colonial Kenya - *I. Sindiga*   175

13. Engineering Education and Related Development – *F. J. Gichaga*   183

PART IV: AGRICULTURAL DEVELOPMENT

14. Population Distribution, Density and Movements Within Kenya's Arable Land - *John O. Oucho*   195

15. Food Production and Population Growth in Kenya - *D. A. Obara*   203

16. Small-holder Food Production: A Geographical Investigation of Maize in South Nyanza - *S.O. Akech*   214

17. Change, Persistence and Development in Small-holder Livestock Production in Western Kenya - *Collette A. Suda*   234

PART V: URBAN INFRASTRUCTURE

18. The State and Low-Income Urban Housing Production and Consumption in Kenya - *G. C. Macoloo*   247

19. Trends of Urban Housing Strategy for Kenya in the Next Decade - *P.M. Syagga*   258

20. Public Transport Modes in Nairobi - *R.A. Obudho and G.O. Aduwo*   271

References   **287**

Appendix: Simeon Hongo Ominde: A Biographical Note - *H. R. Liyai*   **299**

Index   **311**

Foreword

Professor Simeon Hongo Ominde was an utterly intrinsic professional scholar, a pioneer social trail-blazer and an entirely domesticated nurturing provider. He navigated this ship of three masts so well that he was able to reach the shore of his heavenly transition just in time for his allotted "three score and ten years". And he added one more year, seemingly just for the love of it — a great son of Africa who was ever so blissful just to be!

In his seven decades on this Earth, Ominde conquered several high mountain peaks. As a scholar per excellence, he was the first Eastern African to break the glass ceiling on how far up one could climb the academic ladder to the professorial chair, by being appointed the first African Professor in any university in the region and by becoming the first African to be Head of the Department of Geography at the University of Nairobi in 1964. His numerous accomplishments in the advancement of geographical science, particularly in his own specialties in population studies and human geography, are chronicled elsewhere in this book of essays in his honour, *Issues in Resource Management and Development in Kenya*.

In the little book, *The Luo Girl From Infancy to Marriage*, published in 1952, Ominde made a bold statement in an extremely sensitive field, which had essentially been captured by the Christian missionary movement early in the century. While the missionaries imposed a new ethic about social behaviour based on biblical teachings and European and Australian practical interpretations of the latter, Ominde's book showed clearly that the Luo community's social teaching and training, solidly-based on morally inspired social mores and sanctions, produced women graduands who were remarkably stable in their behaviour and social standing, definitely out-performing the Christian, boarding-school trained young women. This statement, consulted afresh today, still reads as a sagacious return to a world of social values of eternal validity

This powerful statement did not mean that Ominde was anti-religion. Ominde's core belief in the divine was profoundly deep — deeper and more resonant in his being than merely the attendance at church services or the periodic performance of religious rituals and ceremonies. His was an ethic of divine devotion, so intense that he always radiated bliss and tranquillity even in the face of evident stress and chaos. In this, he was fortunate that his life partner, the late Mary Ominde, took complete charge of the home-caring, nurturing, and engendering a peaceful atmosphere for family interphasing and growth.

Whenever one discussed issues regarding excellence and productivity in the African academe, Ominde was always wont to inject the need to select, focus upon and pursue the growth points in any of the priority areas of academic concern. He felt that this strategy was the only one open to a continent in a hurry to rejoin the international world of science and scholarship. In this context, he visited me twice in 1964 and 1965 at the University of Cambridge to persuade me to return to Kenya and join in the then fledgeling University College Nairobi, in order to create an African programme of insect physiology and endocrinology. He succeeded, and in July 1965, after successfully defending my Ph.D. thesis in insect reproductive physiology, I joined the Department of Zoology at the University College Nairobi. There, with his collegial help and that of the Principal of the University College, Dr. Arthur Porter, a distinguished man of letters from Sierra Leone, I was able to take my place at the then emerging growth point in the natural sciences in Africa — the exciting new field of insect science.

At the end of one's life on this planet, one wants to produce a kind of balance sheet: has my life in this physical universe made a difference? Has it added any value to the sum-total of human achievement and divine purpose? I think I can be very positive in the case of Professor Simeon Hongo Ominde. He has left an indelible imprint on the human capital development of the Eastern African region — by his founding the Population Studies and Research Institute (PSRI) at the University of Nairobi in 1976 which has rapidly become an international centre for research and postgraduate studies in population and demography. His students and research associates are to be found all over the region and beyond; and the Institute's publications have become invaluable reference points for human geography insofar as Africa is concerned. As to divine purpose, I can only state that his poise and positive nature, in the midst of struggle and chaos in contemporary Africa, is a first-class examples to other souls in this physical universe.

As Walter Stahel has so poignantly stated in his book, *The Limits of Certainty: Facing Risks in the New Service Economy* (London: Kluwer Academic Publishers, 1989), the essence of life cannot be to avoid risks: we take risks by the very fact that we live. The choice must be to live our risky situation in a more or less conscious way ... Ominde did take many risks; but because he was a discerning and judicious shipmaster, he was able to navigate his ship time and time again through treacherous sandbanks to reach the shores of social enlightenment and academic affluence.

THOMAS R. ODHIAMBO
President,
African Academy of Sciences, Nairobi
September 1998

Preface

On a sun-bathed afternoon, on June 24, 1995, Simeon Hongo Ominde was lowered into his grave at his Nyahera home near the lake port of Kisumu, Kenya. This event, which nearly coincided with Ominde's 71st birthday, was to those who have some impression of educational trends in Kenya and East Africa, no ordinary event. The tale was abundantly told in any event for all to note: by the the popular sentiments expressed in the aftermath of his death a fortnight earlier; by the moving testimonies at the several requiem masses held for him; and by the large attendance at the final ceremony, which included not only scholars and the political and adminstrative elite but also ordinary people.

Ominde's position in the Kenyan society, while he lived, has in the first place to be understood within the context of relentless popular aspirations for schooling. For many years in Kenya, the school teacher was everyone's role model by virtue of his being the most visible member of the educated elite. The teacher was seen as the bearer of the responsibility for imparting knowledge to the youth.,and was also perceived as the local community's guide on issues that went into the shaping of social attitudes. The school has been a highly cherished institution in the Kenyan society as the recognised forum for disseminating knowledge and preparing the youth for adult life in an education-demanding social and economic context. And the university was generally seen, especially in the colonial, and the earlier phase of the post-independence periods as the ultimate cradle of knowledge. The university don, in this context, was regarded with the highest esteem.

In the 1940s when there were few educational institutions in East Africa, and when Makerere University College was the hallowed seat of intellectualism, Professor Ominde was among the infinitesimal number of gifted youths who made it with distinction to that institution. His excellence earned him subsequent admission to leading British universities and in 1955 he became himself a don at Makerere University. He was to become the first indigenous holder of the rank of Professor in East and Central Africa with his appointment to the Head of Geography at the University College Nairobi in 1964.

At the University College Nairobi, Professor Ominde settled down as a distinguished scholar and pioneer researcher in the fields of geography and population studies. He was a true doyen among African geographers who rendered more than 40 years of dedicated service not only to Kenyan and African geography in particular. but also to world geography in general.

The death of Ominde, who had formally retired in 1984 but remained in continuous contractual service, was a momentous event for the University of Nairobi's academic community. His passing both deprived scholarship of its distingushed servant and presented a major challenge to his successor academics. In his lifetime he had written profoundly and copiously and otherwise made contributions in the spheres of geography, demography and planning, education and resource management. In the process he had brought together a substantial core of colleagues and students pursuing researches and studies in relevant fields. It was soon after his retirement that Professor Ominde's colleagues thought to write and edit a *Festschrift* in his honour. This initiative could not, unfortunately, be accomplished in time, and its realisation now comes as a memorial to coincide with the fourth anniversary of Professor Ominde's death.

This volume is an attempt to put together what contributors deemed to be a fitting tribute to Professor Ominde. The five parts of the volume are devoted to the core resource areas in which Professor Ominde worked.

The work contains 19 chapters contributed by 23 scholars and scientists. It commences with a prologue, which introduces the reader to Professor Ominde's work and gives insights into his personality and standing and the main landmarks in his career. The second is devoted to the theme of sustainable use of natural resources and presents studies of record on resource development in mountainous environments; environmental stress and conflicts in relation to drainage basins; conjuctive use of ground water and surface water; and the application of systems analysis approaches in medical geographic studies. The next part is concerned with population and human resources and includes state-of-the-art studies on nuptial determinants of fertility; mortality differentials and socio-economic planning; use of models in the study of child mortality; the application of anthropological techniques in demographic field studies; health problems among migrant workers in colonial Kenya and engineering education in Kenya. Part four is devoted to agricultural development and contains authoritative contributions on demographic implications of Kenya's population distribution; food production, and population growth in Kenya's small-holder food production, with special reference to maize and small holder livestock prduction in western Kenya. The last part is concerned with urban infrustructure and under this theme it carries contributions of record on the production of low-income urban housing strategy in Kenya and public transport in Nairobi. The last section is a bibiliography of Professor Ominde's works.

This volume endeavours to accord recognition to Ominde's scholarship through a careful selection for treatment of vital resource management and development issues which, in practice, call for authoritative fact-finding and analysis as a basis for policy making or for further specialised enquiries on environment-related themes. The natural and human resource foundation laid in this study will, we hope, serve as a starting point for a good number of policy-based studies on public planning, management and general governance. The volume as a whole constitutes a critical appreciation of Professor Ominde's contribution to Kenyan geography, demography, population dynamics, education and planning.

We note that the contributors to this volume would make no pretensions of doing full justice to the late Professor Ominde. A full and fair description and analysis of Ominde's achievements would require several volumes and must remain a task reserved for several generations of resource scientists. In his more than 40 years as a professional geographer and demographer, both within and outside academe, Professor Ominde spawned a bibiliography so extensive and varied as to merit distinct recognition as a genuine pace-setter for scholars and other persons. He occupied an exclusive niche in his knowledge of population geography and was unsurpassed in his experience in the application of geographic methods to the solution of practical planning problems.

We feel uniquely privileged, as editors of this volume, to have been accorded this opportunity to convene and guide this posthumous conferment of richly merited accolade on Professor Ominde, which takes the form of learned essays contributed by distinguished social scientists and worthy colleagues of his.

R.A. Obudho
J.B. Ojwang
APRIL 2000

Acknowledgements

We do not take for granted the fact that we have been able to bring out this book. At various stages we had indeed apprehended that the initiative might not succeed. We had the good luck to receive plenty of gracious assistance and vital co-operation, for which we are truly grateful.

we sincerely thank the contributors for agreeing in the first place to support the initiative, for giving of their precious time to prepare original papers for the volume and responding to our many demands, and revising their papers from time to time. We acknowledge with many thanks the help given to us in the organisation of the manuscript by the following persons: Mr. S. O. Owuor, Mr. Peter Abwao, Mr. G. O. Aduwo and Mrs. Alice Atieno. We are indebted to Mrs. Beatrice Ngesa and Ms. Christine Akoth for the preparation of the typescript of the original manuscript. Ms. Elizabeth W. Kung'u assisted us on several occasions with the preparation of typescripts of limited portions, and we record our appreciation.

We acknowledge the invaluable support of the publishers, which took the form of their sustained interest in the work, their encouragement to us, their technical inputs into the editorial process and ultimately, their timely and efficient production of the work.

As is to be expected, we, along with our contributors, take full responsibility for any shortcomings that may be found in this work.

R. A. Obudho
Department of Geography
University of Nairobi.
APRIL, 2000

J. B. Ojwang
Faculty of Law
University of Nairobi.
APRIL, 2000

Tables and Figures

Tables

5.1	World agricultural land	42
5.2	Amount of water needed for irrigation in Egypt	49
5.3	Estimated irrigation potential by rivers in the Lake Basin	53
5.4	Mean annual water discharge at Sudd Swamp in Sudan (in km^2)	57
6.1	Kenya's mean annual rainfall distribution compared to land area	66
6.2	Major drainage areas in Kenya with respective runoff contibutions	67
6.3	Lake water resources	68
6.4	The distribution of boreholes and their estimate in Kenya	69
6.5	Kenya's ecological zones	70
6.6	Distribution of tsetse flies according to drainage basins	72
6.7	Estimated irrigation potential and water demand for each drainage basin in Kenya	72
6.8	The number of ponds of developed grazing block in North-Eastern Province, Kenya	75
6.9	Estimated base livestock population by water units, 1978	76
6.10	Projected livestock population by water management units to 2000	77
8.1	Age at marriage, in western Kenya, 1978	102
8.2	Recorded environmental hazards in western Kenya: Frequency of occurrence by district	104
8.3	Selected indices of fertility levels in western Kenya, 1969 and 1979	106
8.4	Percentage of females of different marital status in western Kenya, 1969 and 1979	107
8.5	Step-wise regression results, 1978	108
8.6	Current fertility rate analysis	109
8.7	Regression results, 1978	110
8.8	Lifetime total fertility rate analysis results in western Kenya, 1978	110
9.1	Mortality differentials by education, Trussell Method (West Model)	118
9.2	Mortality differentials by residence, Trussell Method (West Model)	119
9.3	Mortality differentials by marital status, Trussell method (West Model)	120
9.4	Percentage distribution of births and deaths by region	121
9.5	Percentage distribution of children by duration of sickness and by province	121
9.6	Percentage distribution of children by type of sickness and by province	122
9.7	Mortality differentials by provinces Trussell and Sullivan Methods (West Model)	123
9.8	Percentage distribution of households by distance to social amenities and by province	124
9.9	Percentage distribution of households by access to indicated facilities by province	125
10.1	The probability of dying at the age of 2 yrs for all cases combined	135
10.2	The probability of dying at the age of 2 yrs by education differential, 1979	136
10.3	The probability of dying at the age of 2 yrs by marital status and place of residence, 1979	138
10.4	Infant and child mortality by region using KCPS data, 1984	139
10.5	Infant and child mortality by differentials using KCPS data of 1985	140
10.6	Life expectancies at birth by level of education and place of residence	144
10.7	Life expectancies at birth by marital status	145
10.8	Extremes values	146
10.9	Regression variables for infants and neonatal mortality in selected provinces	152
10.10	Regression coefficients	153
10.11	Analysis of variance of multiple regression	154
10.12	Coefficients of variables	154
10.13	Description of variables used	155
10.14	Infant mortality rate for all cases and by education of mother	156

10.15	Life Expectancies at birth	157
11.1	Sample size, West Agoro Sub-location, Nyakach, Kisumu District	165
11.2	Retrospective marriage, fertility and mortality patterns from the FGP discussion	166
13.1	Commonwealth countries' engineers and technicians needed in 2000, scenarios A,B, &C	188
14.1	Raw and physiological population densities, 1969-1979	198
14.2	Internal migration one year before the 1979 census by province	200
14.3	Growth of urban centres in Kenya by size and number, 1948, 1962, 1969 and 1979	201
15.1	Estimated areas, values and value per hectare for selected commodities, 1983 and 1987	207
15.2	Population and high/medium potential land by province and district	211
16.1	Rotated factor matrix, Quartimax Rotation	223
16.2	Regression results maize yields analysis	224
16.3	Result of regression modelling of maize hectarage	228
16.4	The distribution of adopters for different categories of innovation	231
16.5	Regression results of Model 3: Partial adoption of hybrid seed	231
16.6	Regression Results Model 4: Adoption of farm yard manure and/or commercial fertiliser	232
17.1	Number of households with livestock by district (Siaya and Kakamega)	240
17.2	Relationship between the number of livestock, and family and farm size by district	240
17.3	Use of hired labour by district	241
17.4	Status of livestock by district	242
17.5	Livestock that are easy to tend for by district	243
19.1	Recorded production of dwellings by public and private sectors compared with the formation of new urban centres	264
20.1	Population of Nairobi by race for selected years	273
20.2	Average post-War vehicle growth rates of African colonial urban centres, 1960-1970	275
20.3	Public transport demand in Nairobi, 1985, 1990 and 2000	276
20.4	Growth of KBS fleet and passengers,1962-1990	278
20.5	KBS's trading position and statistical analysis from 1970-1990	279
20.6	The growth of *matatus* in Nairobi and their average daily passengers, 1971-1981	280
20.7	KBS and *matatus*: percentage share of the market in Nairobi	280

Figures

5.1	The 'Zambezi Plan', an ambitious South African water project in the SADCC region	45
5.2	The Nile Basin	46
6.1	Distribution of testse infected areas in Kenya	63
6.2	The geology of Kenya	65
6.3	Kenya's annual water deficit values (in mm)	71
7.1	Development periods of medical geography	82
7.2	Phases of applied systems analysis	84
7.3	Approaches towards an ecologic theory on disease	86
7.4	The relationship between Ruma National Park 'infective cell' and the surrounding human settlement	90
7.6	Stages in modelling process in a TIS development	95
7.7	Structure of a real–time HAT epidemic expert system	96
7.8	The structure of the geographic systems analysis of HAT epidemic	97
8.1	Socio–economic variables in differential fertility analysis	103
10.1	Operation of five groups of proximate determinants on health dynamics of a population	130
10.2	The Stem-Leaf Plot of e(o)	143
13.1	A relation between professional/ technical staff per 1000/population and GNP/ capita	190
16.1	A conceptual framework for a geographic analysis of small-holder maize production	218
16.2	Position of study area in South Nyanza District	220

About the Editors and Contributors

G.O. ADUWO is a Lecturer in Geography at Kenyatta University, Kenya.

S.O. AKECH is a Lecturer in Geography, University of Nairobi (UoN), Kenya.

E.S. ATIENO ODHIAMBO is Professor of History at Rice University, U.S.A.

Elias H.O. AYIEMBA is Associate Professor of Geography, University of Nairobi, Kenya.

Francis J. GICHAGA is Professor of Civil Engineering and Vice-chancellor, University of Nairobi, Kenya.

George Okoye KRHODA is Associate Professor in the Department of Geography, University of Nairobi, Kenya.

Hudson LIYAI is Librarian, University of Nairobi, Kenya.

Gervase C. MACOLOO is Senior Lecturer in Geography, University of Nairobi, Kenya.

Zibeon S. MUGANZI is Associate Professor in Population Studies at the University of Nairobi, Kenya..

Justus I. MWANJE is senior Lecturer in Environmental Studies, Kenyatta University, Kenya.

Dunstan A. OBARA is Associate Professor in Geography, University of Nairobi, Kenya.

Robert A. OBUDHO, co-editor of this volume, is Associate Professor in Geography, University of Nairobi, Kenya. He is editor of *African Urban Quarterly* and Executive Director, Centre for Urban Research.

A.B.C. OCHOLLA-AYAYO is Professor of Anthropology and Population Studies, University of Nairobi, Kenya.

S.A. ODHIAMBO is a student at the Population Studies and Research Institute, University of Nairobi, Kenya.

Jackton B. OJWANG, co-editor of this volume, is Professor of Law, University of Nairobi, Kenya. He is also a member of the Kenya National Academy of Sciences, a Commissioner of the Kenya Law Reform Commission, and a member of the Council of Legal Education.

C.O. OKIDI, formerly a Professor of Environmental Studies, Moi University, is now the Task Manager of the UNEP/UNDP/Dutch Project on Environmental Law and Institutions in East Africa. He is based at the Environmental Law and Institutions Programme Activity Centre at UNEP, Kenya.

J.A.M. OTIENO is Associate Professor of Mathematics, University of Nairobi, Kenya.

John O. OUCHO is Professor of Demography, University of Nairobi, Kenya, currently on leave of absence at the University of Botswana.

F.O. OUMA is a student at the Population Studies and Research Institute, UoN.

The late Isaac SINDIGA, was formerly Associate Professor of Geography, Moi University.and Principal, Kisii College Campus, Egerton University.

Collette SUDA is a Associate Professor at the Institute of African Studies, University of Nairobi, Kenya.

P.M. SYAGGA is Professor of Land Development and Dean, Faculty of Architecture, Design and Deveopment, University of Nairobi, Kenya. He is currently Acting Principal, College of Architecture and Engineering at the University.

Part I

PROLOGUE

Part 1

PROLOGUE

1 Professor Simeon Hongo Ominde's Contribution to Geographical Science

R. S. Odingo

Professor Simeon Hongo Ominde's name is synonymous with the development of geographical sciences in East Africa as its first indigenous Professor of Geography in this sub-region. Appointed to the Chair and Headship of the Department of Geography at University College, Nairobi, in 1964, his work detailing the natural resources and peoples of Kenya and East Africa has had a tremendous impact on the development of the discipline in the region.

Early Training

Professor Ominde's geographical roots are closely intertwined with the growth of the subject to its maturity in Europe and, more particularly, the United Kingdom. The late Professor Samuel James Kenneth Baker, a former student of Professor P.M. Roxby's, who was one of the first Professors of Geography in Britain, brought professional geographical training to East Africa when he was appointed to the Chair of the Geography Department at the then Makerere College in 1947. Professor Baker succeeded in creating an East African geography which is still highly visible today. Makerere College became the centre of teaching and research in geographical sciences for East Africa. It was not until the death of the East African Community in 1977 that Kenya, Tanzania and Uganda began to go their separate ways.

Professor Simeon Ominde was among Professor Baker's students and was the first to make it to professorial rank. Professor Ominde studied at Makerere between 1940 and 1950, then continued his undergraduate and postgraduate studies in geography from 1951-54 at Aberdeen University, Scotland. In 1955, he returned to Makerere University College as a Lecturer in Geography.

In the same way that Professor Baker pioneered geography at Makerere, Professor Ominde pioneered in bringing geography to the service of newly independent Kenya. Professor Baker's main philosophy had been to study the role of man in his environment or the relationships between human societies and their environments, but later he became attracted to population geography. Professor Ominde started by teaching general geography at Maseno School in western Kenya, specialising in human geography during his studies in Scotland. He had a strong affinity for urban geography but opted for population geography at the Ph.D. level. This can be attributed to Professor Baker's influence. Professor Ominde's Ph.D. thesis was entitled, "Land and Population Movements in Kenya," which showed his early interest in the urban aspects of population geography.

Teaching Career

Professor Ominde joined the Geography Department at University College, Nairobi, which Professor W.T.W. Morgan had established in the initial stages of the institution's development. Professor Ominde saw his main task as building a department which would be relevant to the development needs of Kenya, East Africa, and the world. This he poignantly expressed in his Inaugural Lecture: "Geography and African Development," where he stated that intellectual activity should be relevant to national and international development needs and especially to challenges facing the African continent. He was keen to see African geography free itself from narrow intra-disciplinary squabbles. The larger issue was relevance to the "African Development Problematique," while also being alive to developments in related subjects in the developed parts of the world. Professor Ominde emphasised that "the facts of development in Africa require that the role of Geography in the transformation of our natural resources and development of our human resources be correctly appreciated." He also saw the development of geography in post-colonial Africa as "an opportunity to place in perspective the contribution of Western geographical tradition to the understanding of problems of African development and global relations" (Ominde 1965).

Professor Ominde pointed out the growth of different specialisations in the subject of geography. This recognition explains his own inclination towards the interdisciplinary focus. He noted:

> In the field of human geography research workers have now found themselves side by side with other social scientists concerned with such problems as population structures and trends, factors underlying migration, and a wide range of problems of economic development or underdevelopment (Ominde 1965).

As a former school teacher turned professional geographer, Ominde was aware of the link between the subject as taught in schools and at university level. Therefore, he emphasised the need for a secure foundation for the subject in the school system, guided by a dynamic philosophy of development.

Population Studies and Research Institute

In keeping with his research interests in population and human environmental issues, Professor Ominde was instrumental in creating the Population Studies and

Research Institute (PSRI) at the University of Nairobi in 1976. His aim in establishing the PSRI was to provide a forum for a comprehensive multi-disciplinary programme of population studies and research which would not only cater to geographers, but also to economists, sociologists, demographers, political scientists, and mathematicians, among others. Since population is so central to government planning, PSRI was designed to give training at two levels: first, at the Diploma level as an in-service arrangement for those in government; and, secondly, at the Masters and Doctorate levels for professional demographers and population analysts who would engage in teaching and research at various levels. The PSRI was thus established expressly "to bridge the gap in this important area of Kenya's manpower need and to create a self-sustaining programme as a permanent feature of the University's academic programme. It was also established to meet the short and long-term research needs of Kenya in support of the country's development strategy" (Ominde and Henin 1976).

Once it was established, PSRI set out an extensive publications record. This gave Professor Ominde the opportunity to explore many population research themes, as witnessed by his own numerous contributions and those co-authored with colleagues at the PSRI. Because he was convinced that the university needed to avail its tools to planners, Professor Ominde organised population seminars throughout Kenya in collaboration with the Ministry of Planning.

Professor Ominde's Philosophy

Although Professor Ominde was mainly associated with population studies and PSRI, he was first and foremost a geographer at heart. Due to his training at Makerere and Aberdeen Universities, he was at ease with both physical and human geography in a broad sense, as his students and colleagues will attest. He championed the belief that geography and geographical training should face the "African challenge" of development. His efforts succeeded as graduates from his department have made significant contributions to the development process, not only in Kenya, but also in Africa as a whole. The emphasis on the study of natural resources has meant that geography graduates hold their own in any capacity in which they are called upon to service the region.

Professor Ominde's philosophy in geography can be gleaned from his many publications and the public speeches he has given throughout his academic career (see Appendix). In all these one finds his love for human geography and population studies, the link between population growth and economic development throughout Africa. This pan-African perspective can be seen in his founding editorship of the first trans-African population journal, *Jimla Mutane*. The journal had important contributions by scholars from anglophone and francophone Africa. His unquestionable contribution to African demography has strong geographical roots, unlike other population schools.

Professor Ominde's dedication to the improvement of conditions for the people of Kenya was expressed through intense interest in the link between population and environment. He stressed in his first major work, *Land and Population Movements in Kenya* (1968), that "it is becoming increasingly evident that any improvement in the living standards whatsoever is dependent on the rate of growth of the economy as a whole." This conclusion led him to emphasise often the link between population geography and planning.

The two geography departments at Makerere and Nairobi Universities laid special emphasis on the evaluation of African natural resources and the evolution of African economies as they emerged from the colonial period. Professor Ominde went further to emphasise in his teaching, writing, and research the special place of geography in solving natural problems and facing the challenges brought about by political independence. What interested him particularly was the relationship between natural resources and the role which population studies could play in finding constructive solutions to the "development problematique." Some of these ideas were spelled out in detail in the work entitled *"Geography and African Development (1971)"*, which had a strong Kenyan, East African, and pan-African perspective that linked population growth and economic development.

Professor Ominde's Contribution

In addition to the general discussion of the development problems in Africa, Ominde contributed to the promotion of African demography through a series of conferences and workshops as well as *Jimla Mutane*. Some of the issues pursued in these studies include: the sheer size of the population; demographic characteristics of an area, especially the demographic transition; urbanisation; labour force implications of population growth; education and wealth implications of population growth; and, the overall link with natural resources, especially land. Professor Ominde's efforts took into account the gradual evolution of population policies in many African countries. The formation and evolution of a natural population policy is not an easy task as there are many actors at play, and governments have to be convinced that policy will assist rather than hinder economic development.

Professor Ominde underlined the importance of population movements in Kenya — rural to rural, rural to urban, urban to urban, and urban to rural— as being fundamental to the process of economic development. His perspective influenced the evolution of government population policy which has been concerned with four main subjects, namely family planning, urbanisation, health implications and human settlements. Professor Ominde's Ph.D. dissertation entitled "Land and Population Movements in Kenya" was the first exhaustive study of the population of Kenya and the environmental underpinning of population mobility in the country. He did not stop there; working with Professor Henin and others, he made significant contributions to the study of population in East Africa.

His work has not only been in the sphere of population. He has a formidable list of publications on many other aspects of geography. He did not feel threatened by the quantitative revolution in geography. In fact, he encouraged its arrival and made appropriate arrangement for the appointment of staff who had statistical and mathematical skills as one of the best ways of overcoming resistance to the revolution. Although he did not employ abstract quantitative methods and techniques in his own writing, he encouraged the younger geographers who were thus inclined.

As the first indigenous Professor of Geography in Kenya, Professor Ominde encouraged the development of the subject and laid firm foundations on which other geographers have been able to build. His teaching and published works inspired many students, some of whom are now lecturers and professors. The Department of Geography he established has contributed to staff development in other Kenyan universities, both public and private. His long career has left an impressive legacy for Kenya and East Africa as a whole.

2

The Worlds of Ojiji: Luo Women, 1900-1990

E.S. Atieno Odhiambo

The Context

The *Luo Girl From Infancy to Marriage* (1952) marked Simeon Hongo Ominde's first steps in his academic journey. This modest essay, the winner of the Hancock Memorial Prize at Makerere University in 1949, had a cultural poetic umbilical cord to it, for it was in essence an evocative statement about the validity of African indigenous education. It is instructive that Professor Ominde would start his career with a concern for the nature of autochthonous African cultural heritage at a time of deep colonial crisis in Kenya and East Africa generally. These were the years of 'Mau Mau' in Kenya and the 'Kabaka Crisis' in Uganda, and the young Ominde lived through both experiences as a student in Uganda and later as a teacher in both countries. Both movements were in part forceful statements about tradition and against colonially-conditioned change, about the malfunctions of colonial development in the past and its misdirections in the future.

The Mau Mau movement received its moral nurture from the deep cultural symbols of Gikuyu society. It was partly a social contention over values in society, about what was right and wrong with the colonial social order (Lonsdale, 1987: 343-350). The Mau Mau warriors appealed to the notion of Kikuyuness. For succour they appealed to Gikuyu and Mumbi, the founders of Gikuyu society. They bound themselves together through cultural symbols: the oath of secrecy, the bond between blood and soil. This appeal to the soil found an echo in the Kabaka crisis across the border in Uganda, where in 1953 young Kabaka Mutesa contested the legitimacy of British colonial overrule by asserting the ancestral sovereignty of the Baganda over their Territory (Low, 1971: 101-138).

Straddling both borders were the Luo of Kenya. Both movements presented the Luo with the crisis of belonging and of choice. In Kenya some Luos like Achieng Oneko participated in Mau Mau and became part of the essential Mau Mau leadership lore. At the same time, others like Ambrose Ofafa were identified by the Mau Mau as being opposed to the movement and paid the price in life and blood (Miguda, 1987). Likewise in Uganda, the crisis presented itself for the Luo as a choice between those who supported the British and those who empathised with Ganda nationalism or who saw the latter nationalism as being comparable to the nascent Luo nationalism of the same period. By 1954, this Luo nationalism had crystallised into a Luo underground guerrilla movement that the intelligence authorities in Kampala referred to as the *Onegos*. The details of this movement were never explicated, but its potency was recognised in Luoland, where the colonial administration banned the usage of the Luo word 'Onego' (killing), particularly its youthful rendition of the time, *onego nyathi owadgi*—making illicit love.

The crisis of decision and choice was particularly acute for the missionary–educated Kenyans. They were at the forefront of the Christian entanglement with Mau Mau, for in central Kenya, "Churchtianity" had cast Mau Mau as an anti-Christian movement and some of its victims soon ascended to the pantheon of Mau Mau Martyrs through the expedient of the Mau Mau *panga*. Many of these educated persons buried their heads in the sand; silence was their voice during these years. Others pleaded that they were studying overseas—Charles Njonjo and Julius Kiano were exemplars in this— and, therefore, had no input into the colonial crisis by way of commentary. But, howsoever one positioned oneself, the colonial crisis of the post-War years pushed the question of missionary Western education to the forefront of the agenda.

Maseno School, where Ominde had received part of his secondary education, had been the crucible through which Luo neophytes had been forged into would-be elites in the previous half-century. Maseno had for decades been the base for Archdeacon Owen, Carey Francis (*Okaro*) and L.B. Bowers (*Jabawa*), British missionaries who had come to "lighten the darkness" of the "Kavirondo" Luo and Abaluyia. They brought Christianity and Western education and over the years their pupils radiated from Maseno to spread these values in the rural countryside. This education was referred to as "Western education"; the irony, however, was that there was, in the minds of these missionaries, no "other" education among the Luo. The importance of *The Luo Girl* lies precisely in this contention about "the other."

Reading the Text

The Ominde text can be read in various ways. With hindsight, one can read it as a major first step in the enterprise of the production of knowledge for the Luo and by the Luo themselves. At one level, *The Luo Girl* was a challenge to the colonial structure of dominance, to the colonial praxis of hegemony over what constituted knowledge and what did not. It presented the world of the missionaries with an alternative "other." It challenged the book people, *Josomo*, to contend with other forms of knowing, with another world to penetrate. It presented itself as a *secular text* in contrast to the "inspired" Biblical-oral-missionary texts. With hindsight, one can thus read it as an *ideological* statement. It was ideological implicitly, serving as a cultural critique to the "girls' education" that the missionary matrons at Ng'iya,

Rang'ala, Nyakach and Kamagambo girls' boarding schools were offering. At yet another level, it offered different prospects in social education. Social education is above all concerned with the inculcation of values.

Ominde, through this text, posited the values of his grandmothers at Nyahera directly against those of Miss Churchill at Ng'iya (even as, in the same stretch, he sought and married one of Miss Churchill's most polished products, the late Mary Ominde). Most poignantly, this text was a *parade about manners*. It asserted the superiority of Luo pedagogy over the manners that the missionaries sought to inculcate through energetic caning within the school system, but whose visible results were *Amwom Nyar Skul*, the careless school girl who would rush into premarital sex and into pregnancy. In contrast to this missionary failure, Ominde character-sketched the properly brought-up Luo girl, whose sexual etiquette was refined and restrained.

With his opposing strokes (of the pen), Ominde was, to adopt a phrase from B.A. Ogot (1964: 3-4), "reintroducing" the Luo woman into the world . This reintroduction involved pole-vaulting over colonialist anthropology, which had never attributed mind and certainly not autonomy to the African woman in her autochthonous setting. This must be one of the seminal aspects of this work, although it had also remained its least exploited form.

Other ways of reading *The Luo Girl* avail themselves. The text was witness to a process of *recovery*: the recovery of what was embedded in Luo tradition. In this process, Professor Ominde acted as a commentator for his oral informants. The print culture through which the author ushered *The Luo Girl* had its own hallowed history: it empowered the literati. In the previous half-century it had empowered them through content: they knew the Bible and the prescribed syllabi. *The Luo Girl* empowered Luo *tradition*. It was a new message that also amplified and reinforced old values. In doing this, it made a further ideological statement about continuity between the past and the present.

Not least important was the role that Ominde would assume in his society from this time on. Rather than join the bandwagon of some of his contemporaries who cried that they were the "Christians of Two Worlds" and thereby fell into intellectual paralysis when it came to the reaffirmation of the African cultural presence and Africa's civilisation values, this text marked the beginning of engaged scholarship that has been characteristic of Professor Ominde over the years.

Yet another way in which *The Luo Girl* made a statement was in the title itself. The "Luo" on whom the author anchored his study were very much in the process of being "invented" at this same time, notably by "the articulate citizens" of the Luo world: the teacher Samuel Ayany and Shadrack Malo. The text came out at a time when the identity was simultaneously being forged by Luo elites through business, Luo Union, football and cattle trade (Cohen and Atieno Odhiambo, 1988).

The title of the Ominde text made claim to *uniformity* for the Luo community at the moment of its own construction. It is instructive to contrast this period of Luo renaissance with earlier periods five and two decades removed in sequence.

Elsewhere we have argued that it took a period of fifty years for the congeries of the lake shore communities to have a name. At the beginning of this century they were variously known as Nilotic Kavirondo, JaLuo, Wanyifwa or Pagaya. The British anthropologist E.E. Evans-Pritchard, in a study of the Luo in 1936, was very comfortable with the notion of "Luo Tribes," which he identified as Alego, Ugenya, Waholo, Kadimo, Sakwa, Gem, Asembo, Seme, Uyoma, Sagam, Kisumo, Kajulu, Kano, Nyakach, Doho Rachuonyo (consisting of Karachuonyo, Kabondo and Mumbo), Kochia, Kanyada, Kanyamwa, Karungu, Kadem, Kwabwai, Kanyidoto and Kabwoch. He found it necessary to identify the following as non-Luo, mixture of *Mwa* and *Lang'o:* Kagan, Muhuru, Kaksingri, Kasigunga, Gwasi, Kamagambo, Rusinga, Mfan'gano and Kanyamkago. But he did not deem that these tribes would ever belong to a Luo nation. By the time of the Luo Renaissance in the early 1950s, Ominde, Ayany and Malo were now agreed on a Luo nation with a Luo identity that embraced the real Luo and those that Evans-Pritchard had excluded as *Mwa* and *Lang'o*.

This text was making powerful statements about Luo behaviour, perceived of as being under attack and erosion at this time from a whole coterie of forces – European missionaries and administrators, but also *Josomo* (book people), *Jonanga* (civilised people) and *Jopango* (urbanites) among the itinerant Luo.

The World of Ojiji

The word of the *Luo Girl* is a universe of moral order, of certainty about the life-cycle, clearly segmented into infancy, weaning, puberty and initiation, induction into the world of womanhood and work and into marriage. The crucial institution in this world was the *siwindhe*, the grandmothers' dormitory in which girls received education and culture. The key "professor" in this set-up was the grandmother, better known as *pim* in the Dholuo of Ominde's generation. This crucial preceptual setting—of young girls with their *pim* in the *siwindhe*—suffered academic neglect in studies of indigenous education in Africa until David William Cohen evoked it powerfully in the early 1980s. Because it is the crucial node in Luo education and a centre-piece in *The Luo Girl*, it is fitting to quote Cohen at length:

> *Pim* and her charges lived together in the *siwindhe*..... children learned about the past from *pim*. They drew upon her wisdom. They learned about the people, the groups, and the settlements around them. They learned about sexuality and about childbirth..... From *pim*, children learned about health, illness, misfortune, and obligation that would both open and restrict their lives.

Again:

The *pim* instructed the girls concerning their sexuality. *Pim* taught the girls to be tolerant to their future spouses and in-laws during domestic problems. She taught them about responsibilities of the adult woman and she taught them about respect due to husbands in marriage. *Pim* instructed the girls to refrain from sexual relation outside marriage. A successful union was perceived as being marked from the beginning when a young woman was found to be a virgin, *"en kod ringre"*. The girls were taught never to eat at a boyfriend's house. Secret visitations, *wuowo*, with boyfriends were undertaken at night and girls were instructed to be back home before dawn.

Such visitations were only known to *pim* and the girls of the *siwindhe*, not to their parents. They were instructed not to visit their boyfriends during menstruation, *dhi boke*. *Pim* taught the girls to offer persistent boyfriends an experience of lovemaking, but without penetration, through the skilful use of the thighs while making love (Cohen and Atieno-Odhiambo, 1988).

In the fullness of time, the girls emerged from the *siwindhe* prepared to experiment within the world of boy-meets-girl relations. This transitional period bore the true seriousness of the college "dead-week" before examinations for both boy and girl. Evans-Pritchard (1965: 229) captured its centrality in his 1936 study that confirms Cohen's encapsulation forty years later:

Courtship and marriage are intimately connected with the custom of making love to girls *(chodo)* in the *simba* or bachelors' dormitory, the hut nearest to the entrance to a Luo homestead. Youths arrange with girls to visit them there at night, and when a youth is visited by one of his sweethearts the other young men of the home sleep elsewhere. He plays with the girl and has intercourse between her thighs and they sleep together. He must not penetrate her—that is regarded as shameful, and it will be known on the day of marriage. If a girl comes from a distance to visit her lover she may spend several nights with him in the *simba*. A girl may have several lovers in different homesteads who she visits from time to time; she pleases herself in these matters. The young men will not quarrel about her, nor will her father and brothers interfere in her love affairs. She may continue the practice throughout the stages of her marriage to another man right up to the *riso*, the final ceremony of marriage. I was told that a girl does not usually visit the *simba* more often than once or twice a month. Occasionally the lover employs a harpist to entertain the girl and his friends on one of her visits, and there is singing and dancing. Between the songs, youths, and sometimes girls also, stand up and boast of their virtues, of the number of their friends and sweethearts, of their wealth, and of their families and lineages; and when they have finished they throw a gift to the harpist, maybe today a shilling. Other youths then try to outdo them in boasts and generosity. (Evans Pritchard, 1985:229)

They boasted of their virtues. This world of order provided education, but also security. In time, suitors would appear at the homesteads of *The Luo Girl*, bringing with them the startling news (but welcome to the girl's brothers) that they had come to ask for the girl's hand in marriage. Ojiji's rhapsody, sung in celebration of the betrothal of a Gem Rae girl (and aunt to Okoth-Ogendo) in 1931 captures the spirit of that age.

HERALD:	*Nya-Amolo*	Nya-Amolo
	Nya-Amolo, Ojiji Nya-Amolo	Nya-Amolo, Ojiji daughter of Amolo
	Ojiji chwo dwaro!	Ojiji attracts suitors!
OJIJI:	*Chwo angowa?*	Which suitors?
HERALD:	*Chwo angowa Ojiji Nya-Amolo*	Which suitors? Ojiji Nya-Amolo
	Ojiji gin pacho	Ojiji they are at home!
OJIJI:	*Luongna mama!*	Call mother for me!
	Luongna mama mondo osewa	Call my mother, to give me presents
	Ojiji dhiyo!	Ojiji is going
	Luongna dana	Call my grandma
	Luongna dana mondo osewa	Call my grandma, to get me blessings
	Ojiji dhiyo!	Ojiji is going!
MOTHER:	*Ojiji ma nyara*	Ojiji my daughter
	Koro dhiyo	Is going
	Ne achako nera	She's named after my uncle
	Ma wonya Yimbo	From Yimbo,
	Ojiji ma nyara	Ojiji my daughter
	Dhiyo tedo	Will get married
	Ojiji ma nyara	Ojiji my daughter
	Dwa Kelo	Wants to bring
	Dhook!	Cows!
BROTHERS:	*Ojiiji!*	Ojiiji!
	Ojiji dhok mane ikelo	Ojiji the cattle of your dowry
	Osekonyo	have helped out
	Chak to ti wamadho	milk we now guzzle
	Kar Kong'o	instead of beer
	Moo to ti wayudho	ghee we now dip
	Kar Kado	instead of broth
	Dhok to ti wakwayo	cattle we now graze
	Koda roye	and calves also
	Ojiji Okelo dhok!	Ojiji brought cattle!

The Deconstruction of Ojiji

The years since the publication of *The Luo Girl* have witnessed, as it were, a deconstruction and reconstruction of Ojiji's world in various ways. Much of this deconstruction remains uncharted and unstudied and stands out as an urgent research agenda for Luo gender studies in the 1990s and beyond. The field includes the changing conceptions and roles of grandmothers, especially the urban ones; past and present of bride-wealth and marriage; question of divorce and remarriage; reworkings of the roles and responsibilities of *joter* (levirs); urban polygamy among the Luo elites; ambiguities, for the persons involved, in being a Luo woman graduate and a second or third wife and the role that the money nexus plays in this postmodernist domestic arrangement and property relations of co-wives in life and in the death of the wealthy husband and the challenge, for educated Luo women, of being a spinster for life *without* being a concubine to the local notables.

All these areas embody a potency for the mind and for knowledge. Awaiting its own study too is the whole notion of women who work with their bodies, "digging with their buttocks" as the blind Akamba poet Kyeti sang in 1912. Asenath Odaga once provided this author with an evocative Luo vocabulary for the word prostitute: *opamo, ohodho, ochot, mganga, malwa* and *andwayo*, among others. The existing literature tacks this activity as an addendum to the many concerns of the Luo Union.

Agweng' Department

Elsewhere, we have explored yet another dimension, another possibility, that has befallen the children and grandchildren of Ojiji (Cohen and Atieno Odhiambo, 1988: 96-98). Our study *inter alia* confronted the social arena euphemistically referred to by both young and old as the "*Agweng'* Department," the frequent sexual liaisons between young people, really children, inhabiting the same compound or village, *gweng* (hence *agweng'*):

> The Siaya countryside, in the 1980s, is striking for the considerable number of 14-and 15-year-olds who are already unwed mothers. One can get a clear view of them at funerals. They are among the gathering holding their babies, a somewhat uncharacteristic image, given that their own mothers are normally nurturing these uninvited grandchildren; at funerals these mothers-cum-grandmothers are too busy caring for guests to hold the infants.
>
> Kwashiorkor-ridden infants are too eloquent an affirmation of the concept of *amuom nyar skul*, of children begetting children before any training in sexual matters has come their way from the adults of the *dala*. The "children of children" direct our attention to the questions of space and distance, for the fathers of many of these infants are themselves teenagers without a *simba* of their own.

"Sex in the grass"—one way it is known in Siaya—also raises questions concerning intimacy between close cousins, concerns about (*sic*) which are suppressed by adults who fear to expose the identities of fathers of infants produced out of transitory unions between individuals who are too closely related, if not also too young, to marry. The question is also raised as to the sexual *mores* obtaining in the "society of the grass". It is recognised in Kano Koc hogo that lax young women do really exist in the society. They are nicknamed *gwoma*, "those who squat willingly". It is also recognised, poetically, that this quality, this character or persona, is not their fault. One speaks of the "loose loincloth being unprotective" and thereby being the cause; hence the saying, *Afwong'o jang'uono* ("loincloth the generous one"). (Cohen and Atieno-Odhiambo, 1988: 91)

This word *Afuong'o* stands in stark contrast to the punctilious grooming by the *pim* and the strict behaviour code obtaining in the *simba* that constituted the moral of *The Luo Girl*. And if the world of *The Luo Girl* produced its affirmation in the repertoire of Ojiji, this nether world has also, in its time, produced its counter-culture, captured in this song, composed by Richard Onyango Odero of Alego Komenya for the Nyadhi School choir:

SOLO:	*Weche ng'eny*	Mother; Things are happening
	Ka Simba mwalo	At the bachelor's *Simba*!
ALL:	*To igalori agala*	Mother,
	To weche ng'eny	Things are happening
	Ka Simba mwalo	At the bachelor's *Simba*!
SOLO:	*Nyathi Ko*	I am telling you
	Ni min mare	
ALL:	*Mama igalori agala*	Mother; Wake up
	To weche ng'eny	To the things; That are happening
	Ka Simba mwalo	At the bachelor's *Simba*!

Whatever things may be afoot at the *simba* are obviously accessible to the children of these days and to the children of children, in contrast to the discreet nightly visits described by Evans-Pritchard (1965) half a century ago.

The consequences are equally stark. As Festo Ochuka "Wuod-Ahero-Nyando" posited it in 1968, the whole enterprise leads to dead-end options for many of these schoolgirls.

Ochuka sang:

Nyar Sikul	A schoolgirl
Omako ich	Is pregnant
Jodongo laro	Old men jostle
Te te te	Over her
Ma wena!	Please leave this!
Ma wena! Bolingo	Please leave this to me
	Bolingo (It's a washout)!

Conclusion and Beginning

In 1964 the government of the newly independent state of Kenya appointed Professor Simeon Hongo Ominde to be chairman of a commission to prescribe an appropriate system of education for an independent African nation. The muses of the grandmothers at Nyahera, Ominde's home, must have smiled knowingly. Twenty years later, the Mackay Report that ushered into Kenya the 8:4:4 system of education reaffirmed the validity of the social contention of *The Luo Girl* by placing Social Education and Ethics firmly at the centre of the country's value systems. *The Luo Girl* had indeed come of age to be consummated in the mainstream of Kenya's education system into the year 2000 and beyond.

Tinda!
Tinda adongi aromna gi Ominde K'Owiyo!

3 Scholarship in National Development

J. B. Ojwang

Introduction

Those who knew the late Professor Simeon Hongo Ominde could not have mistaken his most elemental characteristic: he was a scholar in every sense. It is only fitting that we should on this occasion reflect upon the theme of scholarship and consider what lessons it brings up in relation to Ominde's work.

What is the place of scholarship in our social progress? Does being a scholar require the same things irrespective of time or place? What kind of scholarship did Professor Ominde practise? What sort of scholastic bequest did he pass on to succeeding generations of intellectuals? How far did he succeed in transmitting his *genre* of scholarship? What kinds of scholastic orientation are today in vogue and how far do they reflect the Ominde legacy?

These questions provide the main reference point in this chapter on Ominde's general orientation as a scholar and researcher.

The Concept of Scholarship

Webster's Third New International Dictionary defines the word 'scholar' as follows:

(a) one who by a long systematic study. . . has gained a high degree of mastery in one or more of the academic disciplines; (especially) one who has engaged in advanced study and acquired the minutiae of knowledge in some special field along with accuracy and skill in investigation and powers of critical analysis in the interpretation of knowledge;

(b) a learned person (especially) one who has the attitude (as curiosity, perseverance; initiative, originality, integrity considered essential for learning. . .)

The dictionary defines 'scholarship' as the character, qualities or attainments of a scholar. . . .'

Scholarship thus entails certain distinct attributes: detailed and methodical study of a subject or phenomenon; command of information relating to that subject or phenomenon; mastery of techniques relating to the subject or phenomenon and the ability to interpret and apply information and knowledge relating to the subject or phenomenon. Scholarship involves not only the identification of the factual characteristics of a phenomenon, but also the application of these to reach certain analytical positions or certain inferences of principle at a more general level. Reasoned generalisations shed light on the broader picture and thus provide knowledge regarding the phenomena in question.

However, such an interpretive process necessarily makes certain assumptions in accordance with the operative social context and on this account the scholar's inferences will invariably reflect the ideals of a particular society. The scholar in Africa is bound to be influenced by the prevailing social conditions; hence his prescriptions will often be different from those which may be preferred in Europe or elsewhere.

African social and intellectual experience has been at the crossroads of Western philosophy, on the one hand, and traditional world-views, on the other. This is partly due to the colonial factor, which took the form of European political control over most African countries for several decades in each case. An important element in this cultural crossroads was the colonial educational system in which Western traditions of scholarship were dominant in the universities and other tertiary institutions of learning. Following the attainment of independence by African countries, the Western academics and researchers in the universities were gradually replaced by African academics. Professor Ominde belonged to the first generation of African professors. It is instructive to reflect upon the interplay between this generation of scholars and the European scholastic heritage. One should appreciate the role of this generation of scholars in the search for new directions in African scholarship.

Received Intellectual Traditions: The European Heritage

African scholars of the first generation such as Professor Ominde were born, bred and schooled in the colonial times when the prevailing rules and methods of scholarship were the same as those approved for the renowned universities of the West such as Oxford, Cambridge and London, among others. Throughout their educational careers they followed curricula that were modelled on British school curricula; for a substantial part of their school life they were taught by British school teachers and at university they were introduced to the well-tested investigative methodologies of the distinguished universities of the West. Thus, to appreciate Professor Ominde's scholastic orientation, one must start from the outlook of the Western scholar and from this standpoint endeavour to perceive his generation's outlook on the scientific study of African phenomena. It follows that the main contribution of Professor Ominde's generation, in relation to scientific inquiry focused upon Africa, is essentially universal and comparative rather than parochial;

innovative and application-oriented; has retained distinct linkages, on the theoretical plane, with the foundations of inquiry evolved in the distinguished centres of learning in the Western countries.

To appreciate Ominde's scholarship in the context of his educational background, we need to bear in mind the guiding features in Western modes of inquiry. The Western scholastic orientation treats scholarship and, indeed, the prevailing broad intellectual standpoints, as part and parcel of a social-evolution package — a package which holds in a relationship of close interplay the economic, social and political dimensions of the life of a people. For a clear depiction of this scenario we quote from the English aristocrat, Sir John Fletcher-Cooke (1966:142):

> . . . representative parliamentary democracy is but one, though no doubt the most important, of a whole series of institutions, patterns and attitudes which should be included in the term "the Westminster model." This whole complex of ideas and institutions represents the Western, and more particularly the British, way of doing things. In addition to representative parliamentary democracy, . . .this complex includes a belief in the rule of law. . . ; support for academic freedom in universities and colleges. . . .This complicated web of ideas and institutions hangs together. To expect parliamentary democracy to take root and flourish in an environment in which these other supporting factors are not present in adequate measure is tantamount to expecting a house to stand up without proper foundations.

Sir John Fletcher-Cooke attributed the concept 'the Westminster model' to a long historical background of nurturing on European terrain, on the basis of dialectical thinking evolved in the Greco-Roman civilisation and tested by practice over a period of time extending beyond ten centuries. It is such a historical context that gave birth to the epistemologies and traditions of inquiry associated with universities, judiciaries, legislatures and investigative commissions, among others.

Those African scholars or judges who were nurtured in the context of such governing ideas and methodologies would naturally have been expected to make this heritage their point of departure as they came to address more squarely the special circumstances of the post-independence period. This heritage, above all else, required (and still requires) the commencement of inquiry from a well-illuminated fact-base and only thereafter proceeding to the puzzles of the unknown. It requires the investigator to invest a considerable amount of time in clearing the ground, in listening to first-hand accounts and in mastering the prevailing state of affairs as a basis for delving into the unknown.

Ominde obtained his secondary school education mainly from British missionary teachers and his university education at the feet of Western scholars in some of the most renowned British universities. It is on record that by 1940, Kenya had only four secondary schools that were open to Africans, namely Alliance, Kabaa/Mangu,

Maseno and Yala (Bogonko, 1992:24). These were feeder centres for Makerere College in Uganda, which was for many years the only institution of learning awarding diplomas and administering the Cambridge School Certificate Examination. Makerere College, first established as an institution of technical education in 1921, became an elite educational centre in East and Central Africa and in 1949 it was renamed the University College of East Africa. From that year Makerere was giving instructions on the basis of British university curricula for the award of degrees of the University of London.

Ominde was taught by some of the most distinguished missionary teachers at both Maseno and Alliance Schools before he proceeded to Makerere where he earned a diploma in 1948. Thereafter, his nurture in Western scholarship took an even deeper level with his admission to the University of Aberdeen in Scotland. While in Scotland, Ominde earned the M.A. degree of the University of Aberdeen as well as the Silver Medal of the Royal Scottish Geographical Society (1954), and in 1955 he obtained a Diploma in Education from the University of Edinburgh. His Western education did not end until 1963, when, as a don already serving at Makerere, he earned a Ph.D. degree from the University of London.

It is clear from such a background of education that Professor Ominde would quite naturally have been expected to make his scholastic legacy in the form of a *mélange*, distinctly marked by the tested principles of Western scholarship. When he entered upon African scholarship, moreover, there was hardly any ready-made investigative orientation and it fell to him and his generation to lay the baselines of academic inquiry.

African Scholars in Search of a New Beginning

While there is hardly any doubt that Kenya's first generation of scholars — such as Professors Ominde, David P.S. Wasawo and Bethwell A. Ogot — had been nurtured in Western traditions of inquiry, they none-the-less had to contend with certain fundamental differences between the Western and the African outlook. In the discipline of history, for instance, a leading British scholar, Professor Sir Hugh Trevor-Roper (later Lord Dacre) had passed the judgement that there could not have been anything called 'African history', since African events merited no better characterisation than as 'a mere phantasmagoria of changing shapes and costumes', and did not carry enough weight to bring them under the classification of 'history'.

Professor Ogot's iconoclastic precedent, taking the form of his famous work, *History of the Southern Luo* (1967) was subsequently followed by distinguished innovations, some by his colleagues such as Gideon Were (1967) and Godfrey Muriuki (1974) and others by his own students such as William Ochieng (1974) and H. Mwanzi (1977). In the end, such a pejorative perspective of African history was entirely discredited.

Ominde, in the same way, did open up new areas of social scientific investigation in relation to African conditions; and a number of his colleagues such as F.F. Ojany, R.B. Ogendo, R.S. Odingo and R.A. Obudho, and his students such as Elias Ayiemba, J.O. Oucho and Z. Muganzi, among others, subsequently took the cue and made highly important scholastic contributions.

The first-generation African scholars had to employ their very specialised knowledge in the traditions of Western scholarship as a vital asset and point of departure from which to undertake a reconstruction and re-assessment of African phenomena.

It is precisely this conjuncture of established traditions and open-ended and novel social issues that accords the first-generation scholars such as Professor Ominde their unique contribution to the evolution of scholarship in Africa. The pioneering status of these scholars and their sensitivity to an evolving intellectual scenario redolent of ideological disharmonies, with the advent of independent statehood and the irresistible clamours for 'Africanisation', gave them occasion to articulate their stand both by their works and through interlocution with learned colleagues.

The late Professor Ominde, in terms of availability to and interaction with fellow scholars on academic issues, had few parallels and the many young scholars with whom he generously communed on the state of learning and the direction of tertiary education in Kenya will readily give testimony on the issues which repeatedly made the agenda. The author, having been one of the many beneficiaries of a deeply academic relationship with the veritable father of academic research in Kenya, to reflects below on Professor Ominde's thoughts and concerns about scholarship in our universities.

Professor Ominde's Thoughts on Scholarship

Professor Ominde was deeply conscious of the delicate institutional linkages that constantly determined the success of a scholar's undertakings. The mission of scholarship was the generation of knowledge. But this knowledge was not necessarily an end in itself because it had an operational significance as it provided guidance for the process of nation-building and it helped in the strengthening of those institutions of society which served as delivery-points for social, economic and political welfare. Professor Ominde saw the interplay between scholarship and institutions as one complex; for instance, scholarship while serving to assure the quality and integrity of institutions depended significantly upon the character and mode of operation of those very institutions to facilitate it. He considered, for example, that the efficient and rational conduct of university management systems was vital to the effective pursuit of learning and the creation of knowledge by able and committed university dons. He was, therefore, greatly concerned when ill-advised policies or odd administrative trends in the university set-up were becoming obstacles in the way of research endeavours by scholars.

It is pertinent to consider critically Professor Ominde's interest in the functioning of administrative, social and other institutions which had a bearing on the activities of scholars. As already noted above, he, unlike a goodly number of socialistic young academics of the post-independence period, did not entertain the pretension of debunking his educational heritage drawn from Western scholarship. He saw this heritage as a major pedagogic resource to be adapted and re-oriented for the purpose of creating an objective understanding of the social reality obtaining in Kenya and in Africa in general. Such a stand was both strategic and realistic, as it provided a practical operations base for the scholar even as it recognised his existing intellectual assets (and possible liabilities). But the critical element in Professor Ominde's position in this regard was his strong belief in the need to recognise the 'continuous' character of social phenomena and trends. He saw no wisdom in investing time in denigrating the historical reality that our society had moved from pre-colonial through colonial to post-independence conditions.

This continuum, as he saw it, necessarily entailed the reality that the present did contain inheritances from the pre-colonial and colonial periods and that post-independence Kenya was inhabited by people whose world-views and social needs bore the imprint of the past. To be of service to such a people, in Professor Ominde's view, the scholar could not avoid employing tools of scholarship which originated from the past. He considered that scholars had an obligation to respect the people's preferences in social institutions and to endeavour to assist them by bringing about a better understanding of the open institutional choices. In this way the people would, in the due course of time, attain a growing measure of institutional stability and such an achievement was a vital element in their social development. If scholars adopted such an attitude, they would also benefit their own calling, as they would thereby attain more objectivity in their methods of investigation and creation of knowledge and a desirable parity would consequently come to exist between scholars and the operative social institutions.

Such a principle may have been partly informed by Professor Ominde's clear understanding of the dynamics of Western societies. As a student in the United Kingdom, he had noticed that there was no fundamental conflict between the higher institutions of learning and the society at large; indeed, a good deal of the institutional improvements that had been made in the economic and political systems of the Western countries were attributable to systematic research outcomes and proposals emanating from the universities.

Professor Ominde keenly observed the changing fortunes of scholarship in Kenya's public universities. In this regard, he considered it highly desirable that all scholars should commit themselves to some scale of time-investment in their academic pursuits and their research work. He considered, on the basis of his own experience, that a scholar's career should begin with a substantial investment in the mastery of the known details, concepts and norms within the relevant discipline, as a basis for self-empowerment towards application and the resolution of the problematic areas. Thereafter, in Professor Ominde's view, the scholar should constantly keep a fair balance between time spent on updating and enhancing his information and that spent on solving new problems and conceiving appropriate prescriptions.

It is to be understood that, against this background, Professor Ominde used to get rather concerned whenever young academics appeared to overlook the basic principle and seemed to be cutting corners so as to expedite their way towards some seemingly popular result. He expected young scholars to be clear-minded on relevant fact-situations pertaining to their subjects of inquiry, properly address the prevailing social reality affecting the investigation and arrive at realistic and constructive innovations. The scholar's enterprise, as Professor Ominde saw it, required methodical inquiry which was the standard formula for arriving at the truth and at a position of knowledge. Although he did not rule out the possibility that a scholar might make a valid finding through sheer luck or serendipity, he considered it idle for a scholar to wait upon chance while foregoing the rewarding opportunity to engage his special skills. The creation of knowledge, as Professor Ominde saw it, required objective investigation and this goal would be defeated if an academic deliberately procured the emergence of a 'social storm' which he then purported to 'investigate'. The scholar, especially in the social sciences, must thus take his subject of inquiry in the condition in which it exists. Professor Ominde was thus naturally rather impatient with academics who strove for the glare of press publicity or resorted to eccentric methods or to propaganda as ways of arriving at particular results. He abhorred caballing among academics as a strategy for pre-empting spheres of practical advantage for privileged enjoyment (*chasses gardées aux médiocrités*).

On many occasions, Professor Ominde expressed concern about the popular image of today's academic—a man of straw who evokes pity and who does all sorts of odd jobs for bare survival! Professor Ominde considered that any academic who gave credence to such a myth would be disingenuous. His reasoning went thus: quality scholarship *will* pay; our country today has such a diversity of specialised and professional activities that the quality researcher, even though he does not advertise himself, will at some point in time be noticed—and in this way somebody will pay him for his good work.

Sustained quality research is not to be found in abundant supply and those who provide it are bound to come to the attention of international organisations, government agencies and foreign governments, among others. Professor Ominde held this perception dear, often typifying it in the vernacular *chiemo gi kalam* (earning a living by courtesy of the pen). He did caution, however, that those who hoped to survive in this way as scholars should avoid the temptation towards irrational haste: they must start off from a well-laid foundation of learning and scholarship and they should, moreover, appreciate that true scholars hardly ever evaluate their worth in purely material terms. Genuine scholars, as Professor Ominde saw it, enjoy great intellectual satisfaction and this must be brought into the equation as part of their affluence and their success.

Professor Ominde's immeasurable love for scholarship explains his continued active service to the university right through to the last moments of his life, despite numerous governmental responsibilities and notwithstanding his failing health. He may not, unfortunately, have felt completely fulfilled in his continuous endeavours to encourage such a kind of scholarship in the universities. Although many promising scholars learned at his feet and he felt justly proud of them, he was rather saddened by the fact that many young academics were not able to devote themselves wholly to scholarship. His earnest hope was that the various universities would be able to take more interest in the research work undertaken by their staff and that they would accord such researchers suitable symbols of recognition.

Professor Ominde on several occasions addressed himself to the apparent conflict between commitment to scholarship, on the one hand, and occupancy of the increasingly attractive administrative positions within the universities, on the other hand. He could cite many specific cases in which senior academics had become heads of major offices in the public universities and in the aftermath their research initiatives had rapidly thinned down. Scholars in this category, as he saw it, in effect had no real opportunity to establish dependable guidelines for their disciplines and as a result they left their successors with hardly any capacity for enhancing their own scholarship.

This concern of Professor Ominde's may be illustrated with a poignant personal example. Having been keenly aware that the University of Nairobi's Faculty of Law had been gravely affected by the re-location of senior staff to management positions, Professor Ominde was most upset when, in October 1988, key administrators of the University made secret arrangements to have the author, who had only recently been elevated to the rank of Associate Professor, translated to a clerical role as the registrar of another university—a university which moreover had no programmes in the discipline of law! It was only thanks to Ominde's initiatives, which went well beyond the call of duty, that the imminent execution of such an act of high-handedness was averted.

Professor Ominde was of the view that the grandiose status currently attached to administrative roles, as compared to purely academic ones in the public universities, ought to be reversed through a reorganised reward system. He apprehended that, under the prevailing conditions, those universities would end up with burgeoning administrative profiles that would increasingly suffocate the conduct of research and the creation of knowledge through systematic investigation.

In many ways Professor Ominde was a mainstream scholar. This was so firstly in the technical sense and secondly in the general sense. In the first category, his writings were mainly in areas of central concern to his discipline, namely geography, population and natural resource management (Ominde, 1968; 1984 and 1988). These works were no doubt intended to coincide with the main agenda of the African student, researcher and policy maker; he was, in effect, seeking to enhance our understanding of those issues which affected us most in our social, economic and political development.

Secondly, the same works exemplified Ominde's familiar perspective on the interplay between academics and the socio-political order. He held the view that the African scholar ought to understand the African society in proper context and assist in stabilising that society by servicing the optimal operation of its main institutions. Such a scholar should recognise that we have come to live in a nation-state which requires the harmonious co-operation of its different elements and thus he should make his contribution towards consolidating the framework of nationhood. Professor Ominde's scholarship had a clear and constructive policy orientation and was of considerable service to public sector planners, especially in matters of economics and national development.

Succeeding Generations of African Scholars

Professor Ominde was not only an originator of important scholastic guidelines, he was indeed a personification of African scholarship. This presents two broad problems for succeeding generation of scholars: firstly, they will need to start from the foundations laid by him and thus be influenced by his guidelines and, secondly, they will be expected to make the difficult personal decisions that go into the making of a true scholar.

The second challenge remains an open one as it must depend so much on the personal capacity and initiative of the individual scholar. The first challenge may be subjected to some analysis here.

In all disciplines, but more particularly so in those which require scientific methods of inquiry, the notion 'knowledge' brings several different elements into a relationship; for instance, the *past* and *present*. On this account, scientific inquiry rests upon *continuity* as its basis for arriving at knowledge, and knowledge rarely takes the form of anecdotal scenarios. Those scholars who seek to generate knowledge in the social sciences and in relation to population movements in Africa, for instance, will find that the baselines of inquiry have been set by Professor Ominde. They will need to invest some time in appreciating the objective conditions which are analysed in his works, and thereafter they will have a basis for the development of their research themes.

It should be noted that the succeeding generations of scholars will be much more privileged in the conduct of research than he was, given the greatly improved methods now available for data processing and information sourcing.

Conclusion

The occasion of this work in memory of Professor Ominde provides a unique framework for reflecting upon the subject of scholarship in the context of post-independence Kenya. The status of a scholar will depend on different factors such as his area of scholarship and the contemporary social relevance of his subject; the innovativeness of the scholar's works and their significance for a particular discipline or professional group; the scholar's contribution to research methodology or tradition the scholar's intellectual impact among a community of scholars and the scholar's contribution to the processes for securing broad national welfare, among others.

Professor Ominde's standing as a scholar entailed an element from most of the foregoing categories. However, the most critical aspect of this standing may be stated as his uncompromised commitment to scholarship, which saw him produce major works of great significance in terms of methodology as well as national policy goals (see Appendix). His special standing as a pioneering scholar also gave him exceptional prominence in the academic community and ensured that he came to play a major role in the servicing of decision-making processes.

Part II

SUSTAINABLE USE OF NATURAL RESOURCES

4 Trends in Resource Development in Mountainous Environments

Peter H. Omara-Ojungu

Introduction

Mountainous environments present one of the most difficult and challenging conditions for human habitation and resource development. Due to effects of altitude, relief, slope angle, aspect, climatic variability and location within zones of high seismic risk, mountainous environments tend to be harsh and vulnerable to the destabilising effects of even slight disturbances. Topographic variations present intimidating challenges to pioneer settlers by raising costs of transportation and communications. In addition, the relatively high frequencies of earthquakes, landslides and intra-mountain valley floods together with anti-mountain tales and myths make mountains so unattractive that one becomes curious to know why in the first place human beings inhabited mountainous areas and what prospects exist for improving and sustaining socio-economic development in these regions.

This chapter documents the evolution of resource use practice and man's impact on mountainous environments. The general trend in mountainous environments is substantiated by case studies from the mountainous regions of Uganda.

The Expansion of Resource Development

Mountainous environments have experienced resource use for many centuries. According to existing records, cattle grazing in the Alps dates back to the 12th century (Scargill, 1975). Skalnik (1978) noted that settlements in mountainous regions of Europe started "through the process of internal colonisation" in the 12th and 13th centuries in the Alps, Carpathians and Caucasus areas while the Inca civilisation in the Andes is known to have existed earlier than the 15th century (McIntyre, 1973).

Most tropical and sub-tropical mountain, slopes have been in human use for over 300 years. While various reasons have been given for the relatively late occupation of mountain slopes, there is a general consensus that resource use in mountainous environments was a response to social stress within traditional lowland settlements. This section will identify the major resource use periods in mountainous environments and describe the resource use features of each period. Emphasis is given to plausible reasons for resource expansion into mountain regions at various time periods.

Mountain resource use periods

From existing literature two major periods of resource use are discernible: a period of subsistence economy and one of market economy. This classification has been used to argue that the present crisis of mountain peasantry and general resource use deterioration were brought about by the shift from a traditional (self-sufficient) subsistence economy to a market-oriented one (Eckholm, 1975; Hewitt, 1976 and Sklanik, 1978). While it is true that today's mountain communities are more market-oriented because of their increasing dependence on developments in lowlands, the claim of a self-reliant economy is difficult to verify even for mediaval mountain communities. By implication, Scargill (1975: 9) suggested that before the peasant crisis in the Alps, intermontane trade existed:

> As early as the 12th century the Alps mountainous areas provided a surplus of cattle for the flourishing towns of the northern and southern forelands. Up to the 18th century the leasing of oxen to Italian merchants constituted one of the major sources of income of the crown.

In the middle ages, mountain communities acquired non-agricultural income from mining, curing, charcoal burning and cottage industries. It seems, therefore, that the periods of subsistence and self-sufficiency and those of a market-oriented economy can be more accurately described as periods of subsistence and less-developed economy and periods of well-developed monetary economy respectively. The period of weak monetary economy corresponds to the pre-industrial period while that of a strong monetary economy corresponds with the industrial and post-industrial periods including their modes of resource allocation and management.

The shift from a subsistence to a market economy has been an inevitable consequence of local pressures and changes in human values in all communities. However, in some mountain regions, the shift was accelerated by the leadership of people from these communities who had different cultures. For instance, tea and coffee were introduced to the East African highlands by European colonisers, consequently changing significantly the nature of human interactions with mountains from one period to another.

The 'pull-push' concept

The causes and nature of resource practices in mountainous environments can be explained by the relationship between 'pull' and 'push' factors. Pull factors retain resource practices within an environment and attract to it practices from other environments. Push factors disperse resource development from an environment into other environments.

In the primordial state, mountainous environments show push factors (restraints) associated with effects of cold climates and extensive snow cover at high altitude steep slopes, rugged topography and poor soils; isolation and inaccessibility; high altitude and decreased oxygen content of air; devastating landslides, floods, snow avalanches, earthquakes and severe natural erosion and traditional anti-mountain myths and tales. Fears associated with these push factors seem to have been more pronounced during the pre-industrial period when technological advance was too poor to overcome mountain vagaries and challenges. These fears are probably the cause of the relatively late human settlement or its absence in some mountain regions.

Initially, human settlement and other resource practices were largely restricted to lowland environments. Pre-industrial expansion from lowlands into adjacent mountain regions probably resulted from the effects of mountain pull factors exceeding the pull to remain in lowlands. Actual emigration from lowlands resulted in push factors which exceeded pull factors. Maintaining resource practices in the mountains must also be the result of pull factors acceding from push factors.

During the pre-industrial period major mountain pull factors include abundance of defensible locations; effective refugium in mountains and abundant opportunities for agriculture, pastoralism, timber and mining of minerals. For instance, Skalnik (1978:11) attributed mediaeval colonisation of European mountains to lowland push factors such as relative overpopulation in the lowlands and lower valleys and feudal interest in [economically] exploiting distant uninhabited and unoccupied territories. He further specified that the Tushebi people of the Caucasus mountains "left their lowland homes because of political pressure, inter-ethnic warfare, feuds and also as a result of religious pressure and persecution from Muslims and Christians (Skalnik, 1978:11)." Similar lowland push factors dominated early occupation of tropical, subtropical and other mountain regions. In Nigeria, the East Africa Highlands, Papua New Guinea, the Philippines and Mexican hills inter-ethnic warfare pushed weaker tribes from lowlands into higher refugia. It is also possible that push factors into the Atlas Mountains, the pre-European Kenyan Highlands and the Ulluguru Mountains of Tanzania included drought conditions in neighbouring lowlands.

These early mountain settlements took place largely through the process of internal colonisation Skalnik (1978:11). During this time isolation and inaccessibility were the principal influences on the lives of mountain people, making many mountain communities relatively more self-reliant and less dependent on external trade. The monetary economy was, therefore, weak. Resource activities were labour intensive with the major practices including subsistence agriculture, pastoralism and logging while mining, carting, charcoal burning and livestock rearing provided additional income. With the advance of industrialisation, mountain resources became increasingly valuable for commercial purposes. Deterrent mountain push factors of isolation and inaccessibility had to be tackled. In Europe and North America, communication routes were opened into mountains. This opened mountain areas to visitors from lowlands; in the Alps and, more recently, the Andes and Nepal, new communication systems facilitated migration out of these mountain regions. This period, therefore, with its institutionalised system of maintaining law and order brought about increased interaction between lowland and mountain communities.

During the industrial period mountain pull factors changed with the requirements and emphases of the time. The pull factors included an abundance of opportunities comprising commercial farming, logging, livestock, bee keeping, mining; sites for hydroelectric development; conservation of water for lowland needs and recreation and wilderness preservation. The lowland push factors included an inadequate resource base for supporting desired rapid economic development and generally high pressure and competition for available lowland resources.

In the Canadian Rockies, New Zealand Highlands, Kenya Highlands and the Indian hill settlements, commercial extraction of resources was primarily due to external colonisation (which involved occupation of a country's mountains by people from other cultures). In the Canadian Rockies and New Zealand Highlands the abundance of wood products, grazing land and minerals provided the initial pull for European settlers. The Indian Hill settlement stations and the "White" Highlands of Kenya attracted European settlement because of the cool climate, absence of mosquitoes at high altitude and relative dryness of adjacent lowlands in case of Kenya.

Developments during the pre-industrial and industrial periods led to the establishment of numerous resource practices in mountainous environments. As new resource practices were introduced, environmental problems mounted and reached crisis proportions in many mountain regions. In brief the crises led to massive farm abandonment by mountain communities, on the one hand, and massive migration into the mountains of lowland urban population, on the other. Two processes are responsible for the contrasting developments: the push to leave mountains for the better life indicated by urban lowland pull factors and the push to get out of the urban setting to the mountain setting. A further examination of the pull-push conditions since the 1970s is necessary to justify the cause of the crises.

In recent times the mountain push factors have included the following increasing incidence of severe erosion, devastating landslide and flood hazards; high population pressure, land shortage and fragmentation in tropical mountains; new expectations resulting from effects of mass media and increased inter-regional contact; increased awareness among some mountain communities of their limited political power and general backwardness; inadequate social services; increasing proportion of the elderly in mountain population structure and relatively marginal economic performance and land use conflicts. The relative impact of these push factors varies from one mountain region to the other. In general the push factors have resulted in massive migration, mainly of the young from the Alps, the mountains of Spain, Lebanon,

Nepal, Peru and Bolivia and to a lesser extent the Elgon and Bufumbiro mountains of Uganda.

While the mountain push factors were in play, lowland urban populations have fallen victim to some mountain pull factors. In recent times mountains, especially those in Europe and North America, have provided urban and lowland populations with opportunities for solitude and recreation; less polluted water and air; relative abundance and naturalness of landscape features including flora and fauna and possibilities for archaeological studies and other forms of research. As a result of these pull factors, large influxes of non-rural populations have occurred into even those mountain regions deserted by rural mountain communities. In the North American Rockies, the demands of urban population have resulted in a change from primary resource extraction to recreation and nature adoration.

The 'pull-push' concept has been used to explain the evolution and resource use in mountainous environments. In the discussions some important theoretical statements have been implied. It is important to summarise and specify the theoretical implications of the pull-push concept as they may provide a basis for understanding human interaction with different environments. The major underlying assumptions of the concept are:

(a) The choice to develop resources of an environment is determined by the balance between effects of pull and push factors in an environment. This assumption can be reduced to the following ratio:

$$\text{Resource Adoption (RA)} = \frac{\text{Pull factors (PL)}}{\text{Push factors (PS)}}$$

(b) When the benefits of exploiting the pull factors in an environment are greater than the costs of persevering push factors in that environment (i.e. $Ra > 1.0$), resource development will take place.

(c) When the costs of persevering push factors exceed the benefits of exploiting the pull factors of an environment (i.e. $Ra < 1.0$), an alternative environment will be sought for resource development.

(d) The choice to develop resources in one of two alternative environments takes place when the ratio of pull to push factors (Ra) of one environment (E1) is greater than the ratio of pull to push factors (Ra) of the alternative environment (E2 i.e. $RaE1 > RaE2$). When RaE1 and RaE2 are equal, both environments (E1 and E2) will be exploited (e.g. through nomadism, transhumance of shifting resource practice).

Human Acceleration of Natural Deterioration

This section examines in detail the consequences of various human interventions on the mountain environment and on humans themselves. Slope stability is of central importance in the use of resources in mountainous environments and the climax of resource use impact is often the tendency for increasing frequencies of rapid mass movement and associated floods. These effects exert a 'push factor' and because of an increasing socio-economic gap between lowlands, especially urban lowlands (and mountain communities), permanent emigration from mountain environments has reached crisis levels in many places.

Although tropical and temperate mountains differ in their types of use (sedentary cultivation in the tropics and sub-tropics and timber harvest in temperate climates) existing literature indicates that mountainous regions have been used for many purposes including human settlement, agriculture, grazing, water supply to lowlands, timber harvesting, hydro-electric power, mining, recreation, tourism and for military practices. These uses have had differing consequences which characterise mounting problems such as increasing human population; forest destruction, soil disturbance and pulverisation, overgrazing, large-scale excavations in mining and built-up areas and increased road density and traffic volume. These developments have upset the delicate equilibrium of most mountain slopes and have inadvertently accelerated severe soil erosion and slope failure due to mass movement.

Human alteration of potentially destructive natural processes is recorded in every mountain region, especially where resource use has a long history. Resource use practice may accelerate erosion due to mass movement by altering vegetation cover, enhancing infiltration capacities and increasing slope over basal support. Existing literature indicates that the effects of forest removal are the most outstanding. However, this assertion plays down the relative and complementary effects of increased pressure due to building; use of heavy trucks serving mines and logging areas and the effects of over-steepening and slope undercutting during mining, building site preparation and road construction.

Resource extraction reduces forest cover and expands, in relative terms, the area of grass cover and bare soils. Many researchers have concluded that forest removal is the primary accelerator of sub-aerial erosion and mass movement. This conclusion is based almost entirely on statements about the role of forests in mountainous regions. Effects of forest removal are well-studied: landslides in logged areas of South-East Alaska are attributed to forest removal while the increased incidence of landslides in Oregon cascades are attributed to logging. Similar findings were made elsewhere. In Nepal part of the Andes and in the East African highlands, increased incidences of landslide hazards and heavy losses of fertile top soils have been associated with rapid population growth (Eckhollm, 1975). Eckhollm Ians attributed the ecological degradation to the rapid destruction of forest caused by extensive cultivation and overgrazing. Because of forest destruction, he argued that the frequencies of landslides, erosion and sedimentation in Bolivia and Peru have reached crisis levels especially as they are accompanied by mass population shifts and farm abandonment. Similar instances of farm abandonment have been recorded in the mountains of Spain and Lebanon where cultivation has denuded the slopes and incidences of mass movement have more than doubled. Hewitt (1976) noted that the number of natural disasters due to mass movement and the degree of damage done to the environment in general have increased in this century. Since nature itself has not become more severe, the origin of these disasters must be sought in human practices, especially those which lead to rapid deforestation, overgrazing and extension of cropping to marginal slopes.

Although ecological degradation has been closely associated with deforestation, forest removal seemingly plays only an indirect role in inducing rapid mass movement. Taking into consideration the effects of relief and climatic energy forest removal may trigger rapid downslope movement by increased infiltration capacity and by reducing the soil anchorage which is otherwise provided by roots.

In contrast to the studies mentioned above, other researchers have found no correlation between forest removal and the incidence of rapid mass movement. For example, it has been found that landslides in the Razorback area of Australia predate the clearing of forests while the sliding in the coast mountains of Southwestern British Columbia occurs in both forested and clear-felled slopes. Other studies have also failed to establish any direct link between timber harvesting and the incidence of rapid mass movement. No attempt has been made to account for the contradictory results from the two groups of research.

Forest removal reduces infiltration capacity which subsequently increases surface runoff and sub-aerial erosion. Selby (1976) found that in South Auckland District, New Zealand infiltration rates into forest soils were three to four times as great as those into grassland soils. Decreased infiltration and increased runoff lead to lower rates of soil saturation. Overburden pressure may also decrease because the conversion from forest to grass results in lower biomass. Consequently, slope stability increases after forest removal. Since forest removal *per se* is not often accompanied by slides, it is logical that the effects of root decay on infiltration and on soil anchorage are either inadequate for stimulating slides or are counteracted by the combined effects of lower overburden pressure and sometimes rapid weed and grass regeneration following sudden exposure of soils to sunlight. Root decay, in any case, is a slow process and effects take long to manifest themselves.

Forest removal accelerates mass movement when other practices complement its effects (e.g. channels left by decaying roots, reduced soil anchorage and possible increased infiltration rates where soil tension cracks have developed). Hong Kong is one of the most significant sites where the effects of acute land shortage have doubled the frequency of mass movement and its resulting damage toll.

Hong Kong's problems range from slow imperceptible creeps of subsoil to rapid earthflows; from isolated boulder falls to wholesale debris avalanche and from surface washouts to deep-seated landslips anaslumps. This ever-rising frequency of the mass movements has been attributed to anthropogenic excavations and cuttings embarkings. He noted that the huge excavations cuttings have resulted from preparation of building sites on ever steeper slopes, road construction and subsoil extraction for reclaiming land from sea.

The drastic effects of road construction, slope undercutting, excavations and heavy trucks are well-illustrated in the Swan Hills of Alberta, Canada where the presence of oil has resulted in access road construction, heavy trucks and drilling. Lengelle (1976:3) noted that in the Swan Hills the "cumulative effect of poorly consolidated bedrock, heavy precipitation and undue careless disturbance of the land has caused intense gullying which increased as time passed. In addition, landslides have been caused by hastily-made roads".

In the foothills of Alberta, increasing resource use pressure has created public concerns about environmental deterioration. Large-scale logging, coal mining, oil extraction and natural gas exploration, highway construction, domestic grazing, recreation activities and hydroelectric power development all converge. The cumulative effects of these converging activities on slope stability can be easily anticipated.

More attention should be paid to the incidence and frequency of small, often imperceptible, tremors and their effects in triggering slides. Most mountain regions lie within heavily faulted and jointed structures where the potential for and frequency of tectonism are high. Human disturbances in conjunction with tremors may combine to account for the high incidence of mass movements.

Case Study: The Mountains of Elgon and Bufumbiro

A final illustration of the human acceleration of natural deterioration may be derived from case studies of the foothills of Mountain Elgon and the Bufumbiro Mountains of Uganda. The study identifies major resource use problems and assesses the cause and severity of landslide hazards and the extent of environmental deterioration in the mountainous regions of Uganda.

The mountainous regions of Uganda, especially that region in the Bufumbiro and Elgon mountains possess almost all of the known symptoms of a high energy and sensitive environment. The regions lie 1,500 metres above sea level and consist of a number of hills and steep slopes rising abruptly to over 300 metres above the surrounding lowlands. Most of the steep hills display a convex-concave profile with slope angles varying from 15 to 45 degrees upslope. These areas receive one of the highest annual rainfalls in Uganda averaging 1,500 mm. Prolonged low-intensity rains are more common than occasional short-lived high intensity rains. Temperatures are relatively cool and rates of evaporation in the hills are lower than in higher mountain slopes where more windy and chilly conditions are prevalent especially in the Elgon mountains. In most of the wetter months, there is an excess of precipitation over potential evaporation, a condition which is responsible for a number of slides. While the soils in the Elgon foothills consist of deep red, sandy clay loams usually showing little tendency to erode, those on the Bufumbiro retain alluvial deposits and sub-angular structures which promote rapid infiltration and minimise sub-aerial erosion. The soils are fertile and in the Elgon foothills they support Arabica coffee and banana plantains. The climatic and soil conditions account for the high density of agricultural settlements and the reduction of slope coverage from luxurious vegetation cover to almost bare ground in both the Elgon foothills and the Bufumbiro Mountains.

Population growth has been enormous in the mountain regions of Uganda, reaching 3.8% in 1948 and 2.9% by the 1969 and 1979 censuses; the present population density at over 320 per square kilometre is one of the highest in tropical Africa. These high densities are reflected by the diminishing sizes of landholding per household, size of buildings and school playgrounds, disappearance of crop-rotation and land fallowing, a firewood shortage, encroachment of agriculture into steeper and marginal slopes and increased outward migration and numbers of landless people. The average farm size in the Elgon foothills is less than 1 hectare as opposed to the Mbale District average of 1.8 hectares and a national average of over 3.0 hectares. In Kabale District average farm size is less than 0.7 hectares. Land shortage and fragmentation were ranked first in the order of seriousness among the perceived resource use problems of the region, especially in Kabale District, where serious land-related conflicts are quite common. On these small (often fragmented) holdings, farmers grow coffee as the major cash crop, banana plantations as the staple food, maize, finger millet, wheat, sorghum, pulses and cabbages. In other mountain slopes of the Ruwenzori even cocoa is grown in addition to coffee and a wide variety of subsistence crops.

Resource development in these mountainous regions has raised the strain on the steep slopes and potential for damage has also increased with the ever-rising human numbers, artifacts and affluence. The strain on the slopes is due to excessive cultivation and soil disturbance; local slope oversteepening due to excavations; increased overburden pressures due to buildings and almost total forest removal. Gravitational energy has consequently been transformed into geomorphic work. Slumping has become more frequent and devastating landslides occurred in 1921, 1926, 1970, 1974, 1981, 1984 and 1986. Substantial losses of life, property and the breach of family ties resulted from those slides. Between 1926 and 1970 about 100 people were estimated to have been killed in over 20 homes affected by the landslides. Since 1980 every rainy season has recorded incidents of devastating slides in the Elgon and Bufumbiro region. Although the frequency of these slides is relatively low, the whole environment and its community are threatened more than ever before. Each slide that occurs creates conditions for subsequent slides and prevention attempts are simple and inadequate involving the capital or technological investment. While terracing is common in many parts of the Bufumbiro Mountains, it is generally lacking in the Elgon mountains. Thus, even within one nation, there is little sharing of techniques and experiences. This gloomy picture must be addressed because as more of the younger generation become urbanised and affluent, homes high up in the mountains will continue to be abandoned to the aged, who are incapable of constructing terraces and taking other conservation measures.

Conclusion

In reviewing trends and conditions in mountainous regions, significant differences can be seen between resource practice in temperate regions and in the tropics. Still, all mountainous regions appear under threat, though the situation is more grave in tropical countries due to problems of population growth, increasing poverty and the subsequent inability to prevent environmental deterioration. Methods for controlling environmental deterioration in mountainous regions exist, but experiences and techniques cannot be shared effectively if the plight of mountain communities is not appreciated and tackled pragmatically. This chapter is a modest step towards attracting attention to the unique problems and experiences of mountain regions.

5 Environmental Stress and Conflicts in Africa: A Case Study of African International Drainage Basins

C. O. Okidi

Introduction

The current dominant international view of Africa is that of a continent beset by environmental and social problems which result in widespread starvation and warfare. While this view is exaggerated, Africa has in recent years been confronted with drought, food scarcity and starvation, partly caused by scarce and maldistributed rainfall. Social conflicts, on the other hand, have been caused by a variety of political and social tensions which may have been aggravated by coincidental socio-economic tensions brought by widespread poverty and the squalor associated with drought. Thus, the two salient problems of Africa — drought/famine and social conflicts — are often concurrent rather than bearing any causal relationship. A critical factor is water shortage which leads to poor agricultural productivity. Africa is generally the driest continent on earth. It has a reticulation of some fifty-four drainage basins, including rivers, which either traverse territorial boundaries or form or part of such boundaries, but which alone cover approximately half of the total area of Africa. Unfortunately, only approximately two per cent of the total African waters are utilised, with the remaining 98 per cent left to fill the oceans, a colossal underutilisation requiring redress (Okidi, 1988).

Africa should, therefore, focus attention on control, apportionment and utilisation of the waters of its drainage basins. It would be expected that after the scarcity of rainfall in the late 1970s and mid-1980s there would be concerted efforts to harness the waters for agricultural and domestic purposes; however, since the quantities of the waters are finite, such measures of control, transfer and utilisation could generate international tension and conflicts. In 1978, for instance, Egypt said it would not allow Ethiopia to harness the Blue Nile for irrigation in the Ethiopian plateau as that would diminish the volume of water reaching Aswan Dam. In return, Ethiopia issued a terse warning that she had the right to exploit her natural resources and to defend that right, even if it meant going to war.

The only way to avoid conflicts in the utilisation of the waters of international drainage basins in Africa is by cooperation and collaboration among the basin states in the management of such waters.

Environmental Management and Stress

Environment may be defined as the total context within which natural resources exist and interact as well as those infrastructures constructed by man to facilitate socio-economic activities (Okidi, 1984:1988). A recent working group of the International Union for the Conservation of Nature (IUCN) in its Commission on Environmental Law (CEL) attempted a tentative definition of the environment as "the totality of nature and natural resources as well as cultural heritage and infrastructures essential for socio-economic activities." While agreeing with the general spirit of the former definition the majority of the Working Group noted that central to the notion of environment was nature encompassing the earth's geosphere, biosphere and associated processes which can be used for socio-economic activities by human beings and other species. Thus, the two definitions agreed on the central tenets of the concept of the environment. First, the environment is the natural context and not any specific resource sectors. Secondly, the various resource sectors such as water, forest, minerals, energy, human beings and air are simply components of the environment. Thirdly, the structures constructed to facilitate socio-economic activities, such as settlements, transport and industrial production are also part of the environment. In fact, it is the latter set of components that make planners and architects environmental scientists.

In pristine settings, the various components of the environment interact, changing one another over time in a form of ionomorphosis, thus maintaining an eternal balance. But human intervention through utilisation of the natural resources as well as the emission of effluents from socio-economic activities have changed the nature of environmental processes. In some cases, the consequence is direct degradation of the environment or its components and in other instances, the result has been depletion or even extinction of some of the natural resources or specific species. These interventions may be generally called environmental stresses.

Identification of the threat to the environment led to the formation of organisations to control the trend. Early organisations especially from the 19th century focused on the different forms of stress on specific organisms and urged their protection (Johnson, 1981:17-23). The environmental movements urged people to manage the natural resources sustainably and to avoid waste of non-renewable natural resources; the critical goal in environmental management is sustainable utilisation of natural resources (World Commision on Environment and Development, 1987).

The underlying notion of sustainable environmental utilisation is the antithesis of environmental stress. It is the concept of sustainability which associates sound environmental management with development. In this context, development is considered as:

> the process by which a country provides for its entire population all basic needs of life, such as health and nutrition; education and shelter; and to provide every one of its population with opportunities to contribute to that very process through

employment as well as scientific and technological construction. Secondly, it is the process by which the national governmental authorities construct and maintain productive mechanisms and infrastructure which diversify and perpetuate the productive base of the country, such as agriculture and industries so as to ensure that the society can overcome the pressures and necessities of the national and related economic system for the present and for all future times. (Okidi, 1984: 92-93)

What makes the relationship between environmental management and development critical is the fact that development requires sustainability in the mobilisation and utilisation of natural resources. Thus, the critical concept for development is environmental conservation because without it sustainability will be jeopardised. Social and political conflicts result in the absence of food and nutrition, safe drinking water and healthy habitats, among others. However, the conflicts may exist in areas with plenty of water and other natural resources, but the situation is exacerbated by maldistribution of natural resources, particularly water as has been evident in Sudan, Ethiopia, Mozambique, Angola and the entire Sahel zone.

It is for these reasons that the suggestion which emerged from the 1970s that environmental conservation was antithetical to development needs was grossly misleading (Kasdan, 1971: 454). Most Developed Countries (MDCs) agree with the thesis of the Brundtland Report that sound environmental management is essential for development. One of the crucial agenda for Africa is how to mobilise the existing natural resources for sustainable development (Mwakwo, 1982 and Ojwang, 1990).

Environmental stress occurs when natural resources and the context in which they exist are threatened either by degradation, reduction of quality, or depletion. Soil erosion, water pollution, depletion of surface or ground water, fishery resources, wildlife, petroleum, oil or mineral reserves and pollution of air, among others, are sources of environmental stress. Similarly, depletion of components of the environment which are not directly used such as the ozone layer also creates environmental stress. The stress on water resources, however, creates a particularly acute problem because water provides life to plants and animals and it is the basis of agriculture. This is why we have to look to Africa's drainage basins where most of the unused waters still exist.

Case Studies of Africa's Drainage Basins

Two problems, drought and famine and the rising costs of imported hydrocarbons for energy have been associated with Africa's development problems during the past two decades. Both of them have focused the attention of some students of development on African rivers because the rivers carry water and they are capable of yielding hydro-electric power.

Of water, Krishnamurthy (1977: 372) said:

Estimates of Africa's total water resources vary from 3,400 billion m^3 of water to 4,600 billion m^3. An analysis of the measured streamflow in African rivers indicates that the total quantity of surface water in Africa's rivers and lakes is in the order of 2,480 billion m^3, the difference between this and the aforementioned estimates being a measure of ground water resources. More than 50 per cent of the total water resources of the continent are in one single basin, Congo/Zaire (1,325 billion) and another 25 per cent in seven other river basins such as Niger (200 billion), the Ogooue, Gabon (149 billion in Gabon), the Zambezi (104 billion at C. Ana), the Nile (84 billion), the Sanage (Cameroon, 165 billion), the Chari-Logone (Chad, 43 billion) and the Volta (40 billion).

There are several other major international rivers such as the Senegal and Limpopo which also flow through areas susceptible to drought and famine. All these waters could be controlled and transferred for agricultural production.

The hydro-electric potential of Africa's rivers is also reported to be vast but largely untapped. Again Krishnamurthy (1977:510) reported that approximately half of the world's potential hydro-electric power is in Africa. He added that despite this high potential, the installed capacity of hydro-power in Africa is only 5.6 per cent of the total and that the ratio of energy generation to the exploitable potential is only 2 per cent. Therefore, there is a considerable reserve of renewable energy which African countries might harness and utilise instead of expensive imported hydrocarbons (UN 1988:28).

Water and power may be relatively abundant but these would not facilitate agriculture unless adequate land is available. A 1982 report by L.A. Odero-Ogwell, Secretary to the World Food Council, said that Africa's agricultural land is still plentiful and compares favourably with the regions of Asia and Latin America (Table 5.1).

Table 5.1 Potential agricultural land: Africa, Asia and Latin America

Region	Potentially cultivable	Presently cultivated (m ha)	Presently cultivated Arable area (m ha)	Average Period %(yrs)
Africa	803.7	193.7	24	3
Southwest Asia	46.0	50.9	110	2
Southeast Asia	324.8	270.5	83	3
South America	914.9	85.2	10	4

Source: Odero-Ogwell, (1982): 6

If water, land and renewable sources of energy are available, then the constraint to agriculture must lie elsewhere. Odero-Ogwell (1982: 6) pointed to the failure of African countries to adopt technological packages particularly irrigation as the main bottleneck. One illustration of this contention is that by 1977 only 1.8 per cent of the cultivated land in Africa was irrigated compared to 28 per cent in Asia and 6.2 per cent in Latin America. Evidently, if Africa is to improve its agricultural productivity, especially in the countries with perennial drought and famine as in Ethiopia, Somalia, Sudan, Mozambique and the Sahel region generally, then there must be more concerted control of river flows and transfer of water to irrigated fields.

Both irrigation and hydro-power production require control of the river by damming and this leads to a change in the flow regime. But while hydro-power generation allows for restoration of the flow regime after the dam fills, abstraction of the basin waters for irrigation reduces or diminishes the quantities flowing drowstream. Both instances may lead to conflicts with irrigation being more severe.

There is a need for basin states to create a forum where every state intending to construct works should inform other states if it plans to change the flow regime of an international river or diminish the quantity of basin waters. But a forum for consultation, functioning only as a mediator, is inadequate due to its *ad hoc* character which does not allow broad and independent planning. What is desirable is a framework for integrated, multi-objective basin-wide planning. This facilitates "a mixture of activities producing marketable goods (electrical energy, agricultural products, industrial goods and transportation services, among others.) and essentially non-marketable services for which public funds are required (UN, 1988: 2). The size or cost of the specific development works is a different issue and informed commentators agree that it is possible to create an institutional and operational arrangement which is sensitive to the exigencies of cost and environmental protection.

There is, therefore, a clear rationale for African countries to mobilise the waters of their drainage basins for irrigation, safe water supply, navigation, fisheries development, flood control and environmental protection. Secondly, there is a necessity for the basin states to arrange integrated and multipurpose basin management within an institutional framework. Such a framework would provide a forum for information exchange, consultation, resolution of possible conflicts and integrated and comprehensive management of the basin and its resources.

The African Drainage Basins

There are 54 international drainage basins in Africa, constituting a complex and widespread reticulation of water bodies. Among them are rivers such as the Nile, Congo, Senegal, Niger, Limpopo and Zambezi. All these were significant in history because they were used either for demarcating colonial spheres of influence or for navigation and colonial penetration. Only in the case of the Nile was the river known also for irrigation which provided the lifeline for the ancient hydraulic civilisation. We have also seen that the quantities of water discharged daily by these rivers into the oceans are colossal.

But rarely have African conflicts linked directly to use of water resources. The problem, however, is that there are a number of actual or planned water use programmes which may lead to conflicts. Given the fact that widespread drought and famine may lead to increased water use such trends may, in fact, exacerbate existing low-level conflicts and violence and lead to acute conflicts.

The Nile, Kagera and Lake Victoria

The Nile is the second longest river in the world after the Mississippi while Lake Victoria is the second largest fresh water lake in the world after Lake Superior (assuming that the waters of the latter can still be considered fresh). Both basins are bordered in different degrees by nine states, namely, Burundi, Egypt, Ethiopia, Kenya, Rwanda, Sudan, Tanzania, Uganda and Zaire. The status of Rwanda and Burundi is unique in that they are brought into the basin by virtue of the Kagera River which drains into Lake Victoria. The entire basin area is estimated at 2.9 million square kilometres which represents approximately one tenth of the African continent.

Lake Victoria sits on the eastern African plateau at an elevation of 900 metres, surrounded by relatively low-lying land averaging 1,100 metres around its shores. The total area of the lake is approximately 68,000 km^2 of which the Kenya's share is about 10 per cent, Ugandan 40 per cent and Tanzanian 50 per cent. Surface water contributed by rivers comes entirely from Kenya and Tanzania. The main rivers from Kenya are the Kuja, Awach (or Kibuon), Miriu, Nyando, Yala, Nzoia and Sio and from Tanzanian the Mara, (which crosses into Kenya) and the Kagera on the southern side, respectively. The Kagera River is significant because it drains the territories of Rwanda and Burundi as well and because it extends the limits of the Nile Basin further to the south-west.

Linked to the lake in Uganda is the Nile, its only drainage outlet from Lake Victoria. Here, at the Ugandan industrial urban centre of Jinja, the exit discharge passes through the Owen Falls Dam which was commissioned in 1954 to provide water security for Egypt and to produce hydro-electric power from Uganda.

The Victoria stretch of the Nile flows from Jinja to Lake Kyoga; between Lake Kyoga and Lake Albert is the Kyoga Nile. It exits from this lake on the northern toe as the Albert Nile and is the only outlet from that lake. It is at Lake Albert that the Democratic Republic of Congo – DRC (formerly Zaire) as a basin state of the Nile makes its contact through the River Semiliki which, flowing from DRC, enters the lake at its southern toe. From Lake Albert to Malakal in the Sudan, the river is known as Bahr el Jebel, part of the White Nile. This is the area of the well-known Sudd of Southern Sudan, where much water is lost through evaporation and

Environmental Stress and Conflicts in Africa

Fig 5.1: The 'Zambezi Plan', an ambitious South African water project in the SADCC region

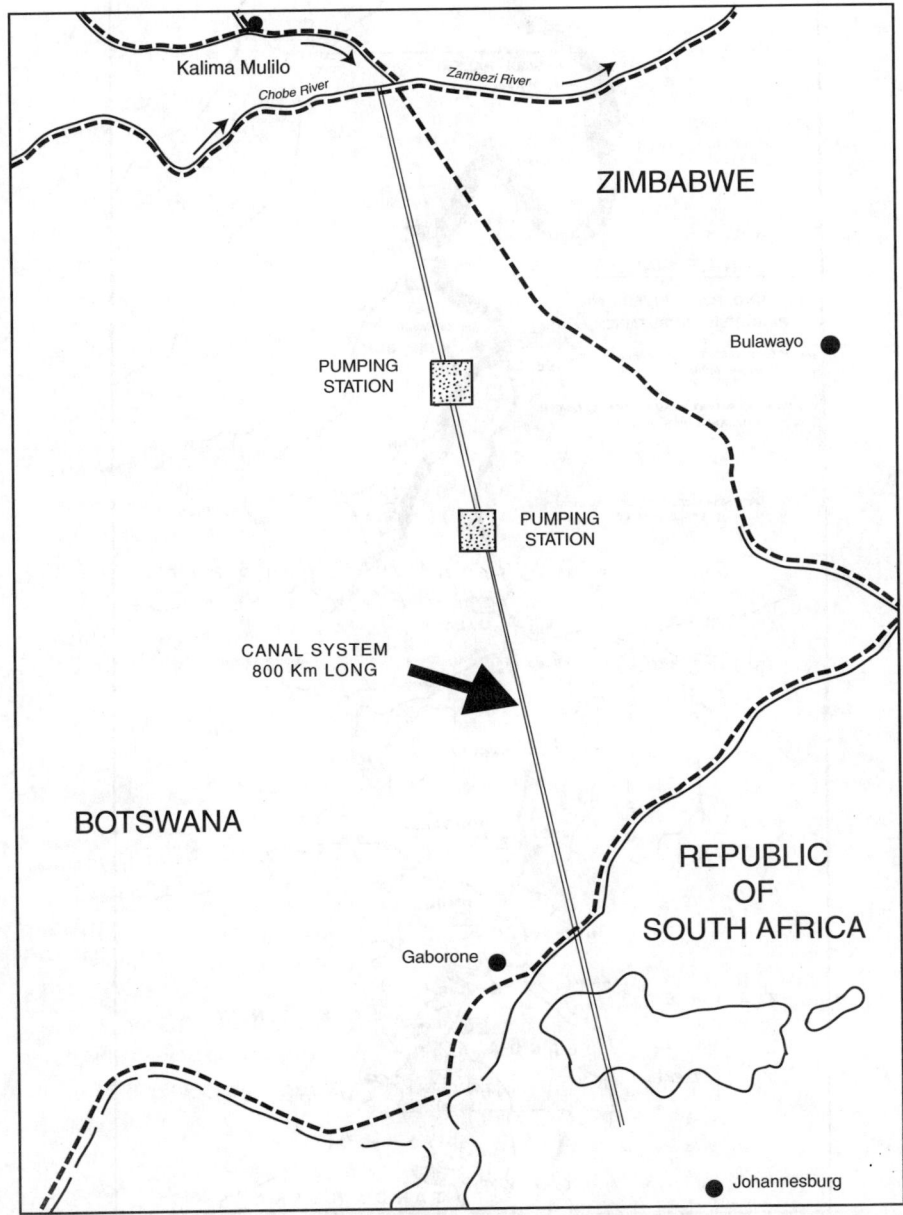

Fig 5.2: The Nile Basin

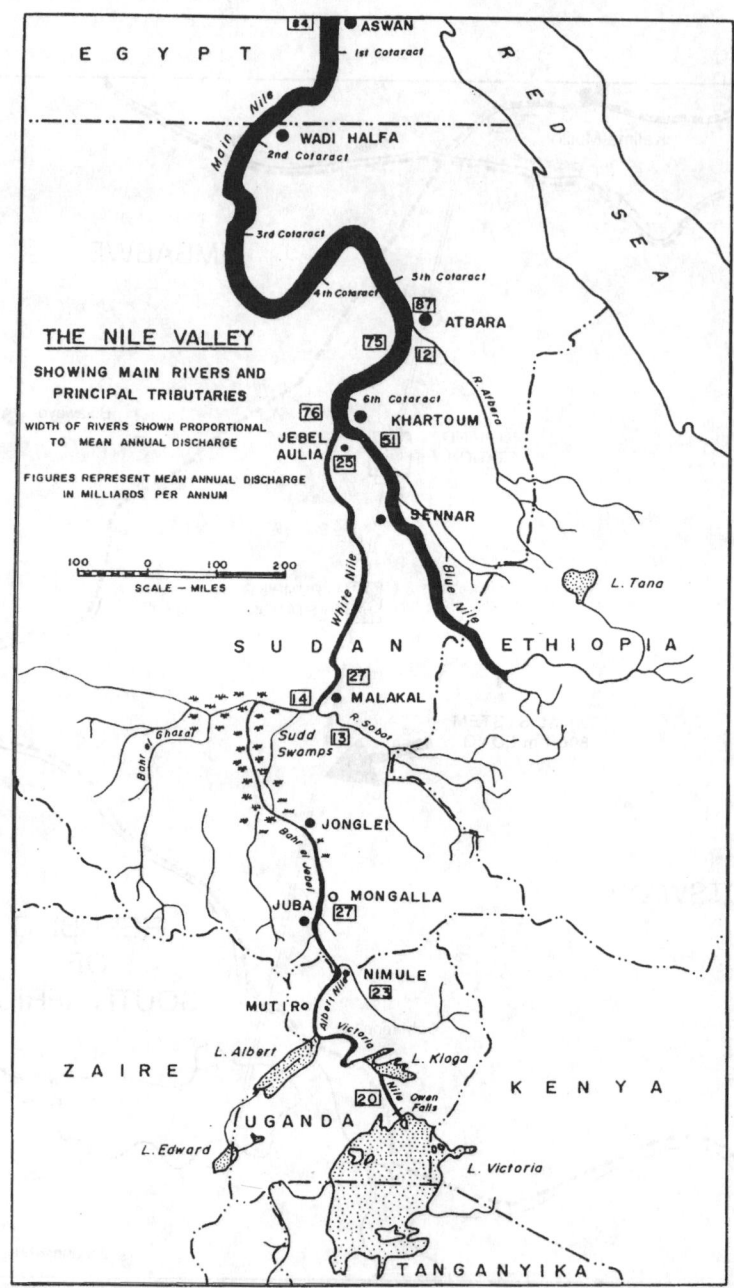

soakage; it is this abundance of water lost through evaporation which necessitated the construction of the Jonglei Canal. Beyond Malakal, the White Nile flows directly northwards up to Khartoum where it is joined by the Blue Nile, which drains Lake Tana in the Ethiopian highlands. Further north it is again joined by the Atbara, also flowing from the Ethiopian highlands. It makes one gentle loop southward then northward crossing the border at Wadi Halfa into Egypt, where it is ushered gently to its full length, estimated at about 4,180 miles from the Jinja exit.

For purposes of international legal and policy perspectives on the Nile basin, there are further geographical-cum-hydrological facts that should be examined. The volume of water each of the riparian contributes to the Nile might be taken into account in the decision of how much water a riparian might properly divert for its national use. In terms of proportions, Gamal Moursi Bard (1959) of Egypt estimates that of the total annual Nile discharge 84 per cent is contributed by Ethiopia and only 16 per cent comes from the "Lake Plateau of Central Africa'. Garretson, (Garretson 1976:259) however, offered the estimate: at the peak of its flood (April - September), the Blue Nile alone supplies 90 per cent of the water passing through Khartoum, but that in the low season (January – March) it provides only 20 per cent. A broad estimate would be that the Lake Plateau of Central Africa contributes between 20 and 25 per cent of the water flowing north of Khartoum while 75 to 80 per cent is contributed by Ethiopia.

To Egypt, a lower riparian dependent on Nile waters for its survival, the contribution from Lake Victoria on an annual basis must be minute compared to that of Ethiopia. However, the Lake Plateau water supply is reliably steady throughout the year because of the storage at Owen Falls Dam. One commentator noted that it was because of the annual flooding due to the Ethiopian contribution that Egypt decided to construct the dam at Aswan to regulate the flow and provide over-year storage for Egypt rather than depend on the remote reservoirs of Lake Tana and the Central African lakes.

Sudan preferred regulation of the flow of Nile waters by a series of smaller dams to the Aswan model which would, in any case, only assure steady supply to Egypt and not into Sudan: also the reservoir would extend into Sudan to flood the urban centre of Wadi Halfa. Besides the flow control, the dam was to be used to generate 10,000,000 kWh of hydro-electric power.

Egyptian interests prevailed and the dam was constructed between 1961 and 1964. The degree of regulation of flow to meet year-round irrigation needs which has been accomplished in fact by the Aswan High Dam needs to be ascertained because it may have a bearing on Egypt's dependence on the waters of central African lakes at present. Also, the volume flowing out of Ethiopia is highly susceptible to variation depending on that nation's plan for future development.

The Sudan contributes no water to the Nile and, in addition to what it consumes for irrigation, there is some volume lost through evaporation and in the Sudd. Apart from precipitation, only the southern and eastern parts of the Lake Victoria basin contribute to the water of the lake. The lake's contribution to the Nile is easily determined by measuring the total discharge at Owen Falls Dam. For purposes of policy in Kenya and Tanzania, the exact proportion of the annual outflow at Owen Falls Dam which is contributed by each country separately needs to be established. This line of analysis should use percentage of volume contributed rather than absolute quantity or volume because when an upper riparian diverts water from an international basin flowing through its territory, the fear of deprivation or injury expressed by a lower riparian is clearest when expressed in term of proportions.

The final question to be considered here is whether the Lake Victoria and River Nile systems constitute one basin. A drainage basin has been defined as, "the entire area, known as the watershed, that contributes water, both surface and underground, to the principal river, stream or lake or other common terminus." (ILA 1966:486).

While the Nile and its tributaries flow directly into the Mediterranean Sea, Lake Victoria drains directly into the Nile, thus contributing water to that one terminus. Therefore, for the purposes of this discussion, the Lake Victoria Basin and the Nile Basin constitute one drainage system with three parts. Regulation of flow through the Owen Falls makes for some semi-autonomy for Lake Victoria's basin which could be managed as a subpart, but it can be argued that hydrologically, a basin may be dammed where it is most convenient. This may explain why the countries around Lake Victoria, especially Kenya and Tanzania, might have unique clusters of interest in the lake which could be poised against those of lower riparians, Sudan and Egypt, in any attempt to work out an up-to-date legal regime for the Victoria and Nile waters. Similarly, it is for these reasons that the Kagera can be legitimately considered a drainage basin since it constitutes a principal river. If there is any kind of conflict it should occur between two or more of the nine basin states.

Pressure Points in the Nile-Victoria-Kagera Basins

A pressure point may be understood as the place or the issue where environmental stress occurs. Simply stated in this context, environmental stress means the degree of use of the water resources which diminishes the quantity or is likely to adversely affect the quality and the construction works which may change the flow regime and, therefore, cause flooding or alter the seasonal flow of water. Activities causing environmental stress within these basins include irrigated agriculture and damming for hydro-electric power.

Irrigated agriculture

Irrigation in Egypt

The question of irrigated agriculture in these basins causing environmental stress revolves around Egypt.

For more than 7000 years, Egypt has enjoyed the uninterrupted use of the Nile for irrigation. According to one commentator, irrigation is vital to Egypt (where no food or economic crops can be grown except under irrigation) and essential to the agricultural economy of the Sudan. Until very recently, irrigation was much less important for the other seven riparians. For Egypt, the study done for the British Government recounted irrigation needs in Egypt with the following illustrative figures:

> Cultivated areas in Egypt amount to 6,150,00 acres (1955), the Egyptians maintain that given water, the area could be extended to 15,000,000 acres. Irrigated cultivated areas in Sudan amount to about 1,400,000 acres (exclusive of basin irrigation); further 5,500,000 acres have been surveyed and could produce crops under irrigation. This total area of irrigation would require water in excess of the amount available in the Nile system, even when all conservation works are in operation (Uganda, 1957:3).

Table 5.2: Amount of water needed for irrigation in Egypt

Water year	Irrigated cropped area (feddans)	Total discharge downstream the Aswan High Dam (milliard M^3)
1968/69	10,740,063	53.116
1969/70	10,732,061	54.852
1970/71	10,747,096	55.364
1971/72	10,742,512	55.955
1972/73	10,863,000	55.285
1973/74	10,976,000	56.295
1974/75	11,333,172	56.245

Reports ten years later showed a general trend in the area under irrigation and the corresponding amount of water essential for the irrigation. Relative to the cultivated areas too much water was being released at Aswan High Dam, which suggests inefficient usage of water.

Egypt vs. Ethiopia

All upstream waters originate from the Ethiopian highlands through the Blue Nile and the Atbara as well as from the White Nile, particularly Lake Victoria. Ironically, there is today no agreement between Egypt and Ethiopia or Egypt and the Lake Victoria Basin countries to secure commitment on the waters reaching Egypt. About 85 per cent of the water flowing down the Nile past Khartoum originates from the Ethiopian highlands via the Blue Nile. Most of the rest comes from Lake Victoria which is also a stable, year round source of water. This is in contrast to the Blue Nile volume which comes down over a short peak period from April to September. Yet Egypt seems to have taken the water for granted asserting that any measures to utilise the water for irrigation would be unacceptable. According to an Egyptian newspaper report:

> Egypt and Sudan were studying with great interest feasibility studies being conducted by the USSR around Lake Tana, where about 85 per cent of the Nile water originates. Egypt will not allow the exploitation of the Nile Waters for political goals, will not tolerate any pressure being brought to bear on it, on fomenting any disputes between itself and its neighbours *Akhbar El Yom* (Cairo) March 13, 1978.

In response, the Ethiopian Government issued a series of statements emphasising that "Ethiopia had all the right to exploit its natural resources." and added that Egypt had gone ahead and built Aswan High Dam which was to depend on security of water from the Blue Nile "without even consulting Ethiopia." But those firm statements from Ethiopia did not deter Egypt. On June 6 1980, while addressing the Second and Third Army officers, the then President Anwar Sadat asked his officers to be ready to fight Ethiopia should the latter interfere with the flow of the Blue Nile. He said to the soldiers:

> If Ethiopia undertakes any action that will affect our full rights to the Nile waters, there is no alternative to the use of force... we will retaliate when something happens but we have to be ready with plans and alternatives to firmly stop any such action. (*Daily Nation* (Nairobi) 7 June, 1980:2).

Clearly this is evidence of an acute conflict, though it could be considered a passing political thought if not backed at the technical level by experts. A similar call, however, was expressed a decade later by a prominent scholar-turned-diplomat Professor Boutros Ghali, then Egypt's Minister of State for Foreign Affairs (and later UN Secretary General), who is reported to have declared that "The next war in our region will be over the waters of the Nile, not politics".

These statements give confusing signals. The Ethiopian leader made an official visit to Cairo in April 1987 at the end of which he and the Egyptian President issued a joint communique in which the first substantive paragraph "reaffirmed the strong political will of the two governments and two peoples to enhance bilateral relations" adding,

> they agreed that as Egypt and Ethiopia are part of the Nile Basin countries, special emphasis must be accorded to the promotion of cooperation and, in particular, in the field of the rational utilisation of the waters of the Nile to the benefit of their two peoples and all the people of the area" (Reported in the *Ethiopian Herald* 14 April, 1987:1).*

It may well be that Egypt is nervous about its future capacity to feed its increasing population and that the security over water is threatened. But two points are not yet clear. First, once the goodwill is expressed as in the above communique why was it not followed up with friendly negotiations rather than the call for war? Secondly, Ethiopia had been unable to feed its own population due to droughts causing crop failure (not to disregard perennial civil strife); how extensive were her irrigation plans which seemed to provoke the Egyptians to nearly irrational contemplation of a war? Were the programmes for irrigation so vast that once started no water would flow down the Blue Nile?

The questions seem to engage considerable international interest. Alistair Matheson (1986:38) reported that Italy had actually designated Alessandro Palmieri as project engineer to commence the first phase of an extensive water project in Ethiopia. Palmieri's view was that while dams would eliminate the area's dependence on seasonal and erratic rainfall it would hardly affect the Nile. In his view, a dam and associated irrigation works would affect only 2 per cent of Blue Nile's annual discharge.

Original studies which proposed the a full works and possible irrigation were done by the United States Bureau of Reclamation in 1964, when the US Government was a close ally of Ethiopia. According to Guariso and Wittington (1985:3)

> The Bureau concluded that there are no lands along the Blue Nile which can be irrigated; the proposed irrigation schemes are located primarily in the plateau valleys at elevations between 335m and 920m chiefly (1) around Lake Tana, (2) on the Sudanese-Ethiopian border, and (3) on the Egypt with an annual water requirement of roughly 6 billion $(10)M^3$.

These reports suggest that there may be additional irrigable areas in other basins. For instance, it is pointed out that while the Diddessa River, a tributary of the Abbay River, drains 34,000 km^2, its waters could irrigate about 53,000 hectares of virgin land.

---

* Reported in *The Ethiopian Herald* (Addis Ababa) 14 April 1982.

It is this range of possible uses that seem to upset Egypt. But surely, it is not to be assumed that Egypt intends that Ethiopia desist completely from using the water for irrigation. Ethiopia has been the subject of rather demeaning worldwide campaigns for food to feed its people. Evidently it should turn to the natural resources of its territory to provide for its people's needs, including the development of irrigated agriculture although Egypt may fear that Ethiopia would consume all of the Blue Nile's water. Guariso and Whittington (1985) have examined this point in detail and found no conflict between Ethiopia's interests *vis-a-vis* those of Sudan and Egypt. They asserted that even if the plans of the US Bureau of Reclamation were fully implemented they would not adversely affect the water supplies of Egypt and the Sudan. Guariso and Whittington (1985:4) drew five conclusions of which three are relevant to this discussion:

i. Full development of the Blue Nile in Ethiopia would effectively end the annual Nile flood.

ii. If Ethiopia were to develop the Blue Nile basin, the amount of water available for agricultural use in Egypt and Sudan would actually increase, because the river could be more easily regulated downstream, thus reducing storage requirements in Sudan and evaporation losses from the Aswan High Dam Reservoir.

iii. There is little, if any, conflict between the riparian states on the broad policy of how such reservoirs in Ethiopia should be operated.

In light of this analysis by Guariso and Whittington, why then would Egypt sound the war alarms, before holding detailed discussions with Ethiopia? Possibly, Egypt has taken the Nile waters for granted and desires the *status quo* despite drought and food shortages in Ethiopia.

Irrigation in Kenya

Kenya is basically an agricultural country even though irrigation has not played a central role in the economy. But if Kenya were to embark on major irrigated agriculture, water would not be the main problem; she has one-tenth of Lake Victoria's 68,000 km^2 as part of her territory. Besides, she has seven major rivers, namely: Kuja, Awach-Kibuon, Oluch, Miriu, Nyando, Nzoia and Yala flowing into the lake throughout the year.

Of all the East African countries, Kenya contributes the largest volume of water into Lake Victoria via the rivers listed above. It is estimated that the rivers contribute about 35-40 per cent of the total water reaching Lake. This is a significant contribution to the storage head for which the Egyptian government agreed with the British colonial government to construct Owen Falls to facilitate year-round control and storage, especially during the floods from the Blue Nile from April to September. For these reasons, Egypt might be nervous about the quantities of water flowing into Lake Victoria from Kenya. Thus, in a recent interview with senior Egyptian authorities in Cairo, Tony Walker (1988:11) was told that: "The day that Kenya decides to use the water of Lake Victoria we'll have less water in Egypt. One litre of water used for their irrigation will be deducted from water received in Cairo". This position is clearly an exaggeration but it is a pointer to the fact that Egypt would be anxious about any massive irrigation scheme in Kenya.

If the utilisation of water in Kenya would concern Egypt, then the planned irrigation within the Lake Basin Development Authority (LBDA) area would worry Egypt as an environmental stress. The LBDA, created through an Act of Parliament in 1979, conceived two broad categories of irrigated agriculture. First, it plans what is called *lake shore irrigation schemes*, confined within 20 kilometres of the lake's shore. The shore area is approximately 157,000 hectares of which 77,900 hectares are in the southern shores while 79,400 hectares are in the northern shores of the lake. The LBDA has surveyed the catchment areas of the various major rivers flowing into Lake Victoria. Such irrigable areas total 200,000 hectares as shown in table 5:3.

Table 5.3: Estimated irrigation potential of rivers in the Lake Basin

River	Location/Area	Irrigable area (ha)
Nzoia	Middle/Lower	65,000
Yala	Yala Swamp	15,000
Sondu/ Miriu/ Kibos	Kano Plains	60,000
Kuja/ Migori	Lower	25,000
Mara	Upper	20,000
Others	Various	15,000
Total		200,000

Source: Lake Basin Development Authority (1987) and Okidi (1988).

The population of Kenya is growing at approximately 3.3 per cent per annum. Yet agricultural production under rain-fed conditions is limited. About two thirds of the entire country is classified as arid or semi-arid requiring irrigation for agricultural productivity. Already there is a study establishing the feasibility of inter-basin transfer from River Nzoia to the Kerio Valley to facilitate irrigation of the semi-desert area of Kenya.

Apart from the planned irrigation within the Lake Victoria basin, a number of irrigation schemes already exist with the largest ones being under the National Irrigation Board. They are at Ahero (1,348 ha), West Kano (1,228 ha), Bunyala (213 ha) and Yala Swamp, where the LBDA, plans to reclaim the swamp and to irrigate over 2,000 hectares.

Kenyans already know the value of irrigation and have drawn on experience outside the Lake Basin, including Mwea, Hola, Perkerra and Bura Irrigation Schemes. Besides, there are a number of irrigation schemes initiated and managed by the Provincial Irrigation Units (PIUs). Within the Lake Basin alone the PIUs total 3,391 hectares. Recently Members of Parliament supported the Ministry of Agriculture's plan to bring an additional 9,900 hectares under new irrigation schemes. How these areas will be distributed has not been specified but it would be expected that some would fall within the catchment of the Lake Victoria Basin. These initiatives in Kenya are occasioned by the erratic climate changes. They may make Egypt anxious, but there is nothing in policy or law to stop Kenya from the essential alternatives for ensuring agricultural productivity. But the stress brought by these irrigation activities create local pressure points.

Tanzanian Irrigation Initiatives

The irrigation initiatives of Tanzania within the Lake Victoria basin, especially using the waters of the rivers or the lake itself have not enjoyed publicity. Only two cases are known: the first is within the Kagera Basin and the second one is for irrigation of the Vembere Plateau in central Tanzania.

(i) The Vembere Plateau

Not much has been heard of the Vembere Plateau project lately. But it was a curious and disquieting issue prompting H.E. Hurst (1952 :26) to ascertain the existence of a major irrigation work planned for Lake Victoria waters:

> I found that the Germans had, before the 1914-1918 War, a project to take water from Smith Sound, a long inlet at the South end of Lake Victoria, over a low country which separates the lake from the land sloping down towards Lake Eyassi. The water would have been used to irrigate Vembere Steppe for the growing of cotton. The scheme which was not a government one, was to start on a small scale with a dam at Manyonga River to store its flood water to irrigate a small experimental area. From this pilot project data would be built on the Manyonga, and hydro-electric station at the dam would supply power to pump water from Lake Victoria. After passing through the turbines the water would irrigate land lower down and finally drain into Lake Eyassi.

The total area to be irrigated in this project was 230,000 hectares, but the precise volume of water to be used was not determined. Rene Dumont, a critical observer, proposed that the project should be activated towards the end of this century and to be commenced early next century.

Two things are important about this project. First, critical observers have found it worth pursuing before the end of this century suggesting it is a worthwhile and feasible project. Second, the fact that a rumour about the project prompted Hurst to travel from Cairo by sea and then up-country, to establish the facts shows the seriousness with which Egypt might view the project as a stress on the water resources rendering the Smith Sound a pressure point. The unanswered question remains: is Tanzania contemplating the implementation of the project? That information is not available to this author.

(II) The Kagera Basin

The Kagera Basin covers approximately 59,800 square kilometres of Burundi, Rwanda, Tanzania and Uganda. It contributes approximately 25 per cent of the annual discharge into Lake Victoria, averaging 184 million cubic meters per second at Kyaka. Approximately 85 per cent of that volume flows from the Nyabarongo and Akanyaru Rivers from western and southern Rwanda (Okidi, 1986). The rest flows from Burundi, which suggests that most of the rainfall in the region is in Rwanda.

The economy of the entire Kagera Basin depends largely on agriculture. But the preliminary studies by the Kagera Basin Organisation (KBO), established by the riparian states in 1977, were not optimistic about large-scale irrigation works. They warned that much of the Kagera River Basin is not economically suited for irrigation either because of poor soil, slope, distance or elevation differences to the dependable water supplies (Okidi, 1986:9-14, 46)

The basin states have, therefore, resolved not to depend on irrigation for agricultural productivity but to use it as a supplement to rain-fed agriculture. Thus, initially seven areas were selected and identified for irrigation but in the end, only three were developed, namely: Bugesera in Rwanda, Rusumo Covette in Burundi and Kyaka/Kakono in Tanzania. This is only 6,500 ha as compared to the original 16,800 hectares which was identified in the earlier plans.

Of special interest here is the range of prospects that Uganda projects. (Uganda is one of the two lowest riparians, along with Tanzania.) Rwanda as an upper riparian which is also the source of most of the Kagera waters, could cause anxiety to lower riparians by excessive *barragement,* and by expanding irrigation so as to diminish the quantities of water reaching the lower riparian. In this case, three factors may mitigate the situation for Uganda. First, it has been observed that soil type as well as the slopes in Rwanda are not suitable for widespread irrigation. Rwanda would have to find other things to do with the water. Secondly, in the profile of Uganda's planned projects under the KBO irrigation is not included.

Apart from a rice scheme, Uganda opted for the intensification of rain-fed agriculture. Thirdly, it seems, in these preliminary stages, that most of the projects identified under the aegis of the KBO might have to be financed through individual country initiatives. The KBO acts as a collective body, though it has not attracted much money. Under present circumstances the KBO, perhaps, still lacks cohesion. Besides it does not appear that there is any impending widespread consumptive utilisation which could diminish the Kagera's contribution of water into Lake Victoria. It is likely that there will be no major combined development work within the Basin while the region remains unstable due to conflict in Rwanda and Burundi.

The Nile in the Sudan

The discussion in this section seeks to outline the position of the Sudan *vis-a-vis* the Nile water controls in favour of Egypt. Ordinarily, the Sudan may be assumed to be a sister state to Egypt for social and historical reasons. However, when it comes to the apportionment of Nile waters the two states have maintained a rather uneasy coexistence. Egypt insisted that any dam construction had to protect her interests, even if the dam was on Sudanese territory. Such was the case with Sennar and Jebel Aulia, which were the only major dams constructed during the Condominium Government (the Anglo-Egyptian Sudan). Moreover, Sudan objected to the construction of Aswan High Dam, but her position was ignored by Egypt. The result was that the dam inundated the Wadi Halfa region displacing about 70,000 inhabitants. Even though Egypt and the Sudan were partners in drought, Egypt ignored Sudan's interests while drawing terms of reference for an international commission constituted to recommend the apportionment of the existing and additional waters. The commission's report in 1926 only discussed the immediate water requirements of the Sudan but concentrated on the ultimate needs of Egypt. The report of the commission was appended to the 1929 agreement.

The Jonglei Canal is one of the projects which might test the long-term relationship between Egypt and the Sudan. The purpose of the project was to build a canal from Jonglei to Malakal and make the Nile water bypass the Sudd area of southern Sudan. In an area approximately 67,900 km^2, the Sudd Swamp's desert heat evaporates more than 50 per cent of the water as shown in table 5:4.

The primary beneficiary of the canal is Egypt which is slated to receive the additional water saved by the canal from evaporation in the Sudd. For the Sudan, the benefits were expected in new dryland suitable for agriculture as well as grazing grounds. But the agricultural experiments have yielded poor results. The sorghum, maize and rice yields were poor (Howell et al, 1988: 437-440). Experiments with pastures for livestock were commenced in the first phase of the Canal, but were interrupted by civil war between the Sudan Peoples Liberation Army (SPLA) and the Khartoum Government. Also halted were projects on fisheries, forestry and water development, although full funding was available from Egypt and the European Community (now European Union). It was the intensification of the civil war that halted the construction of the Jonglei Canal. The work, which started in 1978, attracted a large population of workers including the French company which had

Table 5.4: Mean annual water discharge at Sudd Swamp, Sudan(in Km^3)

Period	At Mongalla	At tail of Swamps	% loss
1905-60	26.8	14.2	47.0
1905-80	33.0	16.1	51.2
1961-80	50.3	21.4	57.5

Source: Howell *et al.* (1988).

undertaken the task. They had done 240 kilometres out of an approximate total of 400 kilometres, when the engineers and other workmen became targets of SPLA attacks. In November 1983 the work was halted.

The conflict which halted the construction of the Jonglei Canal was not caused by the water works nor was it an international conflict of the kind discussed, but the kind of civil war so common in Africa. The significance of this case is that, with increasingly erratic weather and rainfall conditions, there will be a necessity for various forms of water works for irrigation or reclamation of arid or wet lands. Numerous as the internal conflicts are, the Jonglei-type situation may still be witnessed elsewhere. Furthermore, the very fact that such conflicts may impede water works to facilitate increased food production may result in greater instability and aggravated conflicts.

Damming for Hydro-electric Power

Most of the dams for water storage are multipurpose, covering *inter alia*, irrigation, hydro-electric power generation, fisheries and flood control. Not all border dam projects result in inter-state conflicts; for instance, the Aswan High Dam displaced 70,000 inhabitants on the Sudan side as a result of inundation caused by backwater effect. A solution was found in the resettlement of the displaced. Similarly, the planned dam at Rusumo considered a number of alternative elevations and the contracting states accepted the elevation which affected the least number of inhabitants — 2,755 people to be displaced at an elevation of 1,325 feet rather than 25,950 people to be displaced at an elevation of 1,345 feet (Okidi, 1986:22). But it took lengthy deliberations among the basin states to find the solution which also accepted a lower hydro-power output. Rwanda, for instance, as a country without land for resettlement of a large number of people would not accept the displacement of 22,975 people at a dam elevation of 1,345. Instead, she accepted the displacement of 2,220 people at an elevation of 1,325.

In the case of Lake Victoria and the Nile, one question related to the control of the Owen Falls Dam might still lead to conflicts between Egypt, on the one hand,

and Kenya and Tanzania, on the other. The purpose of the Owen Falls Dam was primarily to facilitate year-round storage of water from the equatorial region. Its location was at Jinja, the Nile outlet from Lake Victoria, and it functioned with controlled sluices which are closed and opened according to Egypt's water needs. In other words, during heavy floods from the Blue Nile, the storage function would be engaged by closure of the sluices. On the other hand, they would be opened when the Blue Nile supplies were down.

The dam was also to be used for hydro-electric power generation for Uganda, which was the most important benefit *in situ*. Therefore, the Egyptian storage function of the dam was to be balanced against Uganda's hydro-power needs. Egypt and the British Government (for Uganda) agreed to an initial temporary discharge rate of 600-630 cumecs but later settled for 630 cm, a permanent discharge rate of 505 cumecs.

Certain conditions were expected to follow that rate of discharge and they are significant factors for easing environmental stress arising from the dam construction:

i. Storage in Lake Victoria would be allowed with the effect of (periodically) raising the level of the water up to 1.3 metres above the previously recorded maximum with a range of 3 metres;

ii. Egypt would pay the cost of raising the dam at the Owen Falls to the height required to obtain storage; and

iii. Egypt would pay full compensation for any adverse effects or disturbances of lakeside interests around Lake Victoria(Uganda, 1957:8).

The conditions were duly accepted and included in the Exchange of Notes between the British and Egyptian governments constituting a formal agreement between 1952 and 1953. The Egyptian Government was made responsible for the cost of the Owen Falls Dam, to use Lake Victoria for the storage of water. Secondly, the Egyptian Government would bear the cost of compensation to those whose interests may be affected by the implementation of the scheme. Thirdly, the Egyptian Government agreed that, for the purposes of calculating compensation under the provisions, new flooding around Lake Victoria within the agreed range of three metres was deemed to be due to the implementation of the scheme.

The predicted rise actually occurred steadily from 1961, reaching its highest point estimated at three metres, in 1964, which was described as unusual and unprecedented by the Hydrometeorological Survey Team (UNDP and WMO, 1974: 744). The consequences were manifold: large areas of lake-shore land formerly for agriculture were inundated, sandy fish breeding sites were submerged, leading to disappearance of a number of species such as *tilapia esculenta* and *proptopterus*.

Pierage facilities at Kisumu, Kendu Bay, Homa Bay and Asembo Bay were submerged, forcing the East African Railways and Harbours to seek alternate landing facilities. These are precisely the changes that were anticipated at the beginning of Owen Falls construction. In fact, Uganda (1957:13) had objected to the discharge rate of 505 cumecs arguing that it would be detrimental to the East African Railways and Harbours.

The Hydromet survey should have given a realistic answer to this puzzle. Instead it proposed that the raised level and eventual flooding could have been caused by excessive rainfall in 1961. No attempt was made in the report to ascertain the impact of the controls at Owen Falls. But Kenyan and Tanzanian politicians pointed fingers at the Owen Falls as the cause of the expensive environmental injuries.

Neither Kenya nor Tanzania has taken up this matter to seek redress and compensation from Egypt. It is conceivable, however, that the control of discharges at Owen Falls will continue. Possibly Uganda may propose modification to enhance power generation. The Falls remain a pressure point where conflicts may arise between Egypt and possibly Uganda, on the one hand and Kenya and Tanzania, on the other. Alternatively, conflict may flare up when Egypt seeks restrained use of the rivers for irrigation, or even the Smith Sound project. Therefore, the question of future control of the Owen Falls Dam for the benefit of either Uganda or Egypt could lead to serious conflict.

Conclusion: International Water Resource Management

This section explains some of the common features in problems of management of international rivers and make some recommendations on how to solve them. Management of river basins is desirable; it will lead to the harnessing of water resources for hydro-electric power generation. Unless the management of international drainage basins, particularly the control of flow regimes and abstraction of quantities of water, is done in an orderly manner agreed upon by all basin states, acute conflicts will be inevitable. Both of these points have been demonstrated above as manifestations and consequences of environmental stress.

United States Congressman Lucken has proposed;

> . . .that it is absolutely vital that some agency step in now and either plan one or two routes, one for the emergency that is going to hit us in 20 years, and these people suddenly awaken to the fact that there is not enough water for their population to survive and be prepared for a crisis management of that kind of thing or—I just detest this route—the route that more advanced nations of the world start pressing these people even to the point of suspending financial aid if they don't come to the table and start looking at the water management program immediately (David and Lakayama, 1988:103).

This statement missed the point on several scores. First, on the question of environmental stress, it is not always easy to have an agreement even among the most enlightened of states. The question of acid rain in Western Europe and between the US and Canada are examples of this. For a long time, the US refused to cooperate on the issue while the Western European states agreed to cooperate after serious complaints from the Scandinavian countries in the 1960s and early 1970s.

Moreover, even after an agreement is reached, there is little assurance that the stress will be mitigated. This is clearly demonstrated again by the boundary water problems between Canada and the United States which have been the subject of numerous commentaries. Canada and the United States did not heed the warnings sounded in 1918 by the International Joint Commission (established in 1909) that "conditions exist which imperil the health and welfare of the citizens of both countries in direct contravention of the treaty". By the 1970s Lake Erie was described as a cesspool with a mortuary smell. It is doubtful that the recent efforts at restoring the lakes by the United States and Canada will actually render their water good for human consumption.

In the case of the Nile and Lake Victoria complex, there is the factor of a dominant regional force — Egypt — which has a number of attributes which barely exist in the other African countries. These are a comprehensive water master plan and expertise or technological capabilities. These two attributes may be considered as the prerequisites for peaceful management of water resources.

First, a comprehensive and national water master plan should take into account the national water needs for agrarian, industrial and domestic uses over time. Against these, the master plan should take into account the quantities of water that are available to it, first, exclusively within a national jurisdiction and then determine the water resources of the shared basins. A projection of these resources should be modelled and programmed over time taking into account such factors as the growth of the national population.

In the absence of such studies it is pointless pushing African countries into international negotiations. Only if the master plan can relate the national water needs over time to the national resources and the ways these relate to the water of international character will negotiations be useful. Any agreement based on less information will be short-lived and could be a recipe for conflicts, because a country with inadequate information may call for a review after a few years while a properly informed state may want the previous agreement to last.

Secondly, expertise in water resource sciences is a primary condition for rational management and conflict avoidance. In the case of boundary waters or acid rain in North America, the political organs might have been uncompromising, but the necessary, expertise existed. Once there is a change in political attitudes, scientific capabilities can be called upon to work out remedial measures. It should be noted, however, that the required expertise for water resource management is always in short supply. In a recent Inter-Regional Meeting on River and Lake Basin Development in Africa, the representative of UNESCO (1988:11) observed correctly that:

> ...the field of water resources encompasses a very large variety of socio-economic activities. Water resources development and water projects require, therefore, many different technical and other kinds of manpower, such as civil engineers, hydrologists, geologists, hydrogeologists, agricultural engineers, hydraulic engineers, meteorologists, etc.

UNESCO also observed that rational and comprehensive management of water resources will only be practical when educational institutions are established for water specialisation at secondary and university levels. It is important to emphasise that each of the areas of specialisation should have national cadres trained in research to the highest levels. Indeed, it is not enough to have one person in each area, because quality research thrives on competition and complementarity. The UNESCO list does not include several other relevant specialists in water resource management, such as water and irrigation economists, water quality specialists, *limnologists* and fisheries specialists.

Here again we can refer to Congressman Luckens who suggested that the LDCs (Less Developed Countries) should be offered expertise to assist them in negotiations. Unfortunately, sustainable development cannot be assured with itinerant consultants who have no lasting commitment to the issues. Negotiations must be made with the participation of national researchers who have internalised research information and with lasting commitment to follow-up initiatives. For historical reasons, Egypt and the Sudan have developed cadres of high level expertise and research information in fields related to water resource management. A reasonable approach is to provide scholarships to build up PhD-level and post-doctoral research capabilities in the areas enumerated by UNESCO. Thereafter, development of research will be necessary to promote common understanding of the relevant issues. It would take less than ten years to build up a core research staff in each of the target countries, because several people are available who are already educated at Master's level in each area of specialisation. Numerous well-qualified people are available to pursue doctoral studies in any of the water-related fields.

Once the core of the doctoral-level specialists is made available to national institutions they will bring up additional specialists, thus creating a self-sustaining system. The experts will provide the leadership essential for long-term studies, planning and negotiation.

6 The Conjunctive Use of Ground and Surface Water Resources in Tsetse Infested Areas

George O. Krhoda

Introduction

Water resource development is a vital ingredient in the social and economic well-being of a country. In the past, planning and development of this vital resource in Kenya has concentrated on the use of surface water to the exclusion of ground water. The emphasis has also varied from one region to another, though the increasing use of one may result in loss of the other. Planners have found that surface water resources are easy to explore, plan, and develop while ground water exploration needs skilled manpower and more sophisticated equipment. Many planners have argued that the choice of surface water over ground water is based on the cheap cost of harnessing the former. However, proper accounting of the cost of water supply cannot be complete without a feasibility study of both. Ground water projects have the advantage of reduced costs of storage and low risk of pollution. In addition, their development does not create the ecological hazards that big dams commonly bring which include reservoir siltation, river channel and aquatic changes, water-borne disease, and fertile river valley inundation.

For many years, there has been a peculiar relationship between humans, their stock and the tsetse fly which limited the spread of animal and human populations. But through bush clearing for farming and grazing, the tsetse habitat has been reduced considerably. The traditional tsetse area extended from the coastal belt and stretched inland for a considerable distance carrying *Glossina G. pallidipes, G. longipennis, G. austeni* and some sparsely distributed *G. brevipalpis*. East of Mt. Kenya, the Meru-Embu belt, the main tsetse species are *G. pallidipes* and *G. longipennis* (Figure 6.1). A good section of central and northern Kenya is tsetse-free because of high altitude and aridity, respectively. But towards the Kapenguria-Kitale area, patches of land are infested with *G. longipennis* and *G. pallidipes*. *G. morsitans* inhabit isolated patches west of Lake Turkana and along the Turkwel River Valley. Western Kenya had its first tsetse outbreak in 1902, when the disease was introduced from Uganda through the fishermen of Lake Victoria. *G. pallidipes, G. palpis* and *G. brevipalpis* became very prevalent, with outbreaks intercalated with periods of quiescence. In southern Kenya, *G. longipennis* and *G. pallidipes*,

Fig 6.1: Distribution of tsetse infested areas in Kenya

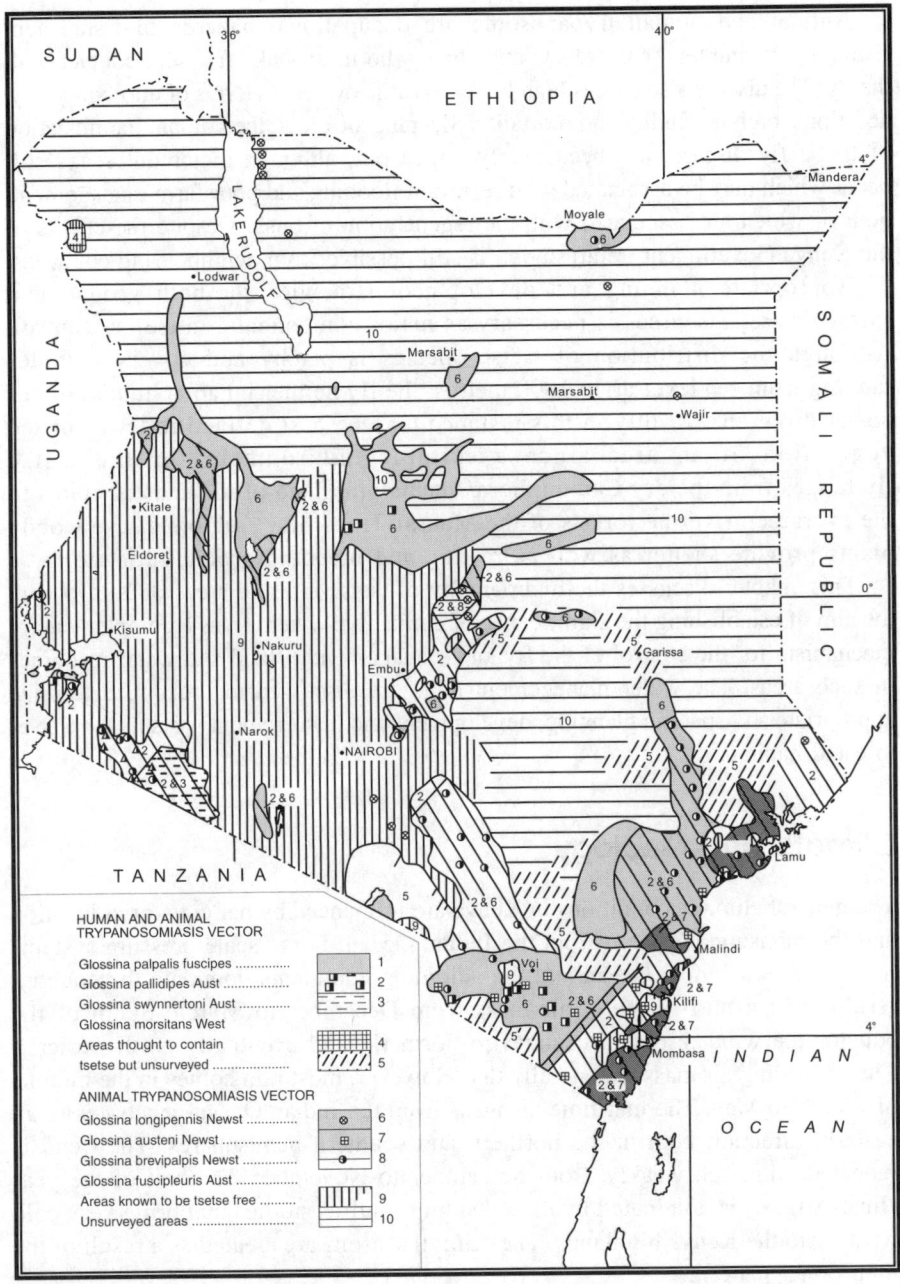

G. swynnertoni and *G. palpalis* came via the Gucha-Ewaso Ngiro Rivers flowing from northern Tanzania.

Animal and human trypanosome are occupational hazards to fishermen, fishmongers, hunters, pastoralists and stock who must make frequent contact with the fly. The disease also causes high losses to cattle owners in terms of unproductivity, abortions, high morbidity and mortality. Sleeping sickness, the human manifestation of tsetse fly disease, has been nearly wiped out, although major outbreaks still occur which may be associated with recurrent flooding and poor farm management, both of which are associated with the regeneration of bush. Despite the efforts of the Kenya Government, reinfestation is still possible from neighbouring countries.

Appropriate planning and development strategies for both ground and surface water resources is necessary as major components in tse-tse control. Although the distribution of tsetse species is patchy and covers altitudes ranging from sea level up to 1800 metres, the fly neither inhabits high altitudes nor arid regions. Aridity in this instance may be best defined in terms of the vegetation of an area since a suitable environment for the tsetse fly ranges from the dry Cammiphora/Acacia grassland of the coastal plains to the evergreen riverine forests of the Maasai-Trans-Mara region. These woody plants provide shelter as well as resting and breeding grounds for the fly.

This chapter discusses the hydrography of tsetse-infested areas of Kenya with the aim of establishing the significance of conjuctive water use as a complementary mechanism for the control of the fly and management of land and water resources in such areas. The water management units are discussed in order to determine appropriate and specific planning, development and management systems for tsetse-infested habitats.

Climate and Geology

The general climatic conditions in Kenya are influenced by her equatorial location and the monsoonal system from the Indian Ocean. Large-scale pressure systems of the western Indian Ocean and the adjoining continents dominate the country, resulting in a north-easterly air mass from December to March. South of the equator the wind direction changes to northerly and eventually north-westerly. This subsiding air mass is generally dry. However, most rain comes in the months of March to May. The maritime air mass from the Indian Ocean penetrates in an easterly direction both in the northern and southern hemispheres. The trend is repeated, although weakly, from September to November. The rest of the year (June-August) is dominated by the subsiding south-easterlies that bring very cold weather to the Kenya highlands. The rainfall seasons are created as a result of the monsoonal pattern.

Fig 6.2: The Geology of Kenya

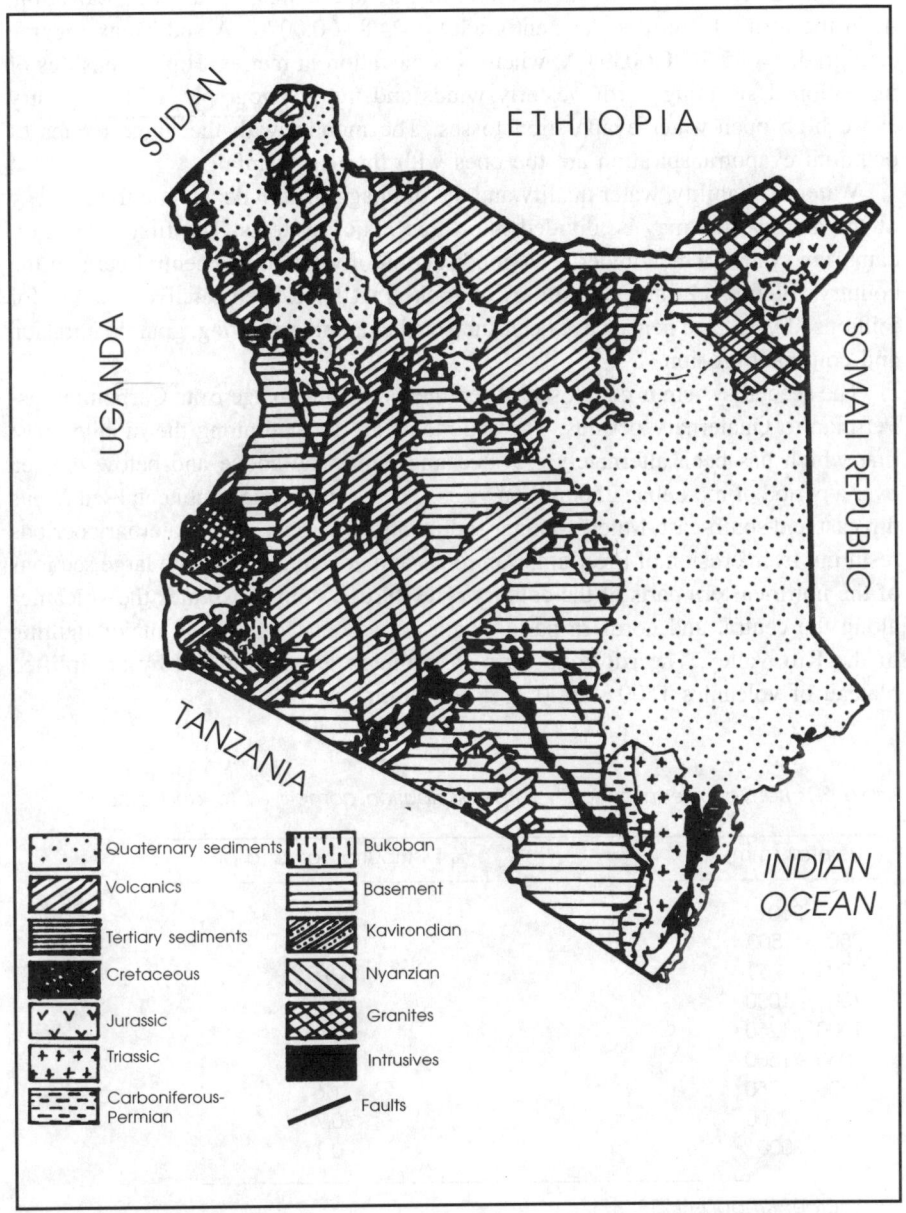

Both rainfall and temperature increase and decrease, respectively, with ground elevation (Table 6.1). Nearly 62 per cent of the total land area of Kenya receives less than 750 mm per annum and this isohyet coincides with the lower limit for the tsetse fly habitat. The maximum and minimum temperatures, T, vary with elevation, A, in the form. Tmin (degrees centigrade) = 24.8 − 0.00705 A and Tmax (degree centigrade) = 35.5 − 0.00594 A, where A is elevation in metres. High intensities of insolation desiccating north-westerly winds and the physiography of the country cause high open-water evaporation losses. The months with the highest rates of potential evapotranspiration are the ones with the least rainfall.

Water availability, water quality and the drainage pattern depend on the geology of the area, which may be divided into three major systems. The first is the Pre-cambrian system which underlies a good section of western and central parts of the country (Figure 6.2) and constitutes of rocks that have been relatively stable for millions of years and hence have been subjected to arching, faulting, granitic intrusion and continual erosion.

The second system is composed of sediments ranging in age from Carboniferous-Permian to Quaternary periods. The sediments are located along the mobile coast line which has been alternately elevated and depressed above and below the sea over a period of more than 250 million years. These marine and continental sediments dip east and south-east. Much younger sediments of Tertiary and Quaternary periods resulting from erosion of Pre-Cambrian rocks and volcanics constitute large sections of the north-eastern parts of the country. The third system constitutes the volcanics along the central and northern parts which are associated with the major faulting of the Rift Valley. The rift zone forms a complex graben flanked by an uplifted plateau of volcanics 1500 - 2000 m in elevation.

Table 6.1: Kenya's mean annual rainfall distribution compared to land area

Rainfall (mm)	% Distribution of land area
< 250	27.0
250 - 500	35.0
500 - 750	18.0
750 - 1000	8.0
1000 - 1250	5.0
1250 - 1500	3.3
1500 - 1750	1.6
1750 - 2000	0.7
> 2000	0.1

Source: Krhoda (1988: 87).

Surface Water Resources

Kenya is endowed with several rivers and lakes which owe their origin to the influence of geologic processes which have given rise to the formation of the Rift Valley with an average width of 64 kilometres. The result of these geologic processes can be observed in the five major drainage basins which include the Lake Victoria Basin, Rift Valley Drainage Basin, Athi River Drainage Basin, Tana River Drainage Basin and Ewaso Nyiro Drainage Basin.

Nearly 50 per cent of the total runoff in the country is generated in the Lake Victoria Drainage Basin which is only 8.4 per cent of the total land area of the country (Table 6:2). The runoff is 5 per cent of the mean annual rainfall of 567mm which represents nearly 487, 394 million m^3 of water. More than 86 per cent of the rainfall is taken to satisfy the demands of evapotranspiration. The fresh water lakes hold about 315 million m^3 of water that is currently under utilised (Table 6.3). The riparian regions of these great lakes are also some of the least developed agriculturally. It is expected that small-scale irrigation projects will be initiated to utilise the lake water. Besides irrigation, fish farming and recreational uses of lakes are other forms of economic activity that may boost the development of the areas around the lakes and also rid such areas of tsetse fly infestation.

Table 6.2: Major drainage areas in Kenya with respective run-off contributions

	Catchment area (km^2)	Mean annual rainfall (mm)	Mean annual runoff (mm)	Mean annual $10^9 \times m^3$	Runoff %	Land area
1. Lake Victoria	49,000	1245	149	7.30	12	8.4
2. Rift Valley	127,000	535	6	0.81	1	21.8
3. Athi River	70,000	585	19	1.30	3	12.0
4. Tana River	132,000	535	36	4.70	7	23.7
5. Ewaso Ngiro	205,000	255	4	0.74	2	35.1
	438,000	510	214	14.85	5	100.0

Source: Krhoda (1987:89)

Ground Water Resources

Ground water and surface water commonly form a linked system. The flow can be in either direction and varies geographically and chronologically. Often the exploitation of one can cause the loss of the other. But most water development projects have been implemented without a careful examination of conjunctive water use.

Table 6.3: Kenya: Lake water resources

Lake	Area (km^2)	Length (km)	Width (km)	Depth (m)	Height above sea level (m)
Victoria*	67493	282	222	46	1134
Turkana	6405	247	32	75	372
Baringo	127	18	6.8	5	975
Magadi	104	26	4	-	579
Naivasha	114-191	16	8	3	1875
Amboseli**	0-114	0-21	0-10	0-0.6	1188
Jipe	39	11.8	3	-	701
Bogoria	34	19	2	-	975
Nakuru**	5-31	2-8	1.6-2	0-6	1757
Elmentaita	18	6.8	2.4	-	1783
On Mt. Kenya					
Hohnel	0.08	0.4	0.2	-	4194
Mitchaelson	0.13	0.4	0.3	-	3965

*Only a small portion is in Kenya
**   Seasonal

Source: Kenya (1973: 4)

Although there are no major ground water basins as such, the distribution and characteristics of aquifers are determined by their geological formations. The Precambrian rock formation, although it has been folded and arched in many places, has been stable over the geological period. Consequently it has been weathered and eroded. Fissures and fault zones have created cracks and conduits on the system and these are responsible for the variable character of the aquifers. Boreholes drilled in Pre-cambrian rock formations have a median depth of 92.0 m, although 20 per cent of all the boreholes have an average depth of less than 45.0 m.

The volcanics occupy the Rift Valley zones and northern and southern parts of the country. They are deeply weathered in many places, fissured and faulted. The boreholes have a median depth of 125 m, although 20 per cent of the cases have a mean depth of less than 75 m. Further towards the north, north-east and coastal zones, the sedimentary rock formations of marine and continental origin are found. The boreholes located there have moderate water yields at depths of 78 m, although the median depth is only 6.0 m. The ground water quality is determined by the genesis of the sediments and distance from recharge zones.

There are relatively few boreholes in the Lake Victoria basin because of the abundant local supply from surface sources while in the Athi River basin there are more people per borehole because of the concentration of people around Nairobi and Thika (Table 6.4). Although the distribution and number of boreholes considered in Table 6.4 may not be a sufficiently representative sample for the country, it demonstrates that there is a relatively high yield that warrants further exploitation to enhance agricultural production.

Table 6.4: Distribution of boreholes and their estimated yield for each drainage basin in relation to population density

Drainage area	Population estimate	Km^2/B/H 1978	No. of B/H	Average yield	Total yield L/min.	Persons/ B/H
Lake Victoria	6,015,049	435	105.8	106.6	42,960	13,828
Rift Valley	1,303,414	771	161.8	139.3	93,610	1,691
Tana River	3,643,691	2,208	31.6	124.0	243,860	1,729
Athi River	3,007,353	328	386.7	88.1	26,080	9,169
Ewaso Ngiro	761,983	587	356.6	74.6	38,200	1,298
Kenya	14,731,490	4,229	135.6		450,700	3,483

Source: Krhoda (1988: 93)

Conjunctive Water Use

The natural availability of water resources in any basin is determined by the hydrological features of the natural supply. The natural supply is limited, however, and in some cases does not coincide with the growing water demand for socio-economic development. Consequently, there is a need for an integrated water resource investigation, planning and development and management in order to utilise the resources. Such an integrated planning strategy involves conjunctive use of water in a given region. The semi-arid and arid areas experience variable but intensive surface hydrological processes that must be harnessed in a manner that reduces the excessive loss of water due to evaporation. On the other hand, each arid area must have recharge either within the region or from external basins. Therefore, planning for both surface and ground water resources fulfils the purpose of optimum use.

It is important to relate the availability of water resources or lack of them to the potential of land-based resources. One of the ways of exploring land potential, especially in those areas inhabited by tsetse, is through the concept of ecological zones. Table 6.5 provides the extent of ecological zones in Kenya. The zones of interest range from zones 4 to 6 in which agriculture is marginal, pastoralism/ grazing/wild life areas become the dominant land use activities and rainfall is low (less than 750 mm p.a.) and variable. The tsetse inhabited area is 25 per cent of the total land area of the country but with more than 42 per cent having rainfall below 750mm per annum.

Table 6.5: Kenya's ecological zones

Ecological zones	Mean annual Rainfall (mm)	Kenya range areas (exact agric. areas) km²
Zones 1 and 2: Agricultural potential 0-81-2-43 ha stock unit*	> 1000	13,535
Zone 3: Marginal agriculturally. High range potential, normally 0-40-2-02 haΣ	(a) 1000 (b) Over 900 (c) 750-900	10,016
Zone 4: Marginal agriculturally. High range potential grazing capacity after less than 4-05 ha per stock unit when developed	(a) 750-1000 (Brachystegia) (b) 650-900	52,924
Zone 5: Medium potential range. Grazing capacity more than 4-05/ha per stock unit	(a) 500-750 (b) 300-650 (c) 300-500	301,569
Zone 6: Semi arid low potential range	Less than 300	115,377
Total		493,421
% tsetse affected		25%

Source: Peberdy, (1972:4).

* 1 stock unit - 450 kg.

In addition, stock carrying capacity is low and water infrastructure is are inadequate in these regions. Tana and Athi River Drainage Basins have up to 50 per cent of the area infested with tsetse (Table 6.6). Although the distribution changes seasonally and with fresh infestations, the area appears to be ecologically suitable for tsetse inhibition. Kenya experiences water deficits (Figure 6.3), which presently are seen in the form of rationing of electricity from hyro-dams (the figures were arrived at by subtracting mean monthly evaporation demands from the monthly rainfall total and accounting for a balance for that year. The isoline values are the annual amount of water that is needed to satisfy the potential evaporative demands.

The fact that deficits occur, however, does not imply that there should be no surplus in the same area. As can be noted, the values are particularly high in the arid and semi-arid regions of the country. Where the values are not excessively high, irrigation measures could be economically undertaken.

Fig 6.3: Kenya's annual water deficit values (in mm)

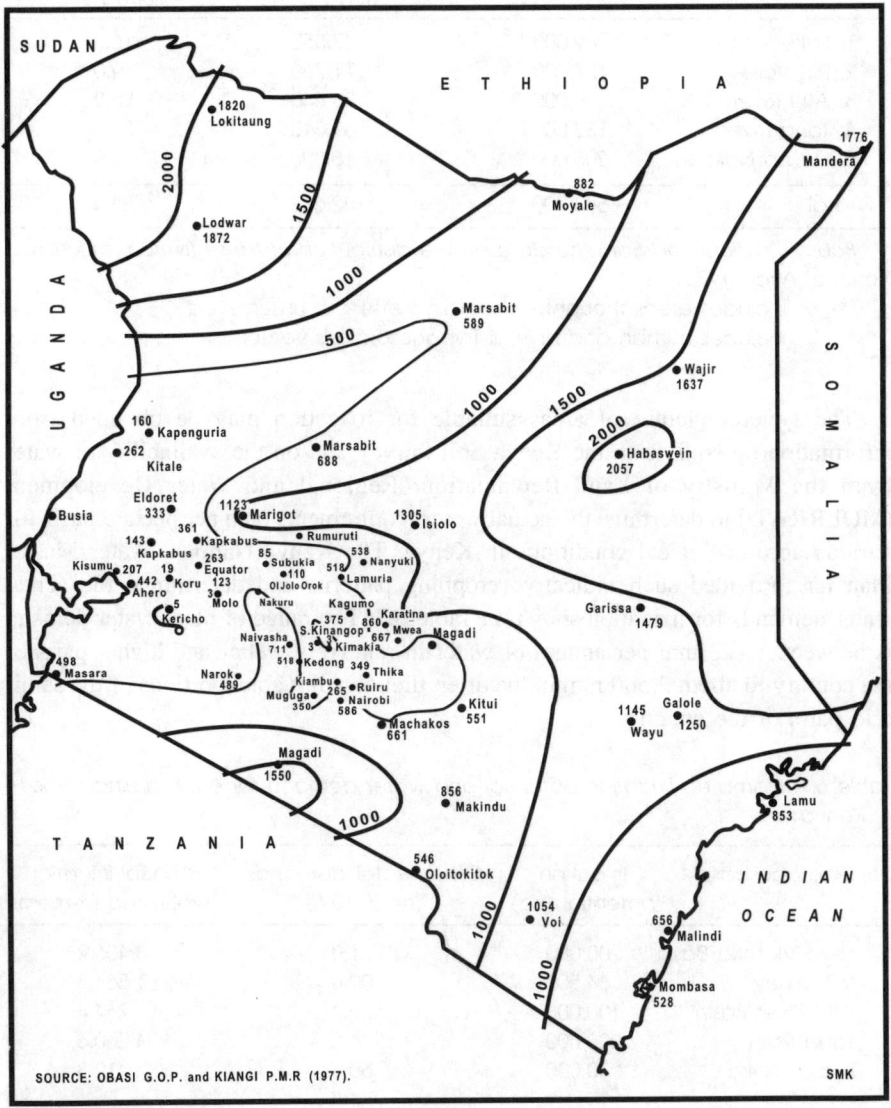

Table 6.6: Distribution of tsetse* flies according to drainage basins

Drainage basin	Total area	Area infested (km^2)	% Total (km^2)**
1. Lake Victoria	49,000	7,852	16.0
2. Rift Valley	127,000	11,760	9.0
3. Athi River	70,000	38,640	55.2
4. Tana River	132,000	69,510	52.6
5. Ewaso Ngiro	205,000	15,300	6.9
Total	583,000	142,612	24.4

Source: Calculated from the data on the distribution of tsetse fly from the Kenya National Atlas, 1973
** Includes areas thought to contain tsetse but unsurveyed
* Includes human and animal Trypanosomiasis vector

The general picture of areas suitable for irrigation may be obtained from information on soils from the Kenya Soil Survey and on the availability of water from the Ministry of Land Reclamation, Regional and Water Development (MOLRRWD) to determine the actual water requirements on a per hectare basis for various agro-ecological conditions in Kenya. The Kenya National Water Master Plan has provided such indicative cropping patterns and determined the actual water demands for irrigation shown in Table 6.7. The range of plant-water demand is between 1,123 mm per annum of water in relatively wetter and higher parts of the country to about 2,500 mm in the drier, though cultivable, portions of the semi-arid parts of the country.

Table 6.7: Estimated irrigation potential and water demand for each drainage basin in Kenya

Drainage basin	Irrigation potential (ha)	Other water demands (m^3 X 10.6)	Annual irrigation water requirements
Lake Victoria Basin	200,000	721.8	3,406.2
Rift Valley	64,500	109.6	1,564.7
Athi River Basin	100,000	530.7	764.4
Tana River	205,000	276.2	4,260.6
Ewaso Nyiro	30,000	60.6	412.3
Kenya	599,500*	1,698.9	10,408.2

Total irrigable land pending soil and water studies for the whole country is 905,000 ha.

Source: Kenya (1978).

It is one thing to identify viable projects in different localities and another to build up a group of development projects that are sustainable on a long-term basis. Sustainability hinges on environmental planning and management of total resources in any given hydrological region in which inputs and outputs may be measured, analysed and modelled for management purposes. Such planning arrangements will inevitably be based on the drainage basin organisation. Both the surface and ground water resources need to be taken into account in order to avoid some of the mistakes that have been made in the past and modern techniques of land use planning based on an ecosystem approach must be employed.

Irrigation development and expansion have been associated with environmental problems. Many people mentally relate irrigation projects to water-related diseases, disruption of families and indebtedness to financial institutions. These observations need not be real for all projects if only careful planning and environmental assessment were made before the projects are undertaken.

The worst problem is water-borne disease. There is evidence that bilharzia, malaria and typhoid are on the increase in irrigation areas. The swamps created by flooding the fields, slow-moving canal water and other swamps within the reservoir area are known to be good breeding grounds for snails and mosquitoes. Snails have been found even at an altitude of 2000 m in the Lake Victoria basin. Research done by the Ministry of Health among mostly school-going children in the Mwea Irrigation Scheme revealed that 31 per cent had bilharzia, 15 per cent had askaris and 6 per cent suffered from hook worms. Evidence coming from other irrigation projects is not different from that coming from the Mwea Irrigation Scheme.

The other main health hazard in irrigation schemes is malnutrition among settlers and their families. Malnutrition often results from settlers growing more cash crops than food crops and it reduces the human endeavour which may be put into productive work. Consequently, the efficiency deteriorates and intended prosperity is hampered by disease.

There are other environmental problems associated with monoculture, aquatic environments and intensive agricultural production. In monoculture, pests and weeds are attracted because of reduced competition, hence farmers use heavy doses of herbicides, pesticides and fertilisers in order to improve productivity. These chemicals persist in the environment and become dangerous for the ecosystem. The chemicals return into the streams as irrigation discharge forms nearly 40 per cent of the total diversion flow. The return discharge contains molluscicides that are used for controlling snails. The return discharge is, therefore, rich in fertilisers and causes eutrophication, thereby enhancing the excess growth of algae and aquatic weeds. The aquatic weeds, in turn, affect the fish ecology and form the swamp habitat valued by mosquitoes and snails.

There is no record of the after-effects of pesticides to downstream ecology. However, it is known that the use of pesticides has increased nearly four-fold in the last ten years. In 1967 the total amount of pesticides used in agriculture had a value of Kshs. 6.3 million and value increased to Kshs. 20.5 million in 1973 and Kshs. 89.1 million in 1975. As labour costs increase, more farmers use herbicides and pesticides to control weeds and diseases which in turn affect many of our river systems.

Irrigation schemes also suffer from flooding and siltation. (Flooding results from extra flow during the rainy season and disrupts the activities of the irrigation scheme.) Besides damage to property, crops and public utilities, flooding washes the fertilisers away from the fields, disrupting snail control and introducing fresh snails into the system. In some cases, flooding causes waterlogging — the ground water table rises and causes salinisation. However, salinisation may also come through inappropriate management of the fields. Nearly 50 per cent of all irrigated lands have been damaged and consequently abandoned. Pockets of saline areas have been reported at Mwea and Perkerra schemes. The results of salinisation include the reduction in yields and consequent loss of soil structure.

Livestock Development and Water Resources

Livestock development is an integral part of the economic development of the rangelands. More than 46 per cent of the 5.2 million hectares of medium- to high potential land is devoted to livestock production. Milk and beef production need to be stimulated by using better breeds and control of disease besides the provision of additional water supplies at strategic points. It is difficult to make estimates on the growth of livestock because a good proportion of producers are nomadic pastoralists who move from one pastoral area to another. Guess-estimates show that there are 18.8 million cattle of which 15.5 per cent are in rangeland, 32.9 per cent in small-scale farms and 51.2 per cent in large-scale farms. By the year 2008 there will be 16.6 million cattle and 14.7 million goats in the country (Kenya, 1978).

Two areas that need attention pertaining to water resource's ability to improve stock-carrying capacity are water supply and forage requirements. The water supply is necessary mainly in the arid and semi-arid lands, which contribute nearly 80 per cent to the total area of the country. Improvements in forage requirements may be targeted to the small holder livestock farmers who may be able to improve fodder production. Land unsuitable for cropping or intensive forestry and where management for stock-raising purposes is best carried through the manipulation of vegetation is called rangelands. According to Figure 6.3, rangelands lie in a two-season rainfall zone in which one season may have a heavier downpour than the other. Nearly 86 per cent (465,000 km^2) of the total area of the country receives less than 750 mm of rainfall per annum. Most of this rangeland was not occupied in the Coast Province because of tsetse infestation.

Between 1945 and 1962 African Land Development (ALDEV) developed 1020 seasonal dams, 332 permanent dams, 308 subsurface dams, 38 rock catchments, and 40 masonry weirs to improve stocking capacity of the rangelands. In addition, ALDEV installed 72 piping schemes, protected 54 springs and drilled 44 boreholes and developed over 200 shallow wells. Since then the stock-carrying capacity at Galana, Kuras and Lamu ranches have improved considerably.

The Livestock Water Development Scheme was started in 1969 with the aim of improving livestock breeding, marketing and management in rangelands. Under this scheme, many boreholes were drilled, equipped, operated and maintained. Additional water pans were constructed in the planned grazing blocks. Phase I of the project spent Kshs. 78 million (1969-1974) for which 74 pans with a total capacity of 1,992 million litres of water were constructed, 2000 kms of access roads were built and 38 boreholes drilled. On phase II of the project (1974-1979) for which Kshs. 118.8 million was budgeted for the construction of 109 pans, 4,100 km of access roads and tracks and 40 boreholes to be drilled, of these were constructed 65 pans holding 2,200 million litres of water, 2400 kms of access roads and tracks and 31 boreholes (only 16 of them successfully completed) were drilled.

Desilting dams and pans is very expensive. In North Eastern Province desilting a pan costs between Kshs. 15,000 and Kshs. 30,000. For the current four year plan, the Province will get Kshs. 110,000,000 for desilting 92 silted pans. Eastern Province is similar, where livestock development has been based on improvement of water supply.

Table 6.8, The number of pans in developed grazing block in North Eastern Province

Name of block	Area (ha)	No. of pastures	No. of pans constituted	No. of pans silted
Garissa District				
Mado Gashe W.	224,688	6	14	3
Mado Gashe E.	282,500	6	13	5
Udhole	702,300	2	10	4
Dadaab W.	323,750	4	11	3
Dadaab E.	N/S	4	nil	nil
Eilein	803,212	4	1	1
Wajir District				
Kalalut	43,125	7	26	23
Sabule	46,982	7	3	nil
Tarba	330,300	4	16	10
Ajawo	436,250	6	22	12
Griftu	414,300	5	22	12
Buna	309,400	5	13	12
Mandera District				
Takabo	N/S	8	12	4
Rhamu	N/S	6	4	1

Source: Compiled by author.

Problems and Prospects of Water Use in Arid and Semi-Arid Areas

Increasing agricultural production for the growing population and for national development must be planned and managed efficiently for the rewards to be meaningful to the community. Kenya is endowed with a sufficient amount of water resources for renewed growth by the year 2000. This chapter has discussed the problems of the geographical distribution of surface and ground water resources and the impact this distribution has on the overall development of the tsetse inhabited areas. Because the high and medium-potential regions are fully utilised, expansion of the agricultural base will depend on the availability of new crop breeds, the control of plant and animal diseases, and the and increase of the carrying capacity of medium- and low-potential areas through proper land use management. The development of total water resources has implications for the rate and direction of renewed growth which cannot be overstated. In this regard, water development projects must increasingly become integrated and made flexible in order to accommodate or generate development in other sectors of the economy. It is possible that the flexibility of any project can only be realised if the alternatives have been evaluated and public priorities taken into account. This requires coordination between different sectors of the Kenya Government.

Renewed growth would require that land use planning and management be totally geared towards optimum production and the elimination of pests (such as the tsetse fly) which inhibit livestock development and debilitate people's productive energy. The agricultural capacity in areas currently infested tsetse flies needs to be increased. In addition, further development of irrigation projects based on the waters of the rivers Yala, Sondu, Kuja-Migori and Mara will be useful in eliminating the habitat of tsetse species along the shores of Lake Victoria. Projects on the river Turkwel, the Kerio Valley and at Lake Baringo will change tsetse habitat into agricultural lands. The irrigated flood plains of the Lower and Upper Tana and Ewaso Nyiro will eliminate the habitats of tsetse fly infestation in these drainage basins.

Table 6.9: Estimated base livestock population by water management units, 1978

Drainage basin	Cattle	Small stock
Lake Victoria Basin	4,646,694	3,088,872
Rift Valley Basin	1,479,466	2,150,058
Athi River Basin	1,399,763	1,715,021
Tana River Basin	1,981,477	1,972,662
Ewaso Ngiro Basin	1,176,860	2,694,327
Total	10,674,260	11,620,940

Source: Kenya (1978)

By increasing agricultural output through the use of underground and surface water resources there is room for increasing livestock production in the present tsetse inhabited areas. Proper range management and sufficient water supply will likely increase production by one-and-a half times in the next two decades (Tables 6.9 and 6.10). In order for the target to be reached, the areas that are presently underutilised because of tsetse infestation need to be rehabilitated and effective tsetse control mechanisms put in place to avoid infestation from neighbouring countries.

Table 6:10 Projected livestock population by water management units to 2000

Drainage basin	Cattle	Small stock
Lake Victoria Basin	7,713,258	3,666,871
Rift Valley Basin	2,133,828	2,658,295
Athi River Basin	1,894,551	2,694,557
Tana River Basin	3,136,118	2,279,932
Ewaso Ngiro Basin	1,705,603	3,370,290
Total	16,583,358	14,669,945

Source: Kenya (1978)

One of the worst losses of water is by flash floods which are prevalent in semi-arid and arid areas. Surface dams and pans have been and continue to be used for storing water for use during dry seasons. Unfortunately, such attempts have succeeded only in part because of high evaporation losses and also due to siltation of pans and reservoirs. Research has been initiated to investigate the rate of siltation of earth dams from various agroclimate and geological provinces. Besides this study, there is an on-going research to investigate the potential of sand river aquifers in semi-arid and arid lands which will open up some nonconventional methods of developing subsurface dams.

As population increases, it will become imperative to manage our water catchment areas appropriately from which perennial rivers are replenished. The total forest cover is about 3.1 per cent of the total area of the country and this figure is decreasing rapidly.

The other threat to our water resources is pollution. The MOLRRWD is charged with monitoring pollution levels. Their experience has not been encouraging, possibly for two reasons: one, the Government of Kenya has been reluctant to enact tough anti-pollution legislation against industries because unemployment is considered of greater national concern than the threat from pollution and, two, the lack of funds to operate systems of policing industries and consistently monitoring air and water quality in order to detect changes over various production levels and weather conditions. Continual international cooperation in tsetse control, water resource development, and research will be an added advantage. Such collaboration will cut the costs of research and enhance cooperation and trade.

Part III

POPULATION AND HUMAN RESOURCES

7

A Systems-Analysis Approach to Medical Geographic Studies: A Case of the Human African Trypanosomiasis Disease in Kenya

Justus I. Mwanje

Introduction

Medical geographic studies date back to the beginning of written history. Classical Greek philosophers such as Thucydides wrote on the plague of Athens which, incidentally, could have been an epidemic of smallpox. Since the 18th century, there has been a steady and logical progression of contemporary trends in philosophical and methodological aspects of medical geography. The focus of these advances has moved from consideration of disease on the basis of an idealist single-factor concept to a more realist multifactorial and multidimensional one over space and time. Indeed, dynamic aspects of commonly-studied diseases tend to vary in scope and extent. This variation could be attributed to the understanding of the relationship between locality and disease occurrence. However, the larger goal of offering geographic solutions to human health problems remains to be attained in a comprehensive sense. Hence, attempts have been made to address this goal, in brief, by introducing the systems analysis approach, a comprehensive methodology for application to the study of medical geography.

It can be argued that geographical variations, in matters of health and disease, like temporal variations, are fundamental facts and of common interest to many scientific disciplines. For instance, disease is viewed by ecologists as maladaptation between organisms, culture and environment interacting in space and time as an agent, pathogen and host. The various approaches associated with this view have been analysed by Pyle (1977). However, to put the picture in perspective, a brief synopsis is presented in the following discussion.

Approaches to Medical Geography

Historically, there has been a change in the purpose of medical geographic studies with regard to the existing health situation and the identification of epidemic trends. In fact, this change can be attributed to the specific research goals adopted. To this end, the evolution of approaches in medical geography has been traced back from the time of Hippocrates to German physicians such as Finke, Schinurrer, Fuchs and

Fig 7.1: Development periods of medical geography

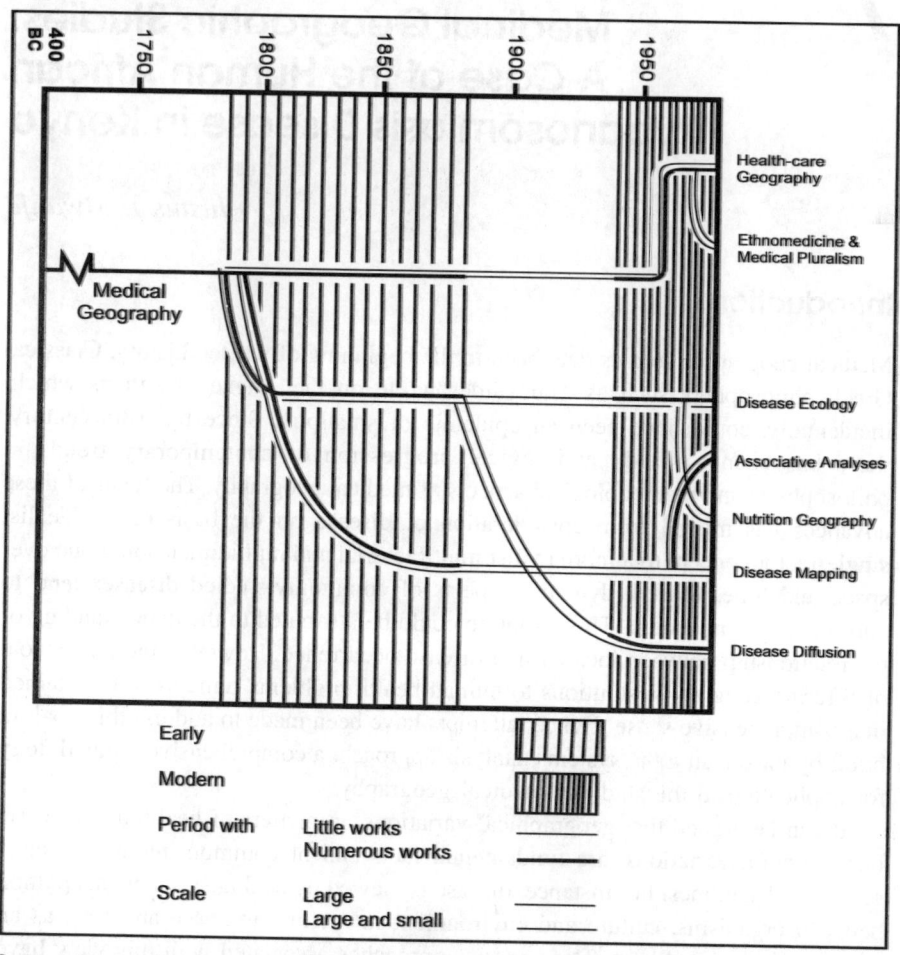

Source: Compiled by author.

Muhry during the late 18th and 19th century and referred to generally as "geographical pathologists", among other terminologies.

On Paul's (1985: 403-409) views on the development of medical geography, it has observed that he has chosen to focus on approaches or content areas rather than on principles. This runs the risk of suggesting that there are many different kinds of medical geography and almost as many kinds as there are workers in the field. What emerges from the critique is that the discipline of medical geography may be facing forces of a "niche specificity" type of compartmentalisation. A vision seems to emerge in Learmonths's (1978) critique to the same paper on the direction of use of systems synthesis, a metaphor capable of stimulating intellectual challenge and rigour. Further, Hunter (1974) chose to consider any such division as merely clusters and not clear-cut. This chapter attempts to forge a unifying theory since the advancement of approaches in a growing discipline medical geography in order to contribute to its conceptual augmentation.

Therefore, the major frontiers of medical geography are integrated into an unified theory that includes the following: historical aspects and the geography of health and disease in particular regions and places geographic epidemiology; geographic pathology or disease ecology and nutrition geography; biometeorological, environmental geochemistry, health and medical services and a health policy.

However, the attempt at developing a unified theory of disease by use of the systems analysis approach should not be construed as a "new" evolution of a package of techniques in medical geography. Nevertheless, this is a contribution towards improving the various spatial approaches in application. It envisages a synthesis of information of geographic nature that may have a bearing on disease landscapes. In addition, the approach can easily accommodate scientific contributions by entomologists, climatologists, ethnobotanists, anthropologists and epidemiologists, among others, on aspects of human disease problems. It is, therefore, an organisational approach for it can integrate vast amounts of information for medical geographic interpretation.

Thus, in the context of systems theory, a medical geographer may play the role of Disease Systems Analyst (DSA) or Health Systems Analyst (HSA) and succeed as a coordinator of multidisciplinary research programmes. Phillips (1981) concurred with this line of thought, although not so explicitly. The tools of systems analysis approach are, therefore, considered in the next section, especially those pertaining to systems modelling and data organisation available to medical geographers in their search for factors and patterns influencing epidemic dynamics. In this regard, policy implications may be synthesised, hence justifying medical geography as an applied subdiscipline of geographical science.

System Analysis Concepts and Approach to Epidemics

A system may be defined as a set of elements interacting among themselves according to some kind of process and with the environment. A system is, therefore, a model of a general nature. The latter, which must be capable of accounting for all the input-output behaviour of a real world system and be valid in all allowable experimental frames, can never be fully known. Thus, the base model of a system is very complex and may require great effort to design and develop.

Fig 7.2: Phases of applied systems analysis

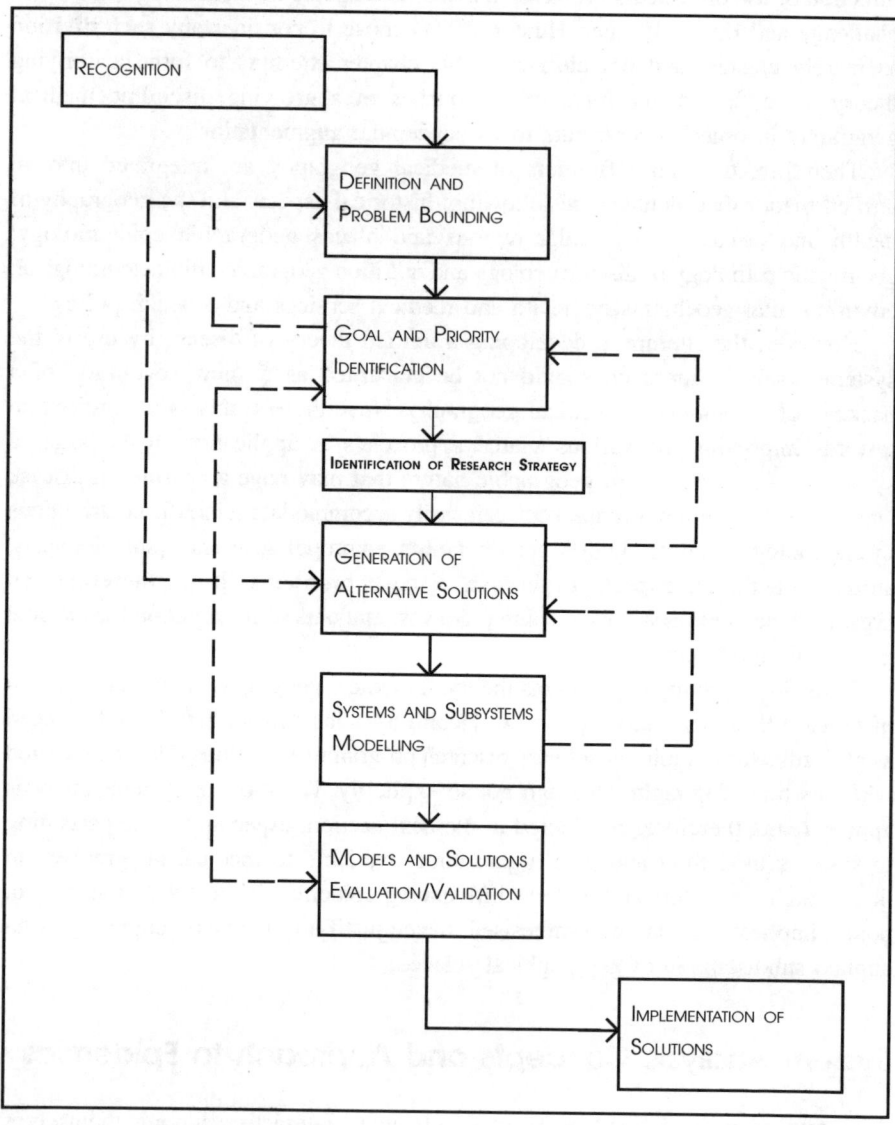

Source: Compiled by the author.

If a system exchanges mass and energy across its boundaries with the environment as a biological organism then it is termed an open system, but if only energy is exchanged, then it is considered a closed system. On the other hand, if neither energy nor mass is exchanged, it is said to be an isolated system. Further, if mass and energy are entering and leaving the system, one is concerned with the rates of exchange or flow rates. Should the flow rates be constant in both space and time, then many systems would attain a steady state of equilibrium.

In general, the theory associated with open systems is termed 'open system thermodynamics' and is the most general formulation. On the other hand, the study of the restricted class of open systems where flows take place at a constant rate is termed 'steady-state thermodynamics.' Indeed, should the flows across the boundary be zero, then the system is 'isolated' in nature and the thermodynamics is termed 'equilibrium' or 'classical'. This chapter considers disease as an open system within the environment where it is endemic.

It is well-known that the design and development of the systems methodology is the occupation of systems analysts; the methodology is basically a collection of tools and models used in an integrated format. The result is an information system on a phenomenon or a problem of interest. Without question the approach gets at the root of the phenomenon or problem, based on the evidence that certain systems properties do not depend on the specific nature of the individual phenomenon or problem (i.e. they are valid for systems of different nature as far as the traditional classification of science is concerned).

Hence, with renewed confidence we can, borrow sophisticated systems theories from fields such as physics, electrical engineering and control systems and apply them in other disciplines. This explains the very large literature on application of systems theory in various fields.

However, systems theory is in a state of flux as paradigms tend to be phenomenon or problem-dependent. The basic goal of systems analysis is to maximise information interpretation at a minimum cost or effort. Given the wide range of its application, the approach is not only possible, but it has also been widely tested and amply proven. However, specific application of this approach to medical geographic phenomena or problems is at an infant developmental stage, to which this chapter seeks to contribute. Indeed the governing systems' paradigms in medical geography are basically evolving.

Where then does a medical geographer performing the role of a coordinator as a DSA/HSA stand? To answer this fundamental question in such circumstances, medical geographers would benefit enormously by applying the systems analysis tools. But while doing so, a great degree of selectivity and caution must be exercised, given that their predictions are likely to have a far reaching impact on human lifestyle within the respective environment. In fact, it is not far-fetched to regard a DSA/HSA as generalised physician without formal training in medicine, but is easy to appreciate that a thorough knowledge of disease theory is necessary for a DSA/HSA to operate. Under certain circumstances, a DSA/HSA may be better informed than a practising physician.

How then would a DSA/HSA set out to build a knowledge base on a given disease? Naturally, the development of epidemiological models would be part of the answer. The other part concerns a thorough understanding of cultural and psychological factors that may contribute towards the prevalence of an epidemic in a given environment.

Fig 7.3: Approaches towards an ecologic theory on disease

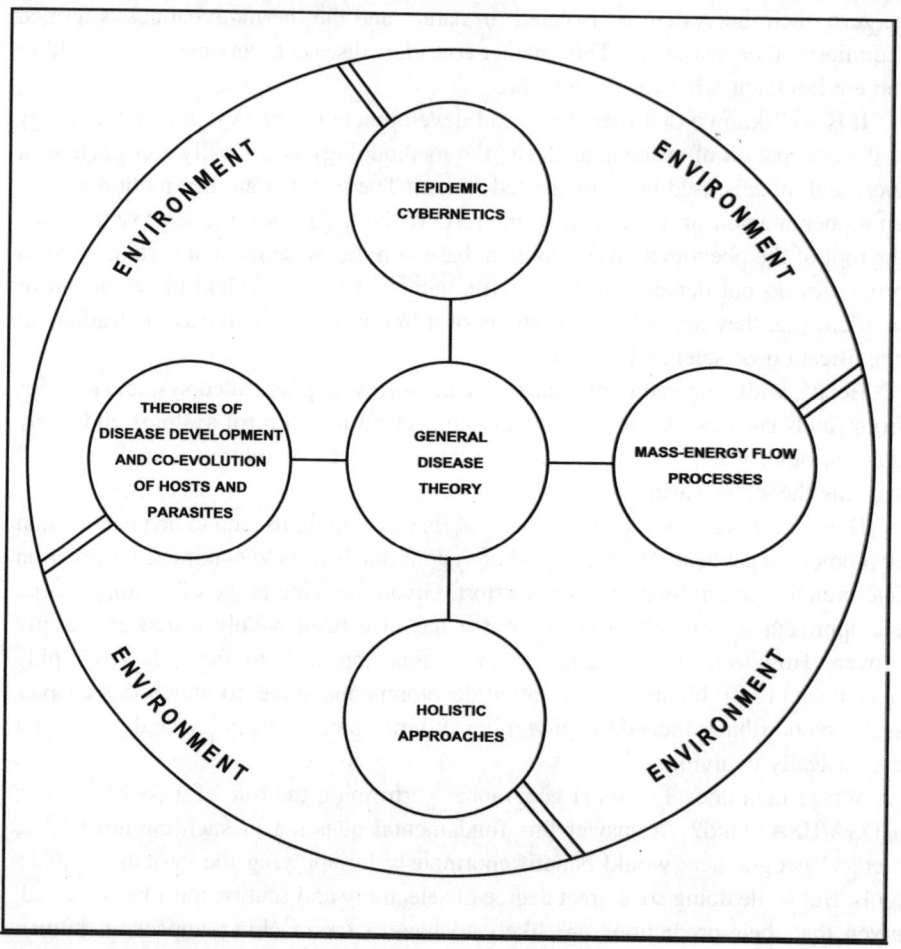

Source: Compiled by the author

In viewing a disease as a complex system, the art of model building demands an authoritative knowledge of the malady under consideration. Such knowledge must, therefore, be built from experimentation and abstraction in a mathematical framework. The experimental stage would have produced what is termed a lumped model, which depicts the real systems' image in which complexity is greatly simplified. However, in order to avoid the plague facing most ecological models, which often tend to have a relatively narrow purpose as they focus ordinarily on simulating only matter-energy flows among a subset of selected ecosystem components, epidemic models must be designed and developed differently. To this end, a DSA/HSA must view disease theory in the context of living systems theory. According to Haug (1982) the theory postulates eight critical subsystems that process matter-energy, nine that process information and two that process matter-energy and information. It states that all information is borne on markers of matter-energy. The markers can be altered in two ways; the matter-energy can be altered by transforming into forms that are more easily transmitted. This is made possible by the input, internal and output transducers or by translating into another code more easily understood within the system. This is made possible by the encoder and decoder in a living system.

Therefore, to develop robust epidemiological models in a geographic framework, the underlying disease processes, matter-energy, and information related to the interactions between critical subsystems must be well understood. Thus, the question of whether or not a disease is a living system must be considered. In retrospect, a DSA/HSA should be able to examine disease theory in a simplified framework, amenable to mathematical modelling. However, this chapter does not discuss these concepts.

However, studies on spatial heterogeneity have focused on the persistence of population in geographic space and time. Consequently, these studies have led to a divergence in views by other researchers. What this indicates is that human beings are still struggling to unravel those complexities that characterise most epidemics.

Therefore, our contribution to the on-going debate focuses on the Systems Analysis (SA) approach which allows probing of epidemic dynamics by the organisation of mathematical models explaining disease processes in a format that mimics nature. From these models, it is expected that threshold epidemic conditions can be isolated. As a result, sensitivity analyses would help to ascertain the impact of spatial heterogeneity, especially on matters of co-evolution, regulation and extinction of parasite-vector systems. This approach does, therefore, allow for an examination of geographic factors of dispersal and immigration which may significantly influence the spread of vectors and epidemic outbreaks. Thus, a DSA/HSA may then be able to access the geographic distribution of disease — causative agents in the field, thereby explaining their prevalence in certain regions. Otherwise, the approach faces the difficulty of classifying interactions between populations as a diversity of biological phenomena. It is proposed that if disease is viewed as a living system and appropriate paradigms are formulated, then the effects of this shortcoming in understanding the underlying complexities would be greatly minimised.

Thus, a medical geographer would often find epidemic outbreaks interesting from both a basic and an applied scientific perspective. However, a DSA/HSA would quickly realise that, although such outbreaks provide an ideal opportunity for model analysis, they are often complex phenomena and are seldom adequately comprehended. Therefore, it is proposed that in order to achieve a comprehensive research methodology for studies on the dynamics of any epidemic, the following systems analysis steps must be adequately formalised:

Step 1: Identify and define with a clear perception the research problem.

Step 2: Exhaustively review the literature pertaining to the research problem. Conceptual linkages between major operational components within the subject matter must be elucidated and existing knowledge gaps identified. Special attention must be paid to methodological achievements.

Step 3: Provide a convincing justification for the commission of the study (with reference to Step 2 above).

Step 4: State the research objectives and their associated scopes along theoretical and practical considerations.

Step 5: Develop the research hypotheses and synchronise them with the requirements of preceding steps.

Step 6: State the conceptual framework pertinent to the research problem and define associated paradigms. Subsequently, develop the conceptual model to assume the linkage role with the foregoing steps.

Step 7: Outline the geography of the area where the research problem is prevalent paying special attention to floral and faunal resource and examine their relationships with humans. In retrospect, opt to evaluate the spatial distribution of health care delivery systems and other infrastructure of significance to the problem.

Step 8: Identify the relevant research methodologies to the various components of the study problem.

Step 9: Take note of the expected limitations to the research project and plan accordingly.

In order to successfully accomplish all the above steps, a DSA/HSA must translate disease system relationships into flow diagrams, mathematical functions, or models. Thus, the aims of modelling should be to achieve epidemic forecasting; an understanding of the biology of the parasite, vector, and humans and policy analysis suitable for the management of the problem.

In the context of the steps and aims outlined above and drawing on Paul's (1985) review, a complete synthesis and integration of the reviewed approaches can be achieved by adopting the systems methodology. Disease systems can, therefore, be decomposed into smaller explainable and manageable components, each of which can be reconstructed into a tractable system of equations along pathways. Hence the approach is suitable to medical geographers studying disease dynamics such as those that are characteristic to vector-borne diseases like Human African Trypanosomiasis (HAT) whose pioneering study was reported by May (1958). The main target of such studies is to eradicate vectors and achieve a reduction in epidemic rates.

Human African Trypanosomiasis

The global spread of vector-borne diseases as in the case of other infectious diseases poses a serious public health problem to the world population. In the case of the HAT disease endemic in Africa, two types exist: *trypanosoma gambiense* (prevalent in West Africa) and transmitted predominantly by the *Glossina palpalis* species of tsetse; and *trypanosoma rhodesiense* (prevalent in East Africa) and transmitted mainly by the *sclossina pallidipes* species of tsetse. The dynamics of the former is beyond the scope of this chapter.

The ecological limit of *Glossina* species lies between 10° N and 35° S, with the human population at risk estimated at 50 million and growing at the rate of at least three per cent annually. The three pertinent components that interact to yield a successful transmission of this disease are considered in the next section.

Parasite

The *trypanosoma rhodesiense* is a protozoan agent that causes the human trypanosomiasis endemic in East Africa. This parasite is regarded as a true zoonosis; the concept of zoonotic processes refers to those processes associated with those zoonotic diseases and infections, (the agents of) which are naturally transmitted to vertebrate animals and man. Heisch et al (1958) confirmed this concept after isolating a parasite strain (i.e trypanosome) infective to humans from the bushback (*tragelaphus scriptus*). Also, Onyango et al (1986) reported the isolation of *trypanosoma rhodesiense* from an East African short-horned zebu ox. The implications are, therefore, that domestic and wildlife animals are confirmed reservoirs for the parasite. However, the key hosts of medical importance are humans.

A comprehensive review of the entire subject of the African trypanosomiases, both human (i.e. sleeping sickness) and veterinary (i.e. nagana caused by the parasite *tryponosoma brucei*), is given by Mulligan and Potts (1970). However, it is important to note that the West and East African *trypanosomes* are morphologically

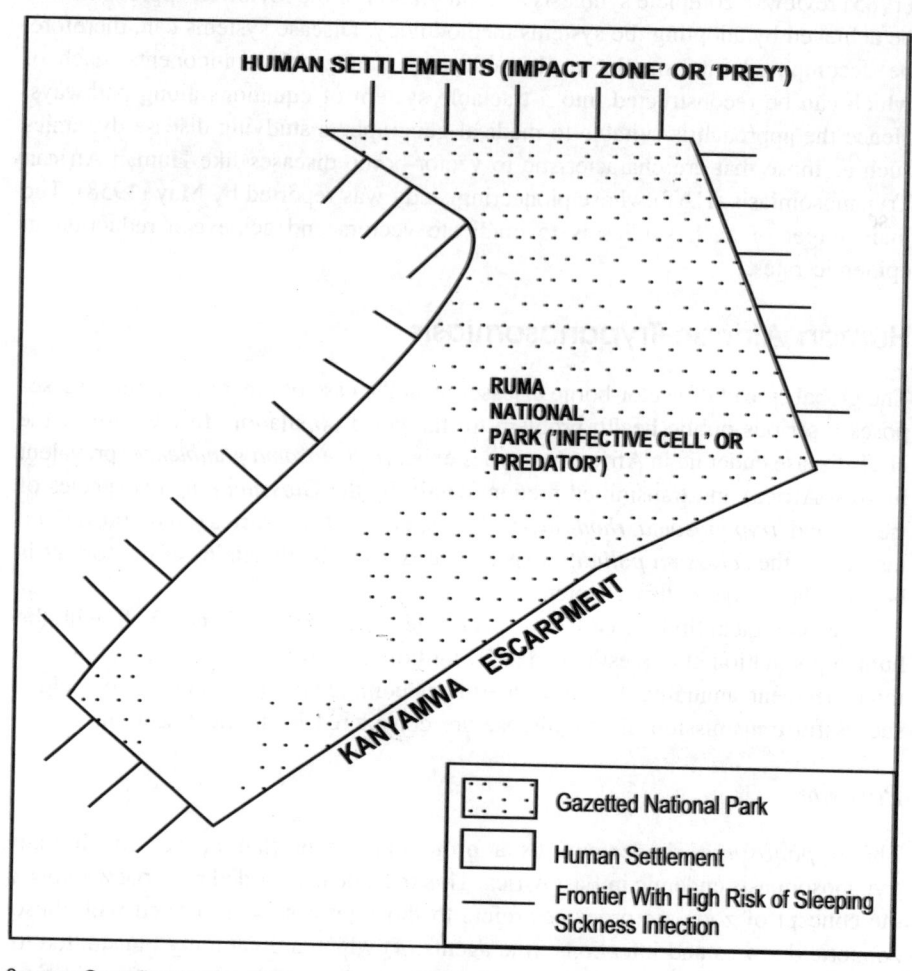

Fig 7.4: The relationship between Ruma National Park 'Infective Cell' and the surrounding human settlements 'Impact Zone'

Source: Compiled by the author.

indistinguishable in stained blood films from *T. brucei*. This identity is seen in both mammalian and vertebrate hosts. It should be readily noted also that little is known about the evolution of the parasite with the varied spectrum of reservoirs resident in the Lambwe Valley ecosystem, the foci of our analyses.

Thus, an attempt to estimate parasite density within the disease endemic areas may prove rather cumbersome. For our purpose, we shall assume endemicity of the parasite in the Lambwe Valley ecosystem and concentrate on understanding the human-vector population interactions, pertinent to the spread of the HAT infection.

Vector

As earlier discussed, the tsetse fly species *G. pallidipes* is an efficient transmitter of the *Rhodesiense* type of HAT. The transmission occurs when a vector acquires a blood meal from infective domestic, wild animals and humans. If humans are not in an infective stage, they contract the infection when exposed to an infected vector. The ecology and physiology of the vector has been extensively studied. The International Centre for Insect Physiology and Ecology (ICIPE) has a specialised tsetse research unit. A synthesis of these works along with field observations, reveals that the breeding system of *G. pallidipes* is complex.

It is evident that three principle parameters, namely soil moisture, temperature and habitat type may be crucial to the average vector survival rate. The latter is a determinant of population success in withstanding the exigencies of the average environment. Thus, an articulated manipulation of these processes and the associated parameters into mathematical models (i.e. N sets of mass-action questions on: growth functions, predator-prey functions and control and feedback functions) would yield a scenario for mimicking vector dynamics over a wide geographic region consequently generating vital information (i.e. estimating vectorial capacity and epidemic threshold levels) that can be used to design population control strategies.

Meanwhile, the control of tsetse population within the Lambwe ecosystem has proved elusive. The basic reason is that the core of the ecosystem is a gazetted wildlife research reserve (Ruma National Park), hence bush clearing, which is a known effective tsetse eradication technique, is impracticable as the biomass is necessary for the survival of wildlife animals. Consequently, suitable tsetse habitats tend to be permanent in this foci from which they disperse easily into the surrounding human settlements. In other words, the park could be viewed as a 'predator' to the surrounding settlements (the 'prey') or 'infective cell.'

In the recent past, attempts have been made to eradicate tsetse flies by spraying insecticides from an aircraft on to mapped habitats. Obviously this method resulted in environmental problems such as the destruction of beneficial non-target lifeforms in addition to the contamination of Lake Victoria-bound streams with disastrous effects. The massive use of this technique has been consequently discontinued.

Therefore, it can be advanced that the control of tsetse flies requires an integrated approach of a carefully-selected combination of strategies along the following paths:

i. *Habitat management:* a systematic clearing of bushes or thickets in tsetse infested areas, as well as potential habitats amenable to inversion of the vector.

ii. *Chemical control:* a selective application of specially-concocted insecticides (i.e. of minimum negative effects to non-target lifeforms such as the beautiful butterflies resident in the area), at specific dosages.

iii. *Biological control:* use of the release of sterile males to compete with non-sterile males thereby significantly reducing productive mating activities with possibilities of reducing the population to tolerable levels. However, this approach requires constant 'seeding'of the tsetse habitats with the steriles, and hence a careful release schedule is mandatory, which may be difficult to maintain on a long-term basis. Another approach would be to develop an efficient natural tsetse predator. The alternative and permanent solution is to de-gazette the game park (i.e. a zero wildlife strategy). However, a recent attempt to do so encountered great public resistance. The park is the only known home of the rare roan antelope (*Hippotragus equinus*) among other life forms.

iv. *Other methods:* use of artificial refuges designed and applied to trap the tsetse flies. The target is to reduce the fly population density by well over 90 per cent, thereby reducing their impact in affected areas.

As observed by Shoemaker (1973), biological control has the advantages of being density-dependent, ecologically -selective, flexible enough to retard the pest's, development of resistance and long lasting. Hence, the tsetse population in the Lambwe ecosystem will continue to persist with a high degree of resilience because of the absence of an effective predator. Therefore, the challenge facing a medical geographer is to develop mathematical models based on Optimal Control Theory for policy design. Perhaps this way there would result a radical reduction in the HAT infections. As for the case of the tsetse menace, such a design would yield vital knowledge on the impact of ecological factors that generate threshold population levels. Consequently, control actions so designed would be more meaningful.

Host

With the rapidly by increasing human population in the study area ($\approx$ 3% per annum), humans continue to exert tremendous pressure on the available land resources. Consequently, humans have invaded habitats around Ruma Park which were previously infested with tsetse flies, an equally rising population. As a result, humans have increasingly exposed themselves to a high degree of interaction with this vector, especially along the fringes of the Park, thus enhancing the likelihood of contracting the HAT infection.

In order to develop solutions to the problem, three important issues must be addressed, namely, an assessment of human population dynamics (especially along the tsetse foci periphery), an estimation of human-tsetse contact probabilities as a consequence of vectorial capacity of infection rates and an evaluation of existing health care delivery systems and relating them to epidemic frequencies. The challenge facing a DSA/HSA is, therefore, to achieve an integration of a wide range of information on the HAT disease in a decision-making framework. In the next section, the underlying theory is considered.

Decision-Making Processes in the Management of Trypanosomiasis

A resource manager in the Lambwe ecosystem would be faced with the problem of developing appropriate resource management strategies bearing in mind that the infective tsetse population is resident in Ruma Park with which humans come into contact. The primary goal of such a manager is would be to increase agricultural productivity in the area as well as achieve near-zero infections of HAT. Definitely, this is a difficult task to accomplish as it is compounded by a rapidly changing human perception of the living environment. All the same, some decisions must be made for an effective implementation of public health policy. These decisions should be based on the developmental stages considered below.

Disease Theory

The development of an authentic disease theory must incorporate epidemic cybernetics, theories of disease development, evolution and co-evolution, energy flow processes and holistic approaches to disease dynamics' modelling. This stage forms the foundation of the systems analysis approach and has three notable phases:

i. *The input phase:* Information is derived from various sources and compiled/integrated into smaller modules for processing in a systems framework;

ii. *The output phase:* Results (i.e. information) generated by systems modules developed at the first stage above are translated into decisions; and

iii. *The systems generator:* The two phases above are driven by the systems generator which provides tools for information processing and management. Here, the electronic digital computer plays a key role.

Trypanosomiasis Information System

A successive development of the disease theory provides the necessary experience to understand the underlying dynamics of the HAT disease. At this moment, the concepts of information systems may be applied to develop a Trypanosomiasis Information System (TIS). However, the design and implementation of a TIS requires accurate and precise data, a rather challenging condition. For instance, morbidity data compiled for national statistics in most LDCs are often marred with inaccuracies. Hence the DSA/HSA would rely heavily on the knowledge gained at stage one and proceed from there to develop the TIS. For a TIS to be authoritative, information on the salient aspects of the disease must be meticulously compiled. These aspects are briefly considered below.

i. *Geographical coverage*: Actual and potential vector habitats are mapped using existing topographic maps and techniques of teledetection (i.e., remote sensing). The epidemic surface should be estimated from morbidity data and vector diffusion patterns synthesised. Ecological variables should be quantified on floral and faunal aspects on both temporal and spatial dimensions. Parameters for ecological models should be generated and the impact of climatic regimes over the region of infestation synthesised.

ii. *Vector Dynamics*: Entomological and ecological literature pertaining to the vector should be synthesised. In the case of *G. pallidipes*, a wide range of literature exists. The efficiency of the vector as a disease transmitter should be well understood.

iii. *Aetiological Aspects*: The role played by the parasite (i.e., *trypanosome*) in the disease ecology should be examined. If possible, information on parasite loads in the vector environment, and the underlying dynamics should be deduced.

iv. *Human Ecology*: The human being's living environment should be synthesised. Their behaviour with regard to interaction with the vector habitats must be examined. Such information is vital in the estimation of probability of susceptibility to infective vector bites. The situation in the study area reveals that the total number of parasite carriers per delineated habitat (j^{th}) is defined as follows:

$$PC_j = \sum_{i-1}^{n} \left\{ PET_{ij} \left(\frac{dN_{ij}}{dt} \right) + P_{fij} \left(\frac{dP_{ij}}{dt} \right) \left(\frac{dA_{ij}}{dt} \right) \right\}$$

where,

n = the delineated study locations within the Lambwe Valley ecosystem.

PET_{ij} = probability of effective transmission of the HAT infection in the jth habitat of the ith location.

P_{fij} = probability of tsetse being infected during one ovarian cycle or age category.

dN_{ij}/dt = density-dependent population (human) growth rate in the jth delineated habitat of the ith location.

dP_{ij}/dt = the rate of change with respect to time of parasites (per unit volume of blood ingested by the vector) from the reservoir agents in the jth habitat of the ith location (Mwanje, 1992).

dA_{ij}/dt = the combined population growth rate for all possible parasite reservoir agents in the jth habitat of the ith location.

Hence, human ecology as influenced by land use systems and environmental conditions should assist in the evaluation of public health policy in the provision of health care to the HAT victims.

Systems Modelling

Systems models need to be developed using a carefully selected menu of qualitative and quantitative approaches for each of the foregoing concerns and then assembled into mnemonic modules. This step is critical towards the success of the systems analysis approach.

Fig 7.5: Stages in the modelling processes in tripanosomiasis information system's (TIS) development

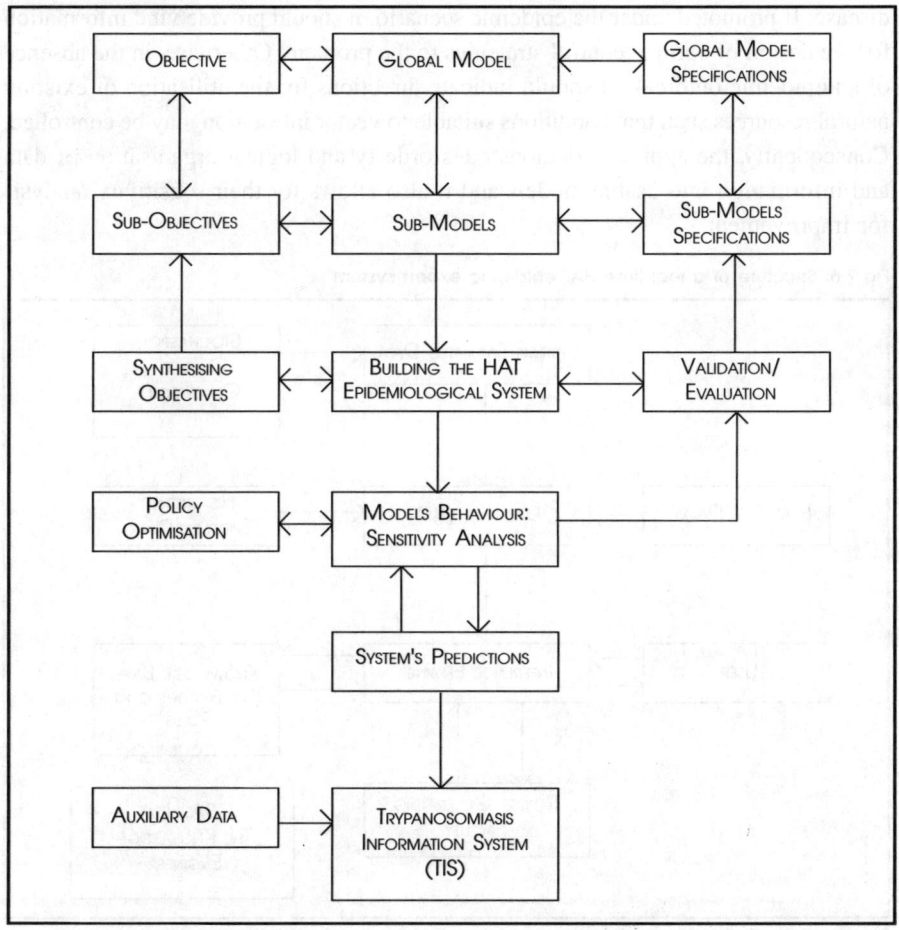

Source: Compiled by the author.

i. *Disease epidemiological model:* It may be necessary to update earlier models and/or modules. The critical sectors are the sensitivity analysis which involves the identification of structures and parameters significant to the generation of answers to the model behaviour which interests the DSA/HSA. If expected behaviour is found wanting, then more effort should be directed to model calibration and validation procedures. It may be necessary to repeat several procedures until satisfactory results are achieved (Mwanje, 1992).

The Human African Trypasonomiasis Information System

The end product of the foregoing procedures is the maturation of TIS. The latter should play a major role in the generation of management decisions on the HAT disease. If promoted under the epidemic scenario, it should provide vital information for the design of optimal control strategies to the problem. Otherwise, in the absence of an epidemic outbreak, it should indicate directions for the utilisation of existing natural resources such that conditions suitable to vector infestation may be controlled. Consequently, the approach demonstrates orderly and logical organisation of data and information into usable models and it also allows for their sensitivity analysis for improvement.

Fig 7.6: Structure of a real-time HAT epidemic expert system

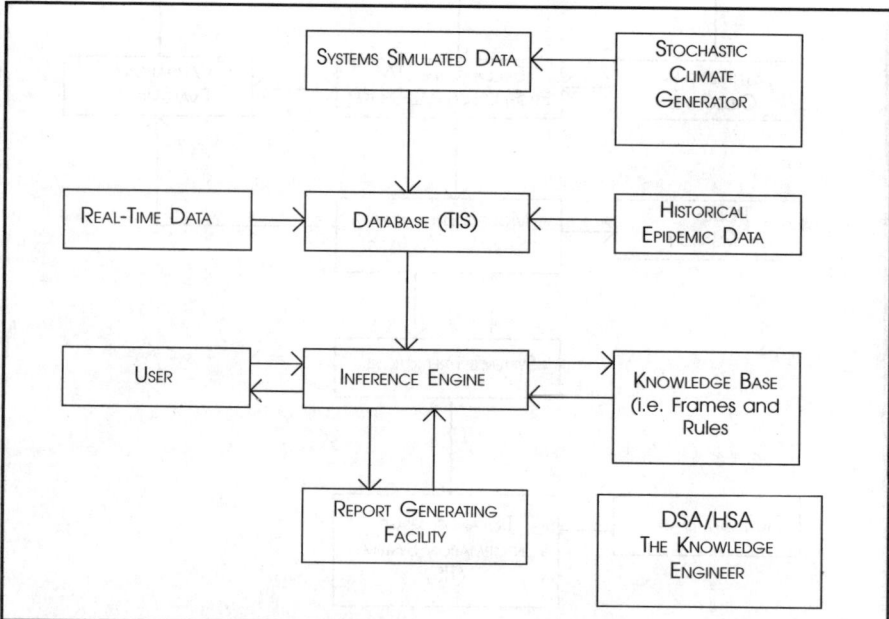

Source: Compiled by the author.

It is easily recognised that each of the procedures explained above determines a specific action in the systems-analysis approach. Further, these techniques are rich as they traverse several disciplines and sub-disciplines of physical, biological, and social sciences. Hence the TIS designer must be adequately and sufficiently sophisticated at coordination. Phillips (1985: 404-407) appears to have had this in mind when he wrote his critique of Paul's review.

The systems-analysis approach promotes good decision-making for such practical applications in disease dynamics. Its framework mitigates serious thought on complex and vast phenomena such as the transmission of the HAT disease, which may not be amenable to solution by simpler approaches, for example, basic field surveys and laboratory experimentation.

An expert system interface

Given a well-developed TIS, the techniques of artificial intelligence show a great potential in medical geography. These techniques have already been applied in ecology integration of qualitative and quantitative knowledge, theoretical development and natural resource management.

Fig 7.7: The structure of the geographic systems analysis of HAT epidemic

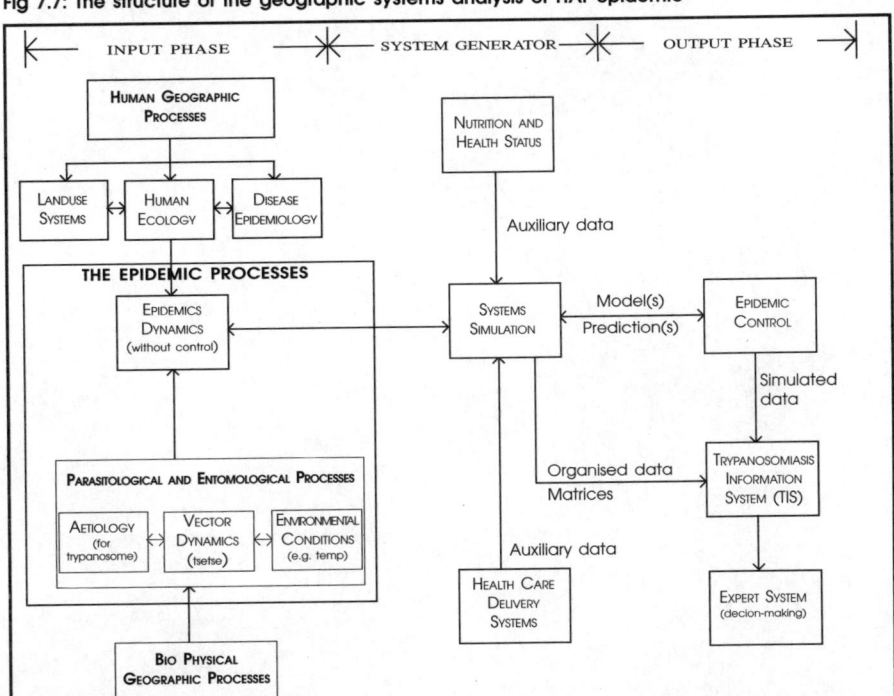

Source: Compiled by the author.

The most versatile branch of Artificial Intelligence (AI) is the 'Expert Systems' which has two key components: a rudimentary understanding of database (in our case, the TIS) structure and a set of rules. The rudimentary structure may be decomposed into what is known as frames. A practical guide on steps to create an expert system is given by Buchanan et al (1983), Hayes-Roth et al (1983) and Genesereth and Nilsson (1987). For the study of of the HAT disease the necessary procedures are as outlined in Figure 7.5.

Conclusion

It is, therefore, feasible to design an expert system for the HAT disease based on the comprehensive TIS developed by a knowledgeable DSA/HSA as a coordinator of a team of researchers. Such a knowledge link between the expert and the non-specialist, field extension and development officers would provide a strategic tool for the management of public health policy on the African continent. Moreover, it would provide an agenda for an integrated utilisation of resources within the Lambwe Valley ecosystem, the case study area, in which human sleeping sickness is prevalent.

8   Nuptial Determinants of Fertility: A Case Study of Western Kenya

Elias H. O. Ayiemba

Introduction

Differential fertility studies incorporating marital conditions as independent variables have increased in importance only during the last decade. The traditional approach to the analysis of fertility determinants which emphasises the significance of nuptial determinants of fertility is rooted in the Malthusian assumption that fertility is confined to marriage. For this reason, marital fertility and the related nuptial factors which determine it are often considered as dependent factors to socio-economic variables which are often assumed to influence fertility directly. The Malthusian analytical framework of fertility determinants does not recognise the possibility that the socio-economic status of an individual could also be determined by changing nuptiality. It is, however, unquestionable that these implied influences occur within the societal institution of marriage or forces which are directly related to nuptial conditions. It is, therefore, an awareness of this among demographers and sociologists that has brought about an emphasis on nuptiality in studies during the last two decades. During this phase of socio-economic and political development in Kenya, fertility research should focus more on factors likely to change substantially the fertility level of a greater proportion of the population.

The significance of nuptial variables in the context of Kenya's current population growth rate of (approximately 3.4% p.a.) is best expressed in the statement that "there appeared to be some evidence from comparison of the Kenya fertility survey with earlier survey results that the total fertility rate had risen by some 18 per cent since 1982 (UNECA, 1979). It should be emphasized that the noted rise in fertility was associated with numerous and stable marriages and a corresponding decrease in childlessness. Whereas the institution of marriage may partly accelerate fertility increase *ceteris paribus*, the same institution can become a significant instrument in society for fertility reduction. In fact, Caldwell (1977: 28) explicitly stated that,

> it is possible to reduce substantially fertility in the contemporary Third World through delayed marriage rather than by any desire to control ultimate family size.

In societies where child-betrothal is commonplace, a practice which is still perpetuated by some ethnic groups in western Kenya such as the Luo, Kuria and suba the customary average age at first marriage is below 18 years. For this reason, there is need for some research into nuptial levels and trends and causative factors in order to establish clearly the relationship between nuptial variables and fertility.

The realisation that the subsistence economy which dominates rural regions of Kenya thrives on a female labour force also augments the need for nuptiality research. Moreover, bride-price or dowry as well as the cultural practices of property inheritance, family immortalisation (through naming of children), polygamy, levirate marriages and land acquisition pivot on the institution of marriage.

From a demographic perspective, medical knowledge and planning for maternal health and child care require adequate information on regional patterns of variables such as age at first marriage. This is because the risk of pregnancy or conception is associated with age at first marriage assuming other variables are constant.

Nuptial levels and trends are a function of socio-economic, cultural and political as well as situational determinants. This realisation reinforces the need for a two-tier system of analysis between nuptiality and fertility. One level of focus should be on the relationship between socio-economic, cultural and political variables including environmental factors and nuptial variables. The fundamental aim should be to identify factors which function through nuptial variables to influence fertility. The second level of focus should regard nuptial variables as an independent factor and show the magnitude and direction of association between nuptiality and fertility. This two-tier system of analysis is crucial for identifying policy measures on fertility behaviour and appropriate instruments for such policy action because the hierarchy of fertility determinants alters their significance with the changing tempo of modernisation processes.

Literature Review and Conceptual Framework

Many studies which focus on the relationship between selected nuptial variables and fertility have regarded marital fertility as a dependent variable. More precisely, such studies have often regarded nuptial variables as insignificant determinants of fertility differences instead over-emphasising socio-economic, cultural and other situational variables which are conceptualised as independent variables in differential fertility analysis models. Such studies have often regarded fertility levels realised outside marriage as too insignificant to warrant any policy measures. Today, this viewpoint is unacceptable because of the rising problem of teenage pregnancies and single parenthood. During the last two decades, however, emphasis has been directed to the analysis of marital phenomena as independent variables in causal analysis of fertility determinants. Such literature may be classified under the following research themes: general studies on nuptial determinants of fertility; age at marriage; marital stability and instability; marriage types (especially polygamy); divorce and separation.

Under the category of general studies on nuptial determinants of fertility, the pioneering work of Hajnal (1974:2) found a positive correlation between an increasing marriage trend and increasing fertility levels in England and Wales. Gabriel (1953: 72) also found some correlation between rising levels of pre-marital pregnancy and fertility in Palestine. Others such as Collier et al (1967); Bumpass et al (1968), Abu-Lughod (1965: 235-253) and Tauber (1960: 264-283) have come to similar conclusions. It should be noted that these studies were disprutive and failed to control the effect of socio-economic forces.

According to the Kenya Fertility Survey (KFS) about 84 per cent of the ever-married women interviewed were still married to their first husbands and only 12 per cent of such women had been divorced or separated, with 5 per cent widowed (Kenya, 1980). This suggested that marital stability among most couples in Kenya contributes to rising fertility levels *ceteris paribus*. Several studies exist on the age at first marriage as a determinant of fertility levels. The UN (1973) summarised the significance of age at first marriage well when it said that,

> of the variables relating to nuptiality age at marriage and the proportions of persons in a population who never marry are the *two* believed to be the most significant accounting for observed variations in fertility levels.

This is supported by Menken (1975: 55-61), Hajnal (1953) and Shafick (1973). The operational hypothesis suggests a positive association between fertility and early age at marriage up to a given age-level, then inversely associates fertility with late age at marriage. It is worth noting that Shafick's study of Cairo (1973), is in particular controlled for child mortality, duration of marriage, religion and education of the mother. For this reason, it yielded better results that clearly portrayed the relationship between age at first marriage and fertility.

The impact of age at first marriage as a determinant of fertility should, therefore, be examined in the context of the following factors: it is a close measure of the level of fecundity; it causes variations in psychological and sociological factors (e.g. it influences perception on sexual frequency control value judgement and adult roles in society and preferences, among others) and it is closely associated with duration of marriage if other factors are constant.

In Kenya, Ahawo (1987) and the KFS (Kenya, 1980a) undertook two significant studies relating age at first marriage with fertility. The relationships observed have not changed much as revealed by the *Kenya Contraceptive Prevalence Survey (KCPS) Report of 1984*. It is, however, evident from Table 8:1 that in western Kenya the Luo, who dominate the districts of Nyanza Province, marry at younger ages relative to the Luhya who dominate the districts of Western Province. Similarly, the Abagusii also have relatively higher age at first marriage. Factors causing these differences are explained in the section discussing ecological factors of the study area. It should be realised that the two cited studies on the Kenyan environment fail to control for the effect of socio-economic forces on fertility levels and trends. The studies also do not isolate the nature of the relationship between socio-economic factors and nuptial factors.

Table 8.1: Age at marriage in western Kenya, 1978

Index of measurement	Nyanza Province	Western Province	Luo	Luhya	Abagusii
Median ages at marriage	16.6 yrs	17.5 yrs	16.2 yrs	16.2 yrs	17.8 yrs
Singulate mean age at marriage (SMAM)	18.9 yrs	19.1 yrs	18.4 yrs	19.0 yrs	20.2 yrs

Source: Kenya (1980a:74).

Further examination of literature on nuptiality reveals that the concepts of marital stability or instability are difficult to define. Nevertheless, a study by Onaka and Yaukey (1973: 457-469) in Latin America found legalised unions to have relatively higher fertility levels than consensual unions which were less stable. More precisely, marital instability, which is apparently high among less fecund women, is positively associated with low fertility levels. Such findings were confirmed by Nerlove and Schultz (1970) for Puerto Rico.

Polygamy depresses fertility. The KFS report noted that polygamy is widely practised in Kenya involving about one third of currently married women. Polygamy was also found to be common among younger women under the age of 25 years, especially among non-educated women. Mosely (1981), analysing KFS data, found that polygamy was causing low fertility levels because of the relatively low pregnancy progression ratio of such women.

Contrary to the above suggestions, few studies have concluded that no significant difference in fertility appears between monogamous and polygamous married women. The rationale for this suggestion is that polygamy is always conditioned by the infertility of the first wife; hence, marrying another wife raises the fertility level of the household. Henin (1979a) and Mosely et al (1981) supported this opinion. Whatever the case may be, polygamy can suppress fertility if reinforced by other cultural practices such as breast-feeding, sex taboos, cultural rituals and the traditional housing system (where every married male resides in his own hut while his wife or each of his wives has her own separate residential hut).

Finally, micro-regional analysis of nuptiality levels and patterns in Kenya associated these levels with fertility as derived from the 1979 census data. Some postgraduate students at the University of Nairobi's Population Studies and Research Institute (PSRI) have also embarked on similar studies that associate nuptial variables with fertility levels and trends, but most of these studies have not yet been completed to provide a basis for critical evaluation.

The Research Model on Determinants of Fertility

This research model has selected only few nuptial variables to be regarded as independent variables in differential fertility analysis. Most of these variables belong to the family of intermediate fertility determinants of Bongaarts (1982). The selected nuptial variables are age at first marriage; proportion single; proportion currently married; proportion divorced and separated; proportion polygamous and proportion in civil or customary unions. In addition, socio-economic variables plugged in the analysis are female educational attainment; employment and income levels; religion; proportion of females urbanised and infant and childhood mortality.

According to the theoretical model as shown in Figure 8.1, socio-economic and situational variables function through nuptial variables to influence fertility. There could be, however, some socio-economic and situational variables with direct impact on fertility levels as demonstrated by the dotted line.

Fig 8.1: Socio-economic variables in diffferential fertility analysis

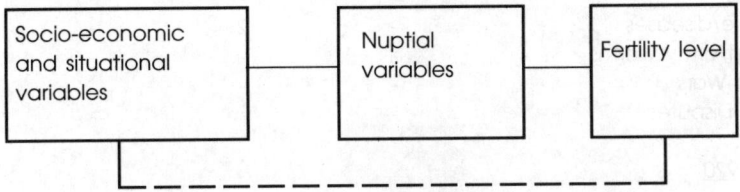

Source: Compiled by the author.

The theoretical framework of the research first correlates selected nuptial variables with fertility levels while holding constant socio-economic and situational variables with nuptial variables; then socio-economic and cultural phenomena are correlated with fertility levels while holding constant effects of nuptial variables. The operational hypothesis is that the selected nuptial variables are better predictor variables of current and lifetime Total Fertility Rates (TFR).

Study Area, Objectives and Methods

Western Kenya was found to be suitable for the study because of several ecological and demographic factors, including the fatct that it is regarded as a melting pot (Ominde, 1963: 26-36; Obudho and Waller, 1976: 1-80). Indeed, this zone experiences constant environmental instability reflected by cyclic patterns of drought, famine, floods, urbanisation and population movements (Table 8:2). The environmental perception created could reinforce the nature of traditional norms or customs which favour a large family size if the environment is perceived to have high risks for survival. Under such circumstances, population policies which favour early marriage are likely to be institutionalised. Moreover, changes in population composition and structure attributed to high mortality rates could reinforce cultural norms which favour levirate marriages due to high incidence of widowhood. Consequently such regions experience high rates of polygamous unions and the associated features of marital instability.

Table 8.2: Recorded environmental hazards in western Kenya: Frequency of occurence by district

Date	Kisii	Kisumu	Siaya	Busia
Before 1920:				
Famine/drought	-	-	1	4
Tribal Wars	27	1	1	5
Livestock Epidemics	35	1	2	1
Floods	-	-	-	-
Plague/diseases	-	-	2	-
Locust/army worms	-	-	-	-
World Wars	-	-	-	-
Land Disputes	-	-	-	-
After 1920				
Famine/drought	-	1	-	5
Tribal Wars	2	1	1	2
Livestock epidemics	1	1	1	3
Plague/diseases	2	3	5	8
Locust/army worms	2	1	2	5
World Wars	1	2	2	2
Land Disputes	1	1	1	1

Source: Compiled by the Author

Furthermore, the frontier populations found in the region are the products of the colonial administration system. Such populations are characterised by high rates of migration which should influence marital characteristics of the study region. Migration may also offer the opportunity for re-marriage for it creates a new socio-economic environment. This probably accounts for the low divorce rate in the study region. Moreover, migration is selective of population characteristics. As a result of this, differentials in sex-ratio could influence the marriage market within the study area. Migration also contributes to changing perceptions of the environment. An individual's perception of ideal family size, norms and age at first marriage cooould be altered significantly.

The regional state of agricultural development in western Kenya is another important factor. Current economic investment projects in the form of large-scale sugarcane farming have also contributed to the nature of population instability in the region. The sugarcane outgrowers receive a high annual income which can be invested in more wives, as the general practice tends to be. In addition, small-scale cash-crop farming affects districts such as Kisii, Bungoma and Kakamega. All these activities could stimulate a desire for early or late marriage or a desire for high standards of living. In traditional economics, territorial expansion encourages polygamous marriage as a means of legitimising land ownership. But, increasing modernisation of the agricultural sector is changing this practice. An individual's psychological behaviour could influence the decision-making process on marital behaviour.

Western Kenya has several urban centres Kisumu, Kakamega, Busia, Kisii, Migori, Homa Bay and Kendu Bay. These are regional nodes of development from which innovative processes such as contraception could be initiated before diffusing to their rural hinterlands. The opportunities generated for different sex roles in these urban centres could influence an individual's desire for early or late marriage, depending on the intensity of rural-urban interaction processes and individual's aspirations are likely to increase with declining distance from urban centres (obudho, 1983).

Focusing on fertility, western Kenya is a region of high fertility (Table 8.3). It is evident that at the provincial level, Nyanza Province on average has a relatively lower fertility rate. The differences between the two provinces (Western and Nyanza) were 24.2 per cent and 13.9 per cent for 1969 and 1979, respectively.

From a geographical viewpoint, the highland regions of the study area constituted the belt of extremely high total fertility rates as Bungoma District had in 1979 an average lifetime TFR of 9.1 births per woman, Kisii District 9.1, and Kakamega District 8.8. (Kisii District now comprises Kisii North, Kisii Central and Kisii South, Bungoma Districts now comprises Bungoma and Mt Elgon Districts; and Kakamega District now comprises Kakamega, Vihiga, Malava – Lugari and Butere – Mumias Districts). Moreover, reference to specific age-groups reveals that a relatively high incidence of current TFR was reported by those aged between 25 and 29 years. This, therefore, suggests that the most important age groups with respect to current fertility were between 30 and 39 years.

It should be emphasised that the level of female infertility is generally moderate in the entire region. The highest proportion of childless women (15-49 years) of 27.0 per cent was observed in Kisii District compared with a relatively low proportion of 21.3 per cent noted in South Nyanza District. (South Nyanza District has now been subdivided into Homa Bay, Migori, Rachuonyo, Kuria and Suba Districts.) This indicates, therefore, that district variations were rather small. The age-groups most affected were 15-24 years, a factor probably caused by high proportion of single women in educational institutions. The proportion of childlessness was, however, noted to increase slightly between 1969 and 1979 for all districts except Kisii. It is realised that, in general, the entire Western Province had a slightly higher proportion of single females compared to Nyanza Province without Kisii District. However, these regional differences were rather small and statistically insignificant. At this juncture, voluntary or involuntary celibacy whether of partial or permanent nature is an important predictor variable of fertility level.

Table 8.3: Selected indices of fertility level in western Kenya, 1969 and 1979

Province and district	Current total fertility rate		Lifetime fertility	Total rates
	1969	1979	1969	1979
NYANZA	6.2	6.5	7.4	7.9
Kisii	7.3	7.3	8.8	9.1
Kisumu	5.6	6.5	7.2	7.7
Siaya	6.0	6.3	6.9	7.0
South Nyanza	5.8	6.0	7.8	7.9
WESTERN	7.7	7.4	8.2	8.6
Bungoma	8.2	8.2	8.7	9.1
Busia	6.5	6.7	7.9	7.7
Kakamega	8.1	7.2	8.2	8.8

Source: Ayiemba (1983 :47).

Universal marriage was a characteristic feature of all districts (Table 8.4). The proportion of currently married women in 1969 ranged between 78.1 per cent in South Nyanza District and 69.5 per cent in Kisii District. In 1979, the proportion was apparently low in all districts as they ranged between 74.9 per cent for South Nyanza District and Kisii District, respectively. In addition, the proportions of divorced and widowed women were observed to be relatively low.

The whole of western Kenya was found to constitute a homogenous zone with respect to socio-economic development. However, three districts tended to be relatively better developed in the agricultural sector. The three districts — Kisii, Bungoma and Kakamega — also had relatively low levels of infant and childhood mortality. Note that with the exception of primary education, the proportion of females aged 15-49 years with other socio-economic characteristics such as employment, income, land, urban residence and religion tended to be too low to have a meaningful impact on nuptiality and fertility. Only two socio-economic and situational variables were found to be important in the study region. These were primary education and the child mortality rate variables.

Table 8.4: Percentage of females of different marital status in western Kenya, 1969 and 1979

Province/ District	of Single women 15-29 yrs.		of Currently married women 15-49 yrs.		of Widowed women 15-49 yrs.		of Divorced/ separated women 15-49 yrs.	
	1969	1979	1969	1979	1969	1979	1969	1979
NYANZA	17.8	21.6	75.0	71.1	4.8	4.4	2.1	2.0
Kisii	23.4	29.4	69.5	64.4	3.8	3.3	2.4	2.1
Kisumu	16.1	19.3	74.9	72.9	4.9	4.6	2.4	2.2
Siaya	14.6	18.6	77.5	73.7	5.0	5.2	1.8	1.8
S. Nyanza	13.0	17.2	78.1	74.9	5.4	5.0	1.8	2.0
WESTERN	21.6	24.7	73.0	69.5	2.3	2.6	2.9	2.2
Bungoma	21.6	24.4	73.1	70.2	1.9	2.1	2.6	2.5
Busia	15.2	19.5	78.1	74.6	3.0	3.5	2.0	1.8
Kakamega	21.5	26.4	71.6	67.5	2.4	2.6	3.2	2.2

Source: 1969 and 1979 census data.

Analysis and Results

A two-stage sampling design was adopted for this research. The first sampling stage was the areal unit sampling frame which enabled 25 per cent of all the 520 sub-locations in the study region to be selected. These sub-locations were categorised on the basis of population density into three categories, namely, high density regions, medium density regions and low density regions.

The choice of households constituted the second stage of the sample frame. On the basis of a list of householders in the selected sub-locations, about 854 households were surveyed in Nyanza Province compared to 411 households in Western Province. These sample sizes represented 0.41and 0.79 per cent of the total households in the provinces, respectively. The application of the F-distribution test supported the conclusion that there was no sufficient evidence to indicate significant differences in the average number of sampled households in both provinces. Furthermore, the research relied on primary and secondary data that covered nuptial, socio-economic and fertility variables. An attempt was also made to graduate such data for response errors, especially with respect to age-sex distribution and fertility response. Data analysis consisted of simple demographic techniques of fertility estimation such as the Brass P/F ratio method for fertility adjustment, Sullivan's (1972) method for mortality estimation and multivariate analysis in determining the magnitude and direction of association between variables.

The application of step-wise regression analysis yielded the results in Table 8.5. It is observed that, with respect to current TFR, the nuptial variable with maximum average parity is widowed women (3.1). It has also the highest value of zero-order partial correlation coefficient (0.8003). On the other hand, the nuptial variable with the least average parity is single women (0.7), though women in polygamous unions yielded the lowest value of zero-order partial correlation coefficient (0.1218).

Table 8.5: Step-wise regression analysis results

Nuptial variables	Average parity (1978)	Correlation coefficients
Single women	0.7	0.1349
Currently married women	2.4	0.5075
Widowed women	3.1	0.8003
Divorced/Separated women	2.0	0.7976
Women in monogamous unions	2.0	0.1308
Women in polygamous unions	1.8	0.1218
	$\Sigma = 12.0 = 2.4929$	
	$x = 2.0 = 0.4155$	

Socio-economic variables	Average parity (1978)	Correlation coefficients
Lower primary education	1.2	0.6869
Upper primary education	0.9	0.4199
Secondary education	0.6	0.5597
Self-employed	1.7	0.2730
Wage-employed	1.7	0.0295
Protestant women	1.8	- 0.9324
Catholic women	1.9	- 0.9696
Urban residence	1.7	0.7103
Land title deed	1.7	- 0.9433
	$\Sigma = 13.2 = - 0.0650$	
	$x = 1.47 = - 0.0072$	

Source: Ayiemba (1983: 255)

Focusing on the results of the listed socio-economic and situational variables, it is noted that all these variables yielded a relatively low average parity of 1.47 compared with 2.0 for nuptial variables. In addition, Catholic women had the highest value of average parity of 1.9 births. Secondary education yielded the least average parity of 0.6 births.

Table 8.6: Current fertility rate analysis

Index of measurement	Average parity		Correlation coefficients	
	Nuptial variables	Socio-economic variables	N- variables	Socio-economic variables
Mean	2.0	>1.33	0.4155	> -0.065
Maximum	3.1	>1.9	0.8003	> 0.7103
Minimum	0.7	>0.6	0.1218	> 0.0295
Mode	2.0	>1.7	0.7976	> 0.2103

Source: Ayiemba (1983: 257)

For the current TFR, there is statistical evidence which supports the hypothesis that nuptial variables are more important than socio-economic and situational variables as predictor variables of current TFR (Table 8.6.) In the case of the lifetime total fertility rate, the situation is definitely reversed Table 8.7. The computed mean parity for nuptial variables is 6.9 and the nuptial variable with the maximum average parity is currently married women (8.2). It is important to realise that the proportion of single women yielded the highest value of positive correlation coefficient (0.4475). This definitely implies that illegitimate fertility is important in the long run and not in the short run. Further analysis shows that the computed mean average parity for socio-economic variables (7.8) is higher than that of nuptial variables (6.88). For the lifetime TFR, there is evidence that nuptial variables are less important than socio-economic and situation variables as predictor variables of lifetime TFR (Table 8.8).

The research in general proved that nuptial variables have relatively greater impact on current TFR or fertility realised within a relatively short period. However, with a longer period, socio-economic and situational variables become more important determinants of fertility levels and trends.

Table 8.7: Regression results, 1978

Nuptial variables	Average parity (1978)	Correlation coefficients
Single women	4.1	0.4475
Currently married women	8.2	0.0628
Widowed women	8.1	0.1213
Divorced/separated women	6.1	0.0715
Women in monogamous unions	7.8	0.0855
Women in polygamous unions	7.0	0.0612
	$\Sigma = 41.3$	$= 0.8498$
	$x = 6.88$	$= 0.1416$

Socio-economic Variables	Average Parity	Correlation Coefficients
Lower primary education	8.1	0.0362
Upper primary education	8.1	0.1726
Secondary education	6.0	-0.0689
Self-employed	7.6	0.2372
Wage-employed	7.5	-0.0225
Protestant women	8.1	-0.9529
Catholic women	8.4	-0.8947
Urban residence	8.1	0.6745
Land title deed	8.3	-0.9569
	$\Sigma = 70.2$	$= -1.7754$
	$x = 7.8$	$= -0.1973$

Table 8.8: Lifetime total fertility rate analysis results in western Kenya, 1978

Index of measurement	Average parity		Correlation coefficients	
	Nuptial variables	Socio-economic variables	N-variables	Socio-economic variables
Mean	6.9	< 7.8	0.1416	< 0.1995
Maximum	8.2	< 8.4	0.4475	< 0.6745
Minimum	4.1	< 6.0	0.0612	> 0.0362
Mode	7.8	< 8.1	0.4472	< 0.6745

Sources (Tables 8.7 & 8.8): Ayiemba (1983: 253-260)

Conclusion

This study had two important objectives, namely, to investigate the strength of the relationship between each of the nuptial variables and total fertility rates and demonstrate that socio-economic and situational factors influence fertility not only directly, but also indirectly through their impact on nuptial variables. These objectives have helped identify the following measures for controlling rapid population growth in the region. Policy makers should pay attention to the following policy issues, namely, age at first marriage education employment and income policy, healthcare, land adjudication and consolidation and family planning. It is hoped that these measures can be applied to other regions in Kenya.

Age at first marriage was found to be generally lower than the stipulated legal age. The research found the tempo of fertility to decline after the age of 26 years. For this reason, it is recommended that the marriage law should set and the minimum age at first marriage at 21 years. Though such a policy would be difficult to implement, putting in place an institutionalised system of punishment and reward to affect couples and parents-in-law through supervision by local chiefs is a possible alternative. Such a policy can be effective if the length of primary education for girls is extended. In other words, an educational policy which encourages higher education for girls is needed. For a long time, female education in Kenya has been secondary to that of boys because of socio-cultural phenomena which place a higher value on male progeny. Educational policy can be reinforced with a strong religious education policy which can reinforce moral restraint among young people and discourage the incidence of polygamy, divorce and separation.

Employment and income policy that focuses on the female population is overdue in Kenya. There is a threshold level in income after which fertility levels begin to decline. In western Kenya this threshold is about Kshs 3,000.00 per month. Employment opportunities increase extra-family activities for women and this reinforces the desire to use contraceptives.

A health care policy that improves survival rates is needed. Childhood mortality is a negative predictor variable of lifetime total fertility rate. So long as the survival rate is still low, women will be reluctant to use contraceptives. It should, therefore, be the responsibility of the Ministry of Health to invest heavily in preventive and curative medicine in order to fight the psychological war on large family size.

Land adjudication and consolidation policy should be intensified to enable the majority of the rural population to discard the "land myth" syndrome. Once land is perceived as scarce, polygamous unions will die out and the age at first marriage will rise with increasing levels of income derived from agricultural investments. (Note that this is a long-term expectation.)

Urbanisation definitely suppresses fertility in the long-run. For this reason, attempts should be made to create more rural service centres that provide urban services with respect to mass media, recreational facilities, accessibility to social clubs for exchange of ideas, etc. Such urban services can provide effective channels for diffusing ideas of modern contraception (Obudho, 1994: 198 -212).

Finally, the current family planning programme needs to be re-evaluated and made more effective through improved delivery service points and skilled personnel. The diffusion of family planning ideas should also focus on the target population at greatest risk namely, the adolescent population and young mothers who still have a long time to attain their desired family sizes.

9 Mortality Differentials and their Implications for Socio-economic Planning in Kenya

Zibeon Muganzi

Introduction

Although mortality is still high in the LDCs as compared to that of MDCs, there has been a dramatic mortality decline in the LDCs which could be attributed partly to socio-economic factors, but mainly to the availability and utilisation of imported medical technology. This technology includes antibiotics used in the treatment of infections and respiratory diseases, immunisation against tuberculosis and smallpox as well as anti-malarial campaigns. Mortality is said to have declined in most countries even where the level of socio-economic development is not extensive.

Nevertheless, despite the impressive measures taken to reduce mortality in the LDCs, there is also increasing evidence that there are still substantial mortality differentials both between and within populations in these countries. These differentials are attributed to the persistence of a multiplicity of factors which are highly interrelated which include infectious diseases, inadequate health facilities, and natural factors such as drought and floods. In addition, there are other numerous socio-political and cultural compounding factors which cannot be directly measured or observed, not to mention the definition and measurement problems which further complicates the interpretation of the results. It is in this context of complexity that the relationship between mortality and socio-economic development continues to engage most governments in the LDCs, amidst limited and unequally distributed resources. This chapter will address the problem as it relates to Kenya.

Historical Background

An understanding of both the determinants and differentials of mortality is essential for a number of reasons:
- its contribution to the total loss of human life is substantial because both it occurs early in life and more so because its level is relatively still high in the developing countries;
- its cause tends to be largely distinct from those which operate at the older ages of childhood and at adult ages;

- its measures provide a useful index of the status of health as well as the standard of living of society;

- its prevention has been and remains a major pre-occupation of the health authorities at both national and international levels; and,

- the degree of success of health programmes could be ascertained on the basis of the observable decline in infant mortality.

Many studies in LDCs have emphasised the inverse relationships between various socio-economic factors and mortality. Some of the variables studied have included income, education, class status and age of mother as well as contextual factors such as piped water, type of housing, accessibility to health facilities and amount of caloric intake.

Studies incorporating these factors and how they relate to mortality have also been completed in Kenya. The Kenyan case has in most cases presented its own peculiarities for the justification of the studies of these relationships. First, the present socio-economic situation is deeply rooted in the colonial past. The colonial government maintained separate levels of development not only for Africans and Europeans, but also for various regions of the country. This imbalance has survived the post-independence era to the extent that there is still an increased maldistribution of services and facilities, especially in the rural areas where more than 80 per cent of the population live. Such services include water, schools, health facilities and roads. The post colonial government effort to correct this imbalance has been slow and expensive, especially with a rapidly increasing population of which more than half is under 15 years.

Closely related to the above regional imbalances is the decision making and policy implementation structure in the country which is hierarchical in nature. Planning takes place at various levels, right from the national to the provincial, district, divisional, locational and sub-locational levels. Bureaucratic delays in the implementation of the desired development programme have led to non-availability of the desired facilities at the local level. This situation is likely to change with the implementation of the District Focus For Rural Development (DFRDS).

Apart from these planning and structural concerns, recent studies in the country show the persistence of large mortality differentials at both micro and macro levels. Indeed, though recent estimates show a steady but gradual decline in infant mortality over the years (from an estimated rate of 184 per 1000 in 1948 to 83 per 1000 in 1980), this trend has not been uniform. Mortality differs not only by region and place of residence, but also by individual characteristics such as age of the mother, education, income, ethnic group and religion. These differentials point out the inadequacies of socio-economic planning, as will be shown later in this chapter.

Literature Review

The history of mortality differentials and their cause and remedies are well-documented in MDCs. Having been able to identify the major factors in child and infant mortality decline in these countries and having implemented successful policies and programmes to affect them, the attention at both the regional and international level is now focused on LDCs where mortality differentials are still dominant and persistent as reflected by income, education, residence, marital status and religion, among others. The major concern is not to identify the differentials as such, which has already been done for some regions, but to understand the factors which underline and perpetuate these mortality differentials among populations, sub-populations and regions.

With the understanding that mortality is affected by a multiplicity of factors, multivariate statistical methods have been employed in identifying the most relevant factors related to mortality differentials. Though some headway has been made, success is still far from sight due to a lack of data on relevant variables. Moreover, problems of definition and measurement have been great obstacles in the analysis of the available data. However, despite these problems, recent and on-going research on this very important issue has led to the discovery of a positive inverse relationship between mortality and such variables as income, literacy, education of mother, ecological and health services. On the other hand, the relation between mortality and some other variables such as politics and culture (which are not easy to measure and define) has been shown to be ambiguous and, therefore, not easy to interpret. Thus, one of the major problems in these studies has been specification.

Preston (1979) using a number of variables such as literacy, income and calories intake per day from cross-sectional data on a number of LDCs for 1940-1970, was able to show through multivariate analysis that there was an inverse relationship between these variables and infant mortality. Preston's study was a macro-analysis pioneer study and has since stimulated interest in this important area. However, this study was not comprehensive since not many variables were included and those left out could have had even high correlations. It, nevertheless, paved the way for more revealing studies that have followed. Caldwell (1979) examined this relationship at micro-level in a study of the Yoruba people in Nigeria, showing that the mother's education is more relevant than other factors. Cadwell, like Preston, also applied multivariate analysis to cross-sectional survey data from 1978, but controlled for most other factors like residence, income and marital status. He found that an educated woman greatly influences the health and care of her children. Relating this to income and residence, he concluded that such a woman is likely not only to have a higher income, but also to live in an urban centre. Behm (1979) using the same analytic techniques and cross-national data for Latin American countries, has confirmed this important finding. Further evidence has been provided by Trussell and Preston (1980) who applied the covariate method on data from Korea and Sri Lanka and showed that mothers' education was important.

More recently, however, the focus has shifted to emphasise not only such socio-economic factors as health service status (i.e. number of doctors, nurses, health centres and hospitals), but also accessibility to drinking water and transportation systems. Flegg (1982), using cross-sectional data, has shown that while income inequality played a major role (indirect relationship) in perpetuating mortality differentials, equally important were such factors as number of physicians and nurses per 10,000 population. His study reflected and emphasised what Rogers (1979) observed in a similar cross-sectional international study that the determinants of changes in mortality in under-developed countries are difficult to unravel. They include improvement in health technology and its availability, education, sanitation, clean water supply and a host of other environmental variables which, however, tend to be strongly collinear, thus making it difficult to isolate their effects.

While the above studies reflect the international scene, several authors have made attempts at studying mortality differentials in Kenya. The study by Anker and Knowles (1980) was more comprehensive than the others, though it was greatly limited by the poor quality of the data at their disposal. The study focused on child mortality differentials at micro (household) and macro (district) level. While Kenya's 1969 census data was used for macro level analysis to show some consistent results, the household survey data for microanalysis proved inadequate and problematic. This was due to the small sample size which made generalisation impossible, in addition to the measurement and definitional problems of the variables used, for example, alternative approximation for such variables as services, nutrition, hospitalisation, malaria-free areas created more interpretational problems due to difficulties in verifying their validity and reliability. However, despite these problems, the authors were able to show the existence of some inverse relationship between such factors as income, education, type of water used and type of toilet and house and mortality.

Mott (1980), using Kenya Fertility Survey (KFS) data, concentrated on infant mortality differentials with less emphasis on child mortality. His regression analysis, however, did not include contextual factors. Nevertheless, his findings were consistent with other findings that infant mortality differed not only by education, age, parity and income but also by region. The work by Henin (1979a) based on the same data and other demographic surveys in Kenya has validated the findings of the other studies, (i.e. persistent infant and child mortality differentials by socio-economic status). Between these Anker and Knowles (1980) were constrained by inadequate data, and Mott's (1980) studies focused only on socio-economic variables. With the availability of more data, not only the KFS, but also the 1979 round of national census and the nutritional surveys, this study aims not only at bridging these earlier studies, but also hopes to shed more light on the observed differentials.

Study Objectives

The objectives of this study were to re-emphasize the need for the study of mortality determinants and differentials as evidence from on-going studies show, especially in the LDCs in general and Kenya in particular; provide an overview of the various sources of data used in the estimation of mortality differentials in the country; critically examine the mortality differentials estimates from such data at both regional and household level and evaluate the implications of such estimates for socio-economic planning and policy formulation, especially within the strategy of DFRDS.

Existing estimates derived from various secondary sources have been used (in this particular case the KFS), which has been used extensively by a number of researchers to derive various mortality estimates. The emphasis here is on the shortcomings of the data as it could affect the resulting estimates and, therefore, their validity in policy utilisation, formulation and implementation. Other sources of secondary data, especially published government reports, were consulted for supporting information relating to such factors as distribution of health services, physicians, nurses and nutrition status. This information is important in helping to account for observed differentials.

Problems associated with the use of secondary cross-section data are well-documented in the literature. The following are some of the factors that should be considered in the utilisation of such data:

i. apart from the reliability and validity of the concepts and definitions used in the surveys during the collection of data, there is the additional problem of under-reporting or misreporting of the number of children who died—a factor influenced either by customary agitations or lack of accurate recall on the part of the respondent.

ii. for certain concepts, care should be taken to clearly define the focus, e.g., in terms of education, is the information sought relating to formal education alone or any other categories? What are the other categories and how are they defined? In terms of health, are we talking of modern health or traditional health as well? Even for marital status care should be taken to sort out divorced women from the separated and the single, for in most cases divorced women report themselves as single for social status purposes.

Educational Differentials

Education has recently been identified as one of the major factors influencing mortality especially in LDCs. As the education of the mother increases, child mortality tends to decline. Women with at least some education influence child mortality through such factors as better nutrition, personal hygiene and proper utilisation of medical facilities. These women, it is argued, break with traditional

117

practices associated with high mortality such as feeding habits and use of traditional medicine.

The effect of education on mortality in Kenya is well documented. Table 9.1 shows the possibility of child survival to a given age, and the mortality rates and life expectancies by education of mothers; the Table 9.1 also shows that a child whose mother has no education can expect to live up to 51 years. The inconsistencies in the estimated qx values as well as in life expectancies may be due to inherent errors in data or poor classification of educational categories. In any case, the pattern is clear though weak that mortality declines as the level of education increases. It has been shown, however, that education can influence mortality to a given point beyond which other factors must operate for further decline to occur.

The above results go along way to support what has been observed elsewhere. Caldwell (1979), among others, has demonstrated the influence of education and child mortality in Nigeria; Preston (1979), using multiple regression based on a cross-sectional study of 120 countries, showed that a rise of 10 per cent in the proportion of literacy is associated with an increase of two years in life expectancy at birth. Other supporting studies include the Philippines study which showed infant and child mortality to decline steeply with education of the mother and one in Latin America which showed that the qx value for 8 countries declined steeply with education. A UN study of 115 countries also demonstrated a higher correlation between literacy and eox for the other factors.

Table 9.1: Mortality differentials by education, Trussell Method (West Model)

Education status	Age	q(x)	+(i)	Level	Average level	e
		2	.1453	3.35	13.85	
No education	3	.1567	5.05	13.89	13.75	51.88
		5	.1795	6.84	13.51	
		2	.1193	2.46	19.44	
1-8 years	3	.1289	4.71	13.04	15.41	56.03
		5	.1401	7.37	13.74	
		2	.1835	2.87	11.84	
9-14 years	3	.1190	4.73	15.97	14.41	53.50
		5	.1400	6.82	15.41	
		2	.1256	2.51	15.01	
15+ years	3	.1950	4.55	12.07	14.68	54.20
		5	.1106	6.93	16.97	

Source: Kenya Fertility Survey data, 1978.

Urban-Rural Differentials

Generally, the results, as indicated in Table 9.2, show more favourable mortality conditions for the urban areas. This is a common phenomenon in the LDCs where the urban centres have claimed the lion's share of health services, schools and social amenities. A person residing in an urban area can expect to live longer than the rural dweller.

Table 9.2: Mortality differentials by residence, Trussell Method (West Model)

Education status	Age	q(x)	t(x)	Level	Average level	e_x
Rural	2	.1352	2.43	14.43		
	3	.1405	4.38	14.76	16.61	54.025
	5	.1552	6.69	14.64		
Urban	2	.1136	2.64	15.76		
	3	.1106	4.73	16.46	16.26	58.15
	5	.1179	7.16	16.57		
Metropolitan	2	.1096	3.05	16.01		
	3	.1184	4.38	16.00	16.32	58.30
	5	.1108	5.78	16.96		

Source: Kenya Fertility Survey data, 1978.

Note: A "Metropolitan" area includes the two major cities of Nairobi and Mombasa, which have more than half of all the urban facilities in the country.

The observed lower life expectancies for the rural areas reflect their high level of malnutrition, epidemic diseases (e.g. malaria, kwashiorkor and cholera), illiteracy, inadequate and unsafe drinking water supplies and poor transportation. These and other associated factors have been shown to be the direct or indirect killers of children in the rural areas.

The urban areas, on the other hand, are endowed with the bulk of these facilities though not equally. For example, Nairobi and Mombasa account for more than half of the hospital beds, hospitals, medical personnel and health centres. The centralisation of government procedures results in extensive delays in the delivery of such essentials as medicine and educational facilities to rural centres. However, as will be shown later, the strategy of DFRDs is meant to correct this anomaly.

Marital Status Differentials

The probabilities of survival to given ages, mortality rates and the associated life expectancies, by marital status of mothers are given in Table 9.3. The results show lower mortality for children of married women than for those of divorced or widowed ones. For example, whereas qx of married women is 0.105, it increased to 0.135 for widowed and 0.171 for divorced women. However, the life expectancy ex given by average mortality atthese ages differs very little between divorced (54.13), widowed (53.50) and married (56.18) women.

Table 9.3: Mortality differentials by marital status, Trussell Method (West Model)

Marital staus	Age	q(x)	t(T)	Level	Average level	e_x
Married	2	.1053	3.91	16.28		
	3	.1285	5.38	15.43	15.46	56.18
	5	.1546	6.74	14.67		
Widowed	2	.1354	4.97	14.42		
	3	.1597	5.94	13.75	14.40	53.50
	5	.1475	6.47	15.02		
Divorced	2	.1717	3.43	12.45		
	3	.0896	5.33	17.73	14.65	54.13
	5	.1737	7.33	13.76		

Source: Kenya Fertility Survey data, 1978.

Note: The low difference between the eo for various levels may be indicative of the increasing influence of other socio-economic factors helping to bridge the gap.

Marital status diminishes considerably in its influence when other factors especially education and place of residence are considered. In any case, this high mortality probability for children of divorced and widowed women may be attributed to the poor conditions children are subjected to when their parents die or are separated great; care is usually taken to reduce their mortality.

Regional Differentials

As we have already pointed out, due to its historical colonial development pattern which emphasised separate development for regions and populations, considerable differentials in mortality exist in the country between and within the regions among different socio-economic strata. Available evidence attests to the fact that these differentials are reinforced by the climatic as well as physical conditions. The regional variation in terms of births and deaths based on KFS is given in Table 9.4.

Table 9.4: Percentage distribution of births and deaths by region

	Nairobi	Central	Coast	Nyanza	Rift Valley	Western	Eastern	Total
Birth Survived 12 months	87.8 (1674)	88.1 (4289)	80.9 (1765)	77.1 (5035)	88.5 (5004)	80.0 (3204)	85.3 (4426)	83.7 (25,397)
Birth Died 12 months	12.2 (232)	11.9 (577)	19.1 (418)	22.9 (1499)	11.5 (648)	20.0 (803)	14.7 (763)	16.3 (4940)
Total Births	6.3 (1906)	16.0 (4866)	7.2 (2183)	21.5 (6534)	18.6 (5652)	13.2 (4007)	17.1 (5189)	100.0 (30,337)

Source: Kenya Fertility Survey, 1978.

Generally, the tabulation shows a higher percentage of deaths in Nyanza, Coast and Western Provinces, a finding which has been observed by a number of researchers in the country. For example, of the total deaths in the sample, Nyanza and Western Provinces accounted for 30 per cent and 16.3 per cent, respectively compared to 11.7 per cent and 13.1 per cent for Central and Rift Valley Provinces. Among the provinces, Nyanza and Western registered higher percentages of deaths compared to the others — i.e. 22.9 per cent and 20.0 per cent, respectively. These statistics are further explained by Tables 9.5 and 9.6 which show the percentage distribution of children by duration of sickness and type of sickness for each province.

Table 9.5: Percentage distribution of children by duration of sickness and by province

Duration	RURAL						URBAN		RURAL	URBAN
	Coast	Eastern	Central	Rift Valley	Nyanza	Western	Coast	Nairobi	Other Kenya	Kenya
1 day	3.6	3.1	4.0	3.7	7.8	8.2	3.4	7.4	4.3	5.2	5.0
2-3	19.2	10.6	7.2	7.1	11.5	13.9	13.2	13.1	15.6	10.6	13.9
4-7	10.6	7.0	8.6	9.8	11.9	19.2	10.8	10.2	15.2	11.0	12.2
More than 1 week	12.0	12.3	12.1	12.7	17.7	21.5	10.4	8.4	15.7	14.3	11.6
Not sick	54.6	67.1	68.1	66.7	54.1	37.0	62.2	60.9	49.7	58.8	57.4
Total =	100%	100.0									

Source: Integrated Rural Survey Report No. 2, 1978.

ISSUES IN RESOURCE MANAGEMENT AND DEVELOPMENT IN KENYA

Table 9.6: Percentage distribution of children by type of sickness and by province

Type	RURAL						URBAN			RURAL	URBAN
	Coast	Eastern	Central	Rift Valley	Nyanza	Western	Coast	Nairobi	Other	Kenya	Kenya
Fever	25.1	16.3	21.8	19.2	26.7	41.0	22.8	22.1	27.2	24.6	24.1
Diarrhoea	7.3	5.3	3.2	5.8	5.2	8.1	0.3	2.0	7.1	5.5	3.2
Fever and Diarrhoea	9.0	2.4	2.1	2.8	7.6	8.3	3.3	3.6	4.4	4.9	3.8
Other	4.0	9.0	4.8	5.5	6.4	5.6	11.4	11.3	11.7	6.2	11.5
Not sick	54.6	67.1	68.1	66.7	54.1	37.0	62.2	60.9	49.7	58.8	57.4

Total = 100% 100.0
Source: Central Bureau of Statistics, 1978

According to Table 9.7 the three provinces of Nyanza, Western and Coast still have a higher percentage of children with an average of 4-7 days of sickness. It is also evident from Table 9.8 that the prevalence of such diseases as fever and diarrhoea is highest in these same regions. For example, Coast Province has a percentage of 25.1, Nyanza 26.7 and Western 41.0 of children suffering from fever compared to 16.3 Eastern Province and 19.2 for Rift Valley Province.

A further examination of the regional variation of the probabilities of child survival is given in Table 9.7 which also shows the life expectancies and mortality rates for the various regions (provinces). As with other factors (i.e. the duration of sickness and type of sickness), Table 9.7 shows that the probability of dying for the infants is higher in the provinces of Nyanza, Coast and Western compared to the other regions of Kenya. The life expectancies also show a considerable range from 68 years for Central Province to 48.43 years in Nyanza Province. These regional variations in probability of deaths for the infants can better be comprehended when viewed in a broader context of the factors associated with child survival where emphasis is on accessibility and the distance covered to tap vital resources.

Tables 9.8 and 9.9 show the percentage of households in each region having access to given facilities and the distances covered to reach them. The great disparity between these regions is clear from the tables. It is evident that whereas one region may be better off in one aspect, it may be lacking in another, a factor which suggests that the observed mortality differentials are not actually accidental.

Overall, it seems Central and Western Provinces have fairly localised amenities in contrast to Coast Province where the facilities are most dispersed. For example, whereas over 95 per cent of households in Coast Province must cover an average of over 2 kilometres to get to a health centre, in Western Province the percentage is only 79. This figure may be deceptive for some health centres, however, close

Table 9.7: Mortality differentials by provinces, Trussell and Sullivan Methods (West Model)

	Age	q(x)	Trussell t(i)	Level	Average level	e(x)	ex	to	Level	eo
NAIROBI	2	.1004	3.17	16.60			.0994	3.17		
	3	.1153	4.31	16.18	16.68	59.10	.1071	4.31	16.66	59.2
	5	.1053	5.46	17.27			.1163	5.46		
CENTRAL	2	.1108	2.25	25.00			.0495	2.25		
	3	.0705	3.65	18.96	20.11	67.78	.0523	3.65	20.22	68.05
	5	.1220	5.33	16.36			.0560	5.33		
COAST	2	.1482	2.76	13.69			.1514	2.76		
	3	.1598	4.66	13.74	13.41	51.03	.1645	4.66	13.52	51.30
	5	.1963	6.82	12.78			.1793	6.82		
NYANZA	2	.1872	2.55	11.65			.1758	2.55		
	3	.1874	4.50	12.43	12.23	48.08	.1915	4.50	12.37	48.43
	5	.2002	6.78	12.62			.2094	6.78		
EASTERN	2	.1114	2.43	15.89			.1172	2.32		
	3	.1346	4.26	15.08	15.40	55.93	.1267	15.53	56.33	—
	5	.1435	6.57	15.23			.1376	6.57		
RIFT VALLEY	2	.0871	2.67	17.48			.0915	2.67		
	3	.0952	4.53	17.38	17.09	60.10	.0985	4.53	17.18	60.21
	5	.1210	6.66	16.40			.0169	—		
Western	2	.1556	2.32	13.29			.1535	2.32		
	3	.1922	4.60	12.20	13.22	50.55	.1668	4.60	13.40	51.03
	5	.1650	7.34	14.16			.1820	7.34		

Source: Computed from Kenya Fertility Survey data, 1978.

they are to the people, they may be ill-equipped to handle even the simplest cases of illness. This is due to lack of supplies which are usually sent from Nairobi and which may take months to arrive at a village dispensary. In this case many people prefer going to hospital than to health centres.

Another aspect relates to the distance covered to get to a source of water. While over 97 per cent of households in both Western and Nyanza Provinces are within 2 kilometres of a water source, only 71 per cent are so placed in Coast Province. Again these figures beg for caution since most sources of water are rivers and streams which in most cases are contaminated.

It is not only the availability but also the accessibility to the given services that is important in their utilisation. According to Table 9.8; for example, in Western Province, only 2 per cent of the households have access to piped water compared

Table 9.8: Percentage distribution of households by distance to social amenities and by province

	Coast	Eastern	Central	Rift Valley	Nyanza	Western	National total
Primary School							
up to 2 km	50.9	77.7	81.7	58.9	76.0	82.9	73.2
over 2 km	49.1	22.3	18.3	41.1	24.0	17.1	26.8
Government Secondary Sch.							
up to 2 km	5.8	13.5	9.3	9.1	7.2	8.4	9.2
over 2 km	94.2	86.5	90.77	90.9	92.8	91.6	90.8
Harambee Secondary Sch.							
up to 2 km	9.7	19.6	30.6	8.7	21.0	19.2	19.1
over 2 km	90.3	80.4	69.4	91.3	79.0	80.8	80.9
Health centre							
up to 2 km	4.7	11.9	14.1	16.4	18.1	21.1	15.5
over 2 km	95.3	88.1	85.9	83.6	81.9	78.9	84.5
Local market							
up to 2 km	17.7	29.2	31.9	24.4	54.5	67.4	38.9
over 2 km	82.3	70.8	68.1	75.6	45.5	32.6	61.1
Co-op store							
up to 2 km	10.9	22.0	27.5	7.3	22.3	13.6	18.2
over 2 km	89.1	88.0	72.5	92.7	77.7	86.4	81.8
Bus route							
up to 2 km	31.5	42.7	50.1	44.7	45.3	52.5	45.8
over 2 km	68.5	57.3	49.9	55.3	54.7	47.5	54.2
Matatu route							
up to 2 km	27.1	59.4	77.8	59.2	57.3	68.3	61.6
over 2 km	72.9	40.6	22.2	40.8	42.8	31.7	38.4
Water source							
up to 2 km	71.0	80.2	98.5	86.8	88.1	97.5	88.1
over 2 km	29.0	19.8	1.5	13.3	11.9	2.5	11.0

Source: Integrated Rural Survey 2, 1978.

to 21 per cent for Coast Province. The high figure for Coast Province reflects to a large extent, the influences of Mombasa metropolitan area.

Still another factor of importance relates to households having access to sanitation. Coast Province is shown to have 75 per cent of its households as having no sanitation. This should be seen within the strong cultural behaviour of the inhibitants of the Coast Province to strongly adhere to certain sanitary

Table 9.9: Percentage distribution of households by access to indicated facilities by

Facilities	Coast	Eastern	Central	Rift Valley	Nyanza	Western
Hospital	21.7	19.9	19.8	32.7	7.2	28.3
Health	54.2	45.5	52.8	35.0	65.3	37.7
Piped water	21.1	42.1	13.4	15.0	2.1	8.3
River water	59.7	31.2	32.5	42.2	28.3	43.8
Other source of water (springs, wells, tanks, etc.)	2.8	17.1	14.2	11.6	57.2	17.6
Pit latrine	89.0	22.5	71.5	32.5	65.9	56.4
No sanitation	9.0	75.0	27.5	64.6	33.0	43.2
Permanent house	11.9	7.7	7.2	9.8	6.2	10.5
Earth (mud) house	87.0	92.3	92.8	89.5	93.5	89.4
Visits to hospital by by 0-4 year old	26.7	28.2	27.6	31.8	32.4	23.6
Presence of malaria	5.6	14.3	28.8	10.7	21.9	0.84

Source: Kenya Integrated Survey 2, 1978.

behaviours. It should also be noted that even the regions seemingly having a higher number of households with pit latrines are no better off. This is because most of the pit latrines are within reach of households and when not taken care of become major sources of food contamination through flies.

The higher incidence and prevalence of malaria has contributed to the higher mortality probabilities for Nyanza, Western and Coast Provinces. Certain features specific to these regions help to explain why they are chronically infested by this deadly disease. Physically these regions are all low-lying lake or ocean-based with low elevation (i.e. less than 300 feet above sea level). This encourages flooding which in turn provides large standing pools for the breeding of mosquitoes, along the coast itself, numerous lagoons and bays provide fertile breeding places which easily spread the disease. Furthermore, these regions, especially Nyanza and Western Provinces, experience some of the heaviest rains in the country due to their geographical location. During floods such as those in the Kano Plains and Busia District, many houses are swept away and large numbers of people are left homeless. Apart from the high human loss associated with such floods, the rest of the population is usually exposed to disease-causing factors.

Implications for Socio-Economic Planning

The importance of the study of human loss, especially in the first few years of life, is well-documented. Such studies have focused not only on the total number who die, but more importantly on how these numbers die differently within and between populations of different regions. This in turn implies the need for the study of the mechanism with which these differentials are maintained across populations.

But, as the UN has appropriately observed, mortality in general and infant mortality in particular is a multi-dimensional complex process that cannot be understood and overcome from one perspective. It occurs within and is affected by a complexity of factors which are themselves interrelated. Thus knowledge relating to specific as well as combined factors is vital for the understanding and reduction of mortality. This study has pointed out selected factors that affect mortality in Kenya. These include education, marital status, place of residence and region of residence. The socio-economic implications of these factors are discussed in the following pages. Before the discussion, however, it must be pointed out that while the Kenya Government will still bear the burden of national planning, more effective localised planning will be handled adequately at the district level within the concept of DFRDs.

The mortality differentials by education of mother imply that not all women have access to education facilities. The low enrolment ratio for females in schools is a well-known phenomenon not only in Kenya, but also in other LDCs. While some districts may not have enough schools to accommodate all those who require places, there is need to encourage the female population to register at all levels of education. The emphasis should especially be for literacy classes for adult women who are mothers to young infants. Thus, there is a need to launch mass literacy campaigns at district and other lower levels so that those who have had no opportunity to go to school will have a chance to learn how to read and write, which will help them contribute positively to the welfare of not only their children but also of themselves.

As has been pointed out elsewhere, education is crucial because it directly influences most other concerns of the mother and the infant. Educated women, for example, will be able to provide not only better health care for the infants but also adequate and better food and clothing. Women who have some kind of education resort to public health and hygiene in taking care of their children. The focus at district level should be on increasing female enrolment as well as providing equal opportunities to both working and non-working women to improve their level of education in order to provide better health care for their children. An educated woman will have a better chance of getting employment, thereby generating income essential for buying vital services such as health care and food. The lower infant and child mortality for married mothers implies that they are able to provide at least some of the services and care needed for their children compared to single or divorced women, for example, the divorced and separated women may not have an economic base to support their children. The problem of single mothers and divorcees is apparently on the rise. The government needs to provide counselling services to women who may not be in a position to help themselves, let alone their children.

Institutions such as the church should be used to accommodate families which cannot hold together after a death, separation or divorce. The government can support programmes which bring together orphans and abandoned children at the local level where such considerations as relatives and land ownership are taken into account. This fits in well with the DFRDs. Since socio-cultural norms and values are closely tied to various ethnic groups who live in particular geographical regions, emphasis should be on the identification of those institutions which could be used to enhance the family as well as individual well-being.

In Kenya, the urban centres claim the lion's share of most public facilities—health, schools and administration as well as major policy-making structures, such as employment opportunities, major industries and transport. The inequality between rural and urban areas has meant increased pressure on the urban facilities as most people flock to the urban centres in the hope of improving their status. The govenments decision to decentralise development from Nairobi to the districts will, it is hoped, bring services closer to the people so that they will not need to make journeys to Nairobi. The government should put emphasis on rural industrial development to create employment, improve small-scale farming to accommodate the majority of the rural population and provide adequate transport to reach as many people as possible to enable easy use of local facilities.

Even more important are the mortality estimates relating to the region of residence. The infants born in Nyanza, Western and Coast Provinces are subject to higher mortality probabilities than those born in other regions. The disparities between regions in terms of essential services, as well as incidence of diseases like malaria for some regions were earlier spelt out. For most regions, a large percentage of the population has to travel more than two kilometres to get to a water source, bus route or health facility. This distance becomes critical in the rural areas, especially for expectant mothers who must walk due to lack of a bus route or inability to pay the high cost of transport. While the government is limited in what it can do with available resources, it must recognise the need to equip and staff the existing facilities, though they are usually far-flung, ill-equipped and understaffed.

Preventive rather than curative medicine should be stressed. With a change of policy, the government could train low-level, cost-effective nurses and nutritionists. While the government cannot presently provide water for everyone in the countryside, it could, however, encourage the population to boil water before using it for drinking or bathing. Local authorities could effectively enforce such rules. The government provision of water through centrally-placed public taps should also be encouraged as everyone is not capable of having piped water in the house. Emphasis should be given to areas least endowed with these facilities because of unequal development in some regions.

Whereas most researchers agree on the high prevalence of malaria in some regions of the country, especially Nyanza and Coast Provinces, it has not been possible to isolate and agree on the contribution of the disease to mortality in the said regions and the country in general. However, the fact that the disease plays a major role is evident from the high percentage of mosquitoes in these regions, a factor attributed to the presence of lots of water in low-lying places. The government should encourage personal hygiene so that the public is educated on the dangers of standing water, children playing in pools and mud and the need for proper nets to prevent mosquitoes from reaching their victims.

Conclusion

We have stressed the inter-relationships of the factors contributing to higher probabilities of infant and child mortality, within populations and between them. No one solution can be suggested as an integrated approach is needed which considers not only socio-economic factors such as education, transport and health, but also stresses the important role of decision-making at the local level. This for Kenya may be realised in the long run on the basis of the DFRDs as is being pursued and implemented by the government.

10 Use of Models in Studying Infant and Child Mortality in Kenya

J. A. M. Otieno, S. A. Odhiambo and F. O. Ouma

Introduction

The study of infant and child mortality is important because its level can be used as a measure of the socio-economic development of an area or a country. High infant and child mortality is associated with high fertility, low socio-economic status and high out-migration. A number of mortality studies in Kenya have now been done at the Population Studies and Research Institute (PSRI) University of Nairobi which was founded in 1976 by Professor S.H. Ominde. The objective of this chapter is to describe models that have been used in studying mortality situations in Kenya and hence to explain the results of the applications of these models, first describing the most commonly-used conceptual framework, namely the Mosley Chen model, demographic techniques and their applications. Description of statistical methods and their applications follows and concludes with recommendations for further research to fill in the gaps.

Mosley-Chen Conceptual Framework

The model or theoretical framework developed by Mosley and Chen (Mosley, 1984) to describe the relationship between socio-economic factors and proximate determinants of mortality has been used in several studies at PSRI. Most demographic researches done to establish associations between socio-economic variables and childhood mortality and morbidity neglected to examine how socio-economic variables operate. By neglecting to look at specific medical causes of death, research could not explain the mechanisms by which socio-economic variables operate to produce observed levels of mortality. In contrast, medical scientists usually focus on disease aetiology or causes of death. For example, if diarrhoea is a cause of death, medical studies will focus on the virus or bacteria that causes diarrhoea. As a result, the intervention would be based on medical technology ignoring the impact of social and environmental conditions where the child lives. A bridge between the two disciplines was, therefore, needed to insure a more comprehensive understanding of morbidity and mortality. Hence the Mosley-Chen model was proposed to incorporate both social and medical science methodologies into a coherent analytical framework of child survival.

Fig 10.1: Operations of five groups of proximate determinants on health dynamics of a populations

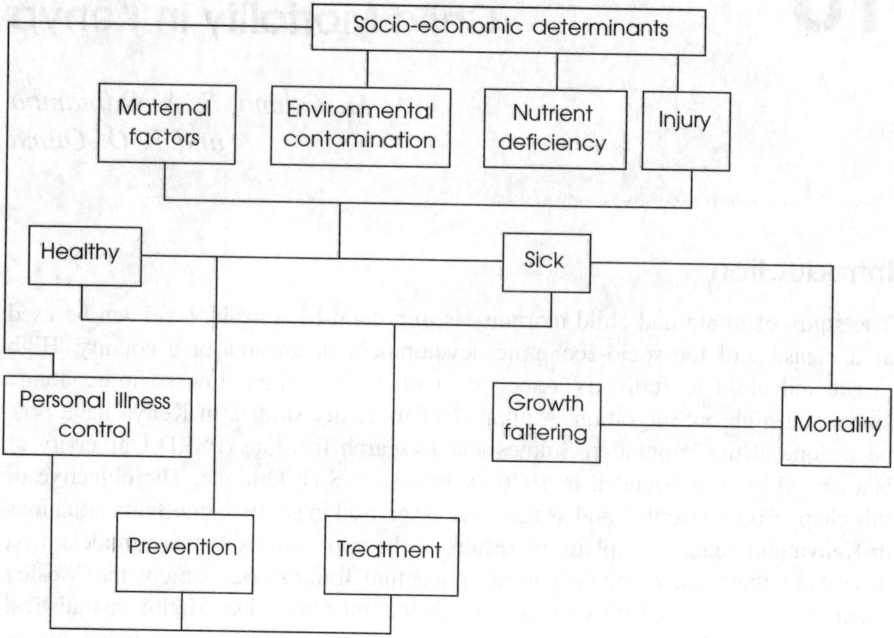

```
                    Socio-economic determinants
         ┌──────────────┬──────────────┬──────────────┐
    Maternal      Environmental     Nutrient        Injury
    factors       contamination    deficiency
         └──────────────┴──────────────┴──────────────┘
              │                              │
          Healthy                          Sick
              │                              │
    Personal illness              Growth              Mortality
       control                    faltering
         ┌──────────┐
    Prevention    Treatment
```

Source: Mosley (1984)

The proximate determinants of the framework defines a set of intermediate (biological) variables through which all social and economic determinants operate to influence infant and child health and survival. In this framework, all biological intermediate variables can be grouped into five categories, namely, maternal, dietary intake, environmental contamination, accidents and personal disease control factors. Broadly, all socio-economic and ecological determinants can be shown to operate through one or more intermediate variables resulting in morbidity, which may be transitory or which result in a permanent residual effect and are cumulative among survivors and/or which ultimately can lead to death. Deeb (1987) gave a brief discussion of what is already known about the interaction of socio-economic variables and relevant intermediate variables that may affect child survival:

Socio-economic variables

The most important socio-economic variable is education. A number of researchers have shown a strong relationship between a mother's education and her child's mortality. Maternal education, however, should not be isolated as one aetiological variable that would automatically improve child survival because the patterns of behaviour and resources associated with maternal education that may minimise the risk of child mortality vary from culture to culture and are not yet fully understood. Also, maternal education is usually closely related to the husband's education and occupation and the household income.

Cultural variables

Mosley and Chen (Mosley, 1984) do not seem to have directly mentioned cultural variables that affect infant and child mortality. An important cultural component that might affect mortality is household structure such as type of marriage and family structure in general. Ethnicity and religion also play an important role.

Maternal factors

Maternal factors such as age of the mother, parity and birth spacing have been shown to be independent risk factors of infant mortality. Higher risks of infant mortality are biologically associated with childbearing at very young ages and older ages with high parity and with short birth intervals. Moreover, studies have shown that children of young, multiparous mothers who have also experienced a short interval are at highest risk of infant mortality. Advanced maternal age is also associated with increased incidence of congenital anomalies while both extremes of age are associated with a higher risk of birth trauma.

Dietary intake

Breastfeeding patterns and food supplementation are considered crucial in determining morbidity during a child's first year of life. Early weaning is also hypothesised to diminish the contraceptive effect of lactation and also to have an adverse effect on nutrition, food supplementation and child care time. This is aggravated among high parity women by the presence of too many children who would compete for the mother's attention and time. This may have a negative effect on each child first by increasing the risk of infection and malnutrition due to early weaning and competition for resources and secondly by the presence of older siblings who might bring home infectious agents.

Environmental factors

The physical setting of the child's household and environment is an important determinant of childhood morbidity and mortality. Infectious diseases are still reported to be among the leading causes that contribute to the toll of mortality among infants and children in developing countries. Environmental factors could act as barriers or could facilitate the transmission of infectious agents. The differential in environmental factors such as availability of clean water, quality of housing and the sewage system are hypothesised to be negatively associated with childhood mortality and morbidity.

Health care behaviour

In most LDCs, respiratory and diarrhoeal diseases and malnutrition rank among the top three causes of death for infants and children. These could be prevented through immunisation programmes or treated with curative services (i.e. antibiotics, oral rehydration and nutritional rehabilitation programmes). Previous studies suggested that the variation in the effectiveness of primary health programmes is usually due to a variety of social, environmental and behavioural variables such as housing conditions, crowding, personal and household hygiene and child care practices.

Moreover, it is widely believed that child care practices are determined by maternal education, cultural beliefs and health knowledge. The lack of access and availability of adequate health care contributes to child morbidity and mortality within individual families. Pre-natal care is thought to be important because it enables women to get general information about infant care and where to get specific medical attention.

Estimating Infant and Child Mortality Under Stable Conditions

There are a number of ways of measuring infant and child mortality using indirect demographic techniques. Probability of dying at age x [i.e.,$q(x)$] is one such measure, as well as life expectancy at birth [$e(o)$]. Other measures are Infant Mortality Rate (IMR) or [$_1q0$], which is the probability of dying between age zero and age one year and [$_1q_4$], which is the probability of dying between age 1 and 5 years.

Calculating $q(x)$

To calculate child mortality, Brass-type models have been used extensively. These techniques require information on Children Ever Born (CEB) and Children Dead (CD), classified by the age of the mother. We also require the Female Population (FPOP) classified by the five year age-groups. The probability of a child dying at age x is $q(x)$ given by the formula:

$$q(x) = K\ (i).D\ (i) \qquad (9.1)$$

for x = 1, 2, 3, 5, 10, 15 and 20

while i = 1, 2, 3, 4, 5, 6 and 7 representing the age groups 15-19, 20-24, ...
    Source: Mosley 1984 and 45-49 age group.

D (i) is the proportion dead in the ith age-group given by

$$D(i) = \frac{CD(i)}{CEB(i)} \quad \text{...(9.2)}$$

K(i) is an adjusting factor which is a function of parity P (i) given by:

$$P(i) = \frac{CED(i)}{FPOP(i)} \quad \text{...(9.3)}$$

According to Brass[3]

$$K(i) = \frac{A_i + B_i P_1}{P_2} \quad \text{...(9.4)}$$

where P_1 and P_2 are parities for the 1st and 2nd age-groups. A_i and B_i are regression coefficients according to Brass model.

According to Sullivan,[4] we have

$$K(i) \frac{A_i + B_i \cdot P(2)}{P(3)} \quad \text{...(9.5)}$$

where A_i and B_i are regression coefficients of the model and P (2) and P (3) are average parities for age group 20-24 and 25-29 respectively.

According to Trussell, we have[5]

$$K(i) = a + \frac{b_i P_1}{P_2} + \frac{c_i P_2}{P_3} \quad \text{...(9.6)}$$

where A_i, B_i and C_i are regression coefficients obtained by Trussell. This model is the one given in manual X.

With regard to the variability of estimates, q(1) and estimates beyond q(10) cannot be taken seriously as precise figures q(1) is an especially untrustworthy figure and estimates beyond q(10) are based on the memory of remote events by women whose responses are not representative of current mortality experience. Thus, the estimates of q(2), q(3) and q(5) have often been accepted as minimum indications of the level of recent infant and child mortality.

Case Studies

Kibet (1981) used Brass's technique to estimate q(2) for the 41 districts in Kenya (at the time of the study) by all cases combined and by the levels of mother's education using the 1979 census data. He found out that infant and child mortality levels are lowest in the highland areas of Kenya and that as one moves towards the lake basin to the west or towards the Indian Ocean coast to the east the levels of early childhood mortality q(2) increases. Thus mortality among the young is low in Central Province and highest in Nyanza Province and Coast Province. The most disadvantaged region in Kenya is South Nyanza District (now comprising Homa Bay, Migori, Rachuonyo and Suba Districts) with an alarming child mortality at age 2 of 216 per 1,000. Nyeri District with a child mortality rate of 49 per 1,000 live births appeared to be better-off than any other district in the whole country.

At the national level, the q(2) values for women with primary education but no more was 64 per cent of that recorded by women with no education while that for women with more than a primary education (i.e. secondary plus) was 37 per cent of that of women with no schooling. At the district level, remarkably similar figures are shown for these two comparisons. Most districts have their q(2) estimates for mothers with no education more than double that for mothers with secondary education or above. In the highest mortality district South Nyanza, the corresponding figures for these two comparisons were 81 percent and 43 per cent, while in the low mortality district (Nyeri) these figures were 62 per cent and 39 per cent, respectively.

Nyamwange (1984) while studying Nairobi's child mortality by administrative wards using Brass's method, found that in-migration was a major determinant of child mortality.

Trussell's technique has been used extensively to study child mortality at district and divisional levels by all cases combined and by the differentials of education, place of residence and marital status. Kichamu (1986) used Trussell's technique to estimate q(x) for the 41 districts in Kenya using both the 1969 and 1979 censuses. He classified the regions into low, medium, high and very high. Nyeri District had the lowest child mortality in Kenya. The mortality declined from 54 to 47 per 1,000 over the ten year period from 1969 to 1979. The Districts of South Nyanza, Siaya, Kisumu, Busia and Kilifi had the highest mortality levels. The levels of differential by education shows that child mortality declines with the rise of the level of mother's education.

Tables 10.1 and 10.2 give the values of q(2) for Kibet and Kichamu.

Table 10.1: The probability of dying at the age of 2 years for all classes combined

Region	1969 Kichamu	1979 Kichamu	1979 Kibet
KENYA	142	122	125
NAIROBI	92	88	93
CENTRAL	86	65	67
Kiambu	92	68	70
Kirinyaga	105	79	82
Murang'a	109	66	68
Nyandarua	61	65	64
Nyeri	54	47	49
COAST	168	114	177
Kilifi	229	195	212
Kwale	170	176	190
Lamu	129	190	200
Mombasa	131	121	120
Taita-Taveta	177	112	116
Tana River	189	172	181
EASTERN	132	95	103
Embu	119	81	83
Isiolo	217	120	127
Kitui	160	141	148
Machakos	128	94	98
Marsabit	189	124	130
Meru	114	71	75
NORTH EASTERN	159	131	135
Garissa	161	126	131
Mandera	137	170	131
Wajir	178	128	129
NYANZA	203	163	174
Kisii	128	96	101
Siaya	244	195	211
South Nyanza	230	200	216
Kisumu	221	183	199
WESTERN	166	145	152
Kakamega	159	136	143
Bungoma	146	135	140
Busia	222	185	198
RIFT VALLEY	101	101	108
Baringo	132	162	171
Marakwet	95	121	127
Laikipia	97	73	77
Nakuru	124	92	97
Kericho	78	86	91
Nandi	96	102	110
Narok	95	87	95
Samburu	85	73	77
Trans-Nzoia	120	106	114
Kajiado	82	80	75
West Pokot	141	174	188
Turkana	95	125	133
Uasin-Gishu	83	86	92

Source: Kichamu 1986 and Kibet (1981)

Table 10.2: The Probability of dying at the age of two by education differential, 1979

Region	None		Primary		Secondary +	
	Kibet	Kichamu	Kibet	Kichamu	Kibet	Kichamu
KENYA	163	144	104	104	61	61
NAIROBI	138	124	101	94	53	52
CENTRAL	94	85	63	61	42	41
Kiambu	103	94	69	66	45	45
Kirinyaga	101	91	73	70	41	41
Murang'a	87	77	65	62	47	46
Nyandarua	93	84	60	57	39	39
Nyeri	79	70	49	47	31	30
COAST	200	182	126	119	75	73
Kilifi	223	202	135	128	66	65
Kwale	200	181	148	140	68	67
Lamu	225	167	108	-	37	21
Mombasa	138	121	119	111	78	75
Taita-Taveta	139	123	110	105	57	57
Tana River	186	175	170	162	119	112
EASTERN	128	118	96	79	51	50
Embu	107	99	74	79	46	50
Isiolo	134	127	101	95	53	48
Kitui	162	147	120	115	92	91
Machakos	136	120	92	88	53	51
Marsabit	135	128	107	105	18	17
Meru	99	89	56	54	35	36
NORTH EASTERN	139	135	120	115	43	44
Garissa	139	133	119	114	45	44
Mandera	151	141	119	142	64	147
Wajir	130	129	107	115	22	22
NYANZA	204	177	158	149	85	84
Kisii	129	106	88	84	54	52
Siaya	237	200	190	178	123	120
South Nyanza	246	215	199	185	107	105
Kisumu	240	200	175	171	94	92
WESTERN	173	155	142	135	86	85
Kakamega	161	143	133	126	78	78
Bungoma	162	147	138	132	87	85
Busia	215	192	180	170	118	114
RIFT VALLEY	125	112	94	88	58	57
Baringo	211	185	113	109	65	65
Marakwet	156	142	108	103	52	51
Laikipia	87	79	71	68	54	53
Nakuru	119	105	91	86	51	56
Kericho	101	88	85	80	46	46
Nandi	122	104	105	99	69	67
Narok	97	86	88	83	39	37
Samburu	75	71	100	100	64	63
Trans-Nzoia	124	109	110	103	75	74
Kajiado	76	67	73	69	55	54
West Pokot	201	185	118	111	43	42
Turkana	137	130	150	138	76	67
Uasin Gishu	107	93	84	80	61	60

Source: Kichamu (1986) and Kibet (1981)

Generally the q(2) values obtained by Kibet (1981) are higher than those obtained by Kichamu (1986). Thus, using Brass coefficients, the q(2) values are higher than using Trussell's coefficients. By levels of education, for the no-education group, Kibet's estimates are higher than Kichamu's. For primary education, Kibet's estimates are also higher, but the difference is now narrowed. For the secondary plus category, the estimates by both Kibet and Kichamu are virtually the same. For the place of residence, the general pattern is that child mortality in the urban areas is lower than that of the rural areas except in Kericho, Nyeri, Kirinyaga, Nyandarua, Nairobi and Trans-Nzoia.

Child mortality for widowed mothers is the highest in general, followed by the divorced/separated. The child mortality for single mothers is lower than those for the married mothers in most cases considered. There are few exceptions to these patterns. Kichamu noticed that high mortality district were also large in size. Coincidentally, most district in Kenya have now been subdivided for ease of administrative purposes. For example Kisii District has now been divided into three districts Kisii, Gucha and Kuria Districts and so has Kakamega (now comprising Kakamega and Vihiga Districts). New divisions, locations and sub-locations have now been formed throughout the country. The DFRDs strategy, introduced in 1984 by the Kenya Government, necessitates analysis of data at sub-district level in order to plan effectively each division and location. Table 10.3 gives the q(2) values for the differential of place of residence and marital status for 1979 data.

Ondimu (1983), applying Trussell's method to Kenya Contraceptive Prevalence Survey (KCPS) data, estimated infant and child mortality at national, provincial and district levels. At the national level, he further estimated the mortality by differentials of education, work status, place of residence, contraceptive use, marital status, type of union and religious identification. Children whose mothers have higher formal education have higher chances of survival compared to those whose mothers have low or no formal education.

In the KCPS, work was defined as doing jobs other than household work for payment in either cash or kind; making things for sale or having business of whatever size in which income is accrued. Infant and child mortality levels were lowest for those whose mothers worked in the past and was highest for those whose mothers had never worked. Infant and child mortality rates were found to be low for those children whose fathers worked in business, moderate for those whose fathers worked on farms and highest for those whose fathers were not working at all. A very interesting finding by Ondimu (1983) was that infant and child mortality in some urban areas was lower compared to urban centres (Nairobi and Mombasa). Children of mothers who had ever used a contraceptive method had a lower mortality experience than those who never applied a contraceptive method.

The Infant Mortality Rate (IMR) was very low for monogamous unions compared to that of polygamous unions. Children of single mothers had the lowest mortality

ISSUES IN RESOURCE MANAGEMENT AND DEVELOPMENT IN KENYA

Table 10.3: The probability of dying at the age of two (2) yrs by marital status and place of residence, 1979

Region	Widowed	Divorced	Single	Married	Urban	Rural
NAIROBI	77	72	71	74	88	-
CENTRAL	95	59	45	52	68	62
Kiambu	26	62	50	57	73	67
Kirinyaga	115	65	50	63	81	79
Murang'a	99	51	48	52	52	66
Nyandarua	130	60	44	52	80	64
Nyeri	75	59	34	37	53	46
COAST	189	131	107	146	118	184
Kilifi	226	169	125	181	135	199
Kwale	179	143	168	118	128	157
Lamu	274	139	83	155	104	65
Mombasa	153	90	87	92	111	-
Taita Taveta	124	99	100	88	99	113
Tana River	161	159	140	146	160	173
EASTERN	114	87	67	77	80	97
Embu	106	60	57	65	82	81
Isiolo	102	151	38	103	133	113
Kitui	106	126	121	118	117	142
Machakos	117	85	67	76	73	97
Marsabit	107	127	69	104	110	145
Meru	125	56	49	55	50	73
NORTH EASTERN	155	163	66	113	146	129
Garissa	108	179	50	109	138	123
Mandera	184	197	68	119	112	52
Wajir	144	119	87	113	150	124
NYANZA	138	131	189	108	150	164
Kisii	114	65	70	78	88	96
Siaya	207	186	152	160	168	196
South Nyanza	201	159	146	174	139	202
Kisumu	150	169	121	157	161	195
WESTERN	173	113	103	122	137	146
Kakamega	199	108	88	114	121	137
Bungoma	90	106	126	112	132	126
Busia	210	153	137	158	163	188
RIFT VALLEY	106	91	75	86	102	75
Baringo	224	116	107	132	93	173
Marakwet	154	106	102	97	47	122
Laikipia	64	74	64	62	71	74
Nakuru	131	83	72	77	86	94
Kericho	66	85	60	75	113	84
Nandi	106	90	81	85	117	102
Narok	98	90	125	76	87	87
Samburu	67	69	62	65	97	53
Trans-Nzoia	132	90	59	92	108	103
Kajiado	58	65	83	61	87	67
West Pokot	179	141	109	146	114	176
Turkana	113	66	99	95	140	123
Uasin Gishu	105	107	62	71	105	82

Source: Compiled by authors.

experience followed by separated, widowed, married and divorced, in that order. Protestants had the lowest mortality experience followed by other non-sects, Catholics and Muslims, in that order. Children born of Kikuyu parents experienced the lowest infant and child mortality rates followed by Meru-Embu, Kamba, Kalenjin, Kisii, "Other tribes", Luhya, Mijikenda and Luo, in that order. Ondimu also showed that regional variations of infant and child mortality exist in Kenya.

The life expectancy ranged from 61.77 years for Central Province to 45.91 years for Nyanza Province. The same experience is also evident in the case of districts. Nyeri had the lowest mortality whereas South Nyanza and Kisumu district had the highest.

Tables 10.4 and 10.5 show infant and child mortality estimates by regions and differentials using KCPS data.

Table 10.4: Infant and child mortality by region using KCPS data, 1984

Region	q(2)	IMR	$_4q_1$	e_o
KENYA	115	92	63	52
COAST PROVINCE	131	115	88	48
Kilifi	139	124	97	45
Kwale	102	127	101	45
Mombasa	122	89	60	53
Taita-Taveta	231	105	76	50
CENTRAL PROVINCE	39	43	20	64
Kiambu	51	53	28	62
Kirinyaga	29	64	38	59
Murang'a	55	38	16	66
Nyeri	31	35	14	66
NYANZA	145	125	98	46
Kisii	68	78	49	56
Kisumu	206	135	108	44
Siaya	150	132	105	45
South Nyanza	136	135	108	44
WESTERN	167	117	89	47
Busia	189	87	58	54
Bungoma	184	107	78	49
Kakamega	60	92	63	53
RIFT VALLEY	94	83	55	55
Kericho	127	103	74	50
West Pokot	18	97	68	51
Nakuru	72	60	34	60
Baringo	168	70	42	58
Trans-Nzoia	46	50	26	63
Uasin-Gishu	80	62	35	60
Nandi	93	93	64	52
EASTERN	98	80	51	55
Embu	27	64	37	59
Meru	65	76	48	56
Machakos	151	84	55	54
Kitui	122	102	73	51
NAIROBI	46	75	-	57

Source: Ondimu (1987)

Table 10.5: Infant and child mortality by region using KCPS data, 1985

Differential	q(2)	IMR	$_4q_1$	e_o
EDUCATION				
NE	121	107	78	45
1-4 years	133	98	69	51
5-8 years	105	74	46	57
9+	52	40	17	65
WORK STATUS OF MOTHER				
Currently	105	82	54	55
Past	137	92	66	52
Never	109	96	67	52
WORK STATUS OF FATHER				
Own farm	100	91	61	53
Other farm	84	88	59	53
Own business	72	79	50	56
Other business	96	80	51	55
Not working	110	105	76	50
PLACE OF RESIDENCE				
Metropolitan	95	82	53	55
Other urban	100	71	44	57
Rural	119	96	67	52
CONTRACEPTIVE USE				
Ever users	86	69	42	58
Non users	123	103	75	50
TYPE OF UNION				
Polygamy	128	118	90	47
Monogamy	92	82	53	55
MARITAL STATUS				
Single	66	73	45	57
Married	98	89	60	53
Separated	27	76	48	56
Divorced	183	108	80	49
Widowed	56	81	52	55
RELIGION				
Catholic	118	102	73	50
Protestant	113	90	61	53
Muslim	106	109	81	49
Others	57	96	67	52
No religion	92	110	82	49
ETHNICITY				
Kikuyu	56	49	25	63
Luo	175	145	117	42
Luhya	146	108	75	49
Kamba	140	83	55	55
Meru-Embu	62	75	47	56
Mijikenda	126	115	87	48
Kalenjin	94	87	58	54
Others	87	94	65	52

Source: Ondimu (1987).

At divisional levels, using the 1979 census data and Trussell's model, a number of studies estimating child mortality have been made for Nyanza, Western and Coast Provinces. The districts surrounding Lake Victoria — South Nyanza, Kisumu, Siaya and Busia — have a striking mortality pattern. Mortality seems to rise as you go towards the lake with the exception of Siaya District which has the opposite phenomenon. In Siaya District, Bondo Division has the lowest (relatively) child mortality while Ukwala Division has the highest. In Busia District, Hakati Division has the highest mortality and in Kakamega District, Mumias Division has the highest mortality. We should note that the three divisions, namely Ukwala in Siaya District, Mumias in Kakamega District and Hakati in Busia District border each other.

In Coast province, we also have a very interesting phenomenon. In Taita-Taveta District, the divisions bordering national parks have higher child mortality than those at the centre. In Kwale District, Kuyoh (1990) found that Kinango Division had the highest mortality values among all the divisions. Kubo and Msambweni Divisions had almost the same levels whereas Matuga Division had the lowest values of q(x). In all divisions without exception, child mortality by education status of the mother follows the same pattern. Child mortality is lowest among mothers with secondary plus education. For place of residence, in general, most divisions have the patterns whereby urban mortality is lower than rural mortality. In South Nyanza District, the exception is Migori Division while in Siaya the exception is Bondo Division.

Mutai (1987) has studied Kericho's child mortality by locations. He found that in general, the urban areas had higher infant and child mortality than the rural areas. For marital status, divorced mothers had the highest child mortality followed by married then widowed mothers, while the single mothers had the least.

Construction of a Life Table

Other measures of infant and child mortality are based on the life table functions. These are IMR which is defined as the probability of dying between age 0 and 1 ($_1q_0$). Life expectancy at birth is denoted by e_0 and the probability of dying between age 1 and 5 denoted by $_4q_1$.

The first step in constructing a life table is to calculate the mortality levels corresponding to the values of q(x). The calculation of mortality levels involves the use of linear interpolation. Once this is done, the mean mortality usually x = 2, 3 and 5 is taken. The mortality level determines the appropriate value of l(x) which is the number of survivors at age x.

After getting the value of l_x, other values of the life table can be obtained as follows:

$$nPx = \text{......... the probability of surviving between age x and x+n}$$

$$= \frac{l_{x+n}}{l_x} \quad \text{..} \quad (10.7)$$

$_n d_x$ = ... number of persons who die between age x and x+n
$= 1_x - 1_{x+n}$ .... (10.8)

$_n L_x$ = the number of person's years lived between age x and x+n

where

$_1 L_0 = 0.3 * 1_0 + 0.7 * 1_1$ .... (10.9)

$_4 L_1 = 1.3 * 1_1 + 2.7 * 1_4$ .... (10.10)

$_5 L_x = 5 * (1_x + 1_{x+5})/2$ for x = 5, 10, 15, ..., 70 .... (10.11)

$L_{75x} = L_{75} + * \log_{10} l_{75+}$ .... (10.12)

T_x = the total person's year lived from age x
$= T_{x+n} + {}_n L_x$ .... (10.13)

$_n q_x$ = probability of dying between age x and x + n
$= 1 - {}_n P_x$ .... (10.14)

e_x = expectation of life at age x

$= \dfrac{T_x}{1_x}$ .... (10.15)

Application of Statistical Techniques

An effective analysis of quantitative empirical data is a two-stage process, with the initial stage involving surveying and exploring the data using some quick and non-rigorous methods and the second stage consisting of model building and hypothesis testing using the procedures of classical inferential statistics. The first stage is called Exploratory Data Analysis (EDA) and the second is called Confirmatory Data Analysis (CDA).

Exploratory Data Analysis

EDA is detective in nature. There has been developed a large and diverse array of intuitively sound and theoretically supportable tools for explaining data prior to the application of confirmatory tools. Two basic EDA display measures are the stem-and-leaf and the box-plot.

A stem-and-leaf, display is a basic organisational procedure of the EDA. Like a simple sorting of data values, the display organises the values in numerical order and facilitates the study of the shape, spread and distributional characteristics of empirical values. Unlike a sort and histogram, however, the stem-and-leaf display retains information on individual data and provides both graphical summary and convenient calculation of median and mid-spread. The plot is divided into two components—the leading digits called the stem and the trailing referred to as the leaf.

Figure 10.2: The Stem-Leaf Plot of e_o

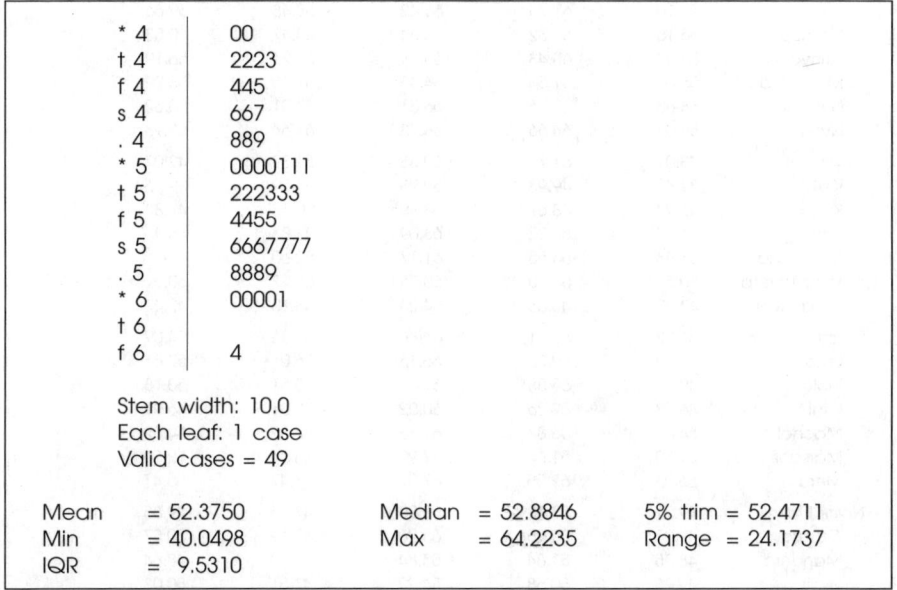

Source: Compiled by the authors.

Perhaps a better way of displaying the distribution of a set of numbers is the box-plot. A box-plot is obtained by plotting the lower and upper quartile values of a batch of numbers and drawing a box to identify the length of the spread. Instead of plotting actual values, a box-plot displays summary statistics of the distribution. It plots the median, the 25th, and the 75th percentile, the mean and the k per cent trimmed mean, for some k and the values that are far removed from the rest (extremes) as well as other measures of dispersion like the variance, standard deviation and skewness. An asterix in or on the box represents the location of the median; crosses at the end of the vertical line drawn outwards from the lower and upper quartiles are the best data points that lie within 1-5 spread from the quartiles. Numbers lying outside these crosses are called outliers. Outliers more than 3 times the mid-spread away from its ends are called "far out" observations. Box-plots are particularly useful for comparing the distribution of values in several groups.

Rapemo (1991) has used the EDA technique to examine mortality estimates he had obtained using the demographic techniques that have been described above. He picked the life expectancy at birth, e_o and the probability of dying at age 5, [q(5)]. A summary of these values have been given in Tables 10.6 and 10.7.

Table 10.6: Life expectancies at birth by level of education and place of residence

Region	NE	PE	SEC+	Urban	Rural
KENYA	48.04	54.07	63.67	-	50.00
NAIROBI	51.55	56.68	64.98	57.13	-
CENTRAL	57.10	61.75	66.42	58.48	59.66
Kiambu	56.15	60.32	66.64	58.47	60.58
Kirinyaga	54.50	60.43	65.34	57.24	56.19
Murang'a	56.01	61.56	64.43	51.18	58.29
Nyandarua	58.68	61.76	66.31	57.01	60.52
Nyeri	59.02	64.66	68.73	61.64	63.98
COAST	43.01	51.76	60.33	51.39	42.07
Kilifi	39.40	49.93	59.93	49.29	39.86
Kwale	42.94	48.67	58.89	50.68	42.83
Lamu	41.67	52.38	63.09	41.83	46.45
Mombasa	51.43	53.68	61.17	52.60	-
Taita-Taveta	50.69	53.50	60.78	53.27	50.91
Tana River	43.23	45.36	54.24	43.10	43.42
EASTERN	51.52	58.21	64.86	57.36	54.09
Embu	55.10	59.02	65.16	56.01	57.81
Isiolo	49.33	55.88	61.04	50.51	50.13
Kitui	46.73	52.26	60.02	52.31	46.99
Machakos	54.67	56.87	64.63	57.82	54.52
Marsabit	51.50	51.64	57.90	58.14	52.50
Meru	56.23	62.55	67.07	63.47	55.47
NORTH EASTERN	49.21	53.13	58.90	47.15	49.85
Garissa	49.31	55.05	65.26	47.19	50.57
Mandera	48.35	51.84	53.84	46.69	48.64
Wajir	49.75	50.58	54.29	47.51	50.08
NYANZA	43.85	47.26	58.48	46.24	44.12
Kisii	54.17	57.91	65.04	55.83	54.91
Siaya	40.86	43.15	52.20	48.41	40.16
South Nyanza	39.27	42.29	53.46	46.44	39.49
Kisumu	39.08	44.31	58.05	44.77	39.87
WESTERN	45.92	49.34	57.32	49.50	46.68
Kakamega	47.40	50.10	58.02	52.31	47.79
Bungoma	48.78	50.10	57.66	50.52	48.98
Busia	40.94	44.61	53.72	45.29	40.65
RIFT VALLEY	53.07	56.90	63.63	56.01	53.93
Baringo	43.40	52.86	62.92	55.07	40.49
Marakwet	48.37	54.95	55.66	63.75	50.49
Laikipia	57.43	60.29	66.79	59.55	58.79
Nakuru	54.98	56.61	64.32	56.01	55.49
Kericho	57.28	58.64	64.93	52.91	57.36
Nandi	54.39	55.68	61.61	54.62	54.02
Narok	57.07	58.08	61.90	59.13	56.72
Samburu	59.07	56.75	55.98	47.05	60.03
Trans Nzoia	53.53	55.03	62.37	55.91	52.96
Kajiado	51.15	60.25	64.49	58.15	60.86
West Pokot	41.24	53.34	63.75	51.74	41.94
Turkana	49.44	51.15	54.75	47.26	49.91
Uasin Gishu	56.38	58.19	63.60	55.64	65.97

Source: Compiled by the authors.

Table 10.7: Life expectancies at birth by marital status

Region	Widowed	Divorced	Married	Single
NAIROBI	56.64	56.88	61.23	59.53
CENTRAL	54.56	58.73	64.08	59.66
Kiambu	57.91	58.95	63.31	62.75
Kirinyaga	49.51	56.16	61.37	61.49
Murang'a	58.49	59.18	63.08	61.81
Nyandarua	55.06	55.57	64.02	64.37
Nyeri	61.10	62.52	66.27	64.48
COAST	43.70	47.36	48.65	48.65
Kilifi	42.06	43.01	43.03	44.00
Kwale	45.01	45.29	47.42	47.21
Lamu	39.74	47.76	50.62	48.66
Mombasa	49.76	52.24	57.48	53.49
Taita-Taveta	51.81	50.48	58.08	52.84
Tana River	43.35	46.08	49.07	46.53
EASTERN	50.11	53.24	59.24	58.17
Embu	54.16	56.71	61.39	58.38
Isiolo	50.46	52.80	53.77	54.92
Kitui	46.66	47.55	52.23	48.59
Machakos	54.75	53.40	59.79	58.42
Marsabit	50.64	52.27	55.66	59.55
Meru	53.86	55.25	62.83	62.94
NORTH EASTERN	48.19	46.50	52.36	55.91
Garissa	48.46	47.85	53.51	55.43
Mandera	51.51	42.21	52.81	58.38
Wajir	47.53	49.36	52.76	55.95
NYANZA	41.08	46.45	49.61	48.85
Kisii	54.35	54.69	59.09	57.57
Siaya	36.74	39.88	42.32	44.65
South Nyanza	37.52	43.44	44.59	40.95
Kisumu	39.51	41.97	43.37	45.60
WESTERN	43.96	48.57	51.99	50.19
Kakamega	47.14	48.63	53.13	53.33
Bungoma	50.06	48.92	53.95	48.79
Busia	39.68	42.94	46.26	43.10
RIFT VALLEY	52.33	53.83	57.98	56.07
Baringo	47.35	48.28	51.74	48.59
Marakwet	43.81	55.62	56.19	52.31
Laikipia	57.91	55.98	62.12	59.58
Nakuru	53.19	54.75	59.53	57.01
Kericho	55.15	58.84	59.89	59.70
Nandi	51.65	53.13	58.50	55.48
Narok	56.18	58.05	59.46	51.14
Samburu	56.06	57.44	61.27	58.46
Trans Nzoia	52.03	55.75	54.82	54.82
Kajiado	58.84	60.21	62.78	58.68
West Pokot	41.43	44.10	48.44	46.95
Turkana	49.17	54.42	56.90	50.12
Uasin Gishu	55.50	55.88	60.65	58.33

Source: Compiled by the authors.

Table 10.8 Extremes values

5 highest	Region	5 Lowest	Region
64.223	Nyeri	40.050	South Nyanza
61.041	Kiambu	40.767	Siaya
60.843	Nyandarua	42.093	Kisumu
60.809	Central Province	42.101	Busia
60.507	Kajiado	42.128	Kilifi

Source: Compiled by the authors.

In the stem-leaf plot, the vertical line has separated the stem and the leaves. Each leaf represents one case (region) whose estimate has been rounded to the nearest ones. The stem gives the range (in this case the first digit of each value) and the leaves give the second digit. The leaves are divided into five parts: (1) '*' representing leaves of 0 and 1; (2) 't' representing leaves of 2 and 3; (3) 'f' for values of 4 and 5; (4) 's' for leaves of 6 and 7; and, (5) '.' for leaves of 8 and 9.

In the first row, there are 2 cases lying in the range of 40-41; the exact values are 40.04978 and 40.76659. The second row has 4 cases whose exact values are 42.09298, 42.10138, 42.12798, and 43.43599. The plot gives all these values in this order. They are the life expectancies for the of South Nyanza, Siaya, Kisumu, Busia, Kilifi and West Pokot Districts, respectively.

The distribution shown on the stem-leaf plot appears to be bimodal, with the major peaks around the life expectancies in the 50s. If the aggregate distribution is visualised as the sum of the two component distributions, a mode of e_o's of early 50s is observed which is where the life expectancy for Kenya lies. The general outlook is, however, one leaning towards "high" life expectancies. This shows that many regions in Kenya have made some improvements in their mortality conditions.

Along with stem-leaf display, the extremes which are the five lowest and five highest observed e_o values are shown in Table 10.8.

Time Series Analysis

There have been a few seasonality studies in Kenya connected to mortality. Bunyasia (1985) in his study on the seasonality patterns of causes of death in Kenya, concentrated on the monthly occurrence of deaths by cause between 1975 and 1979. The study used vital, climatic and morbidity conditions statistics and established that the seasonal death curve is low at the beginning of the year and rises as the year advances. The major peak was identified as occurring between June and August while the secondary peak occurs in April, May, September and October. The curve was observed to be at its lowest during the months between November and February.

Kichamu (1986), in his study on the estimation of mortality in Kenya using vital registration, dedicated a section for the seasonality of vital events. The study found that there is a time lag of about 3-4 months from the rainfall peak in April and the peak of deaths in August. At the same time, the study also showed that another peak occurs in April and it concluded that there is an association between rainfall and seasonal variations of death. This association, the study argued, is a result of high morbidity associated with rainfall originating from use of dirty water and infectious and parasitic diseases, which are common during rainy seasons. Ojwang (1992) also found the seasonal variations of deaths using Civil Registration Demonstration Project Data and identified the various climatic, socio-cultural, socio-economic and medical factors which correlate with seasonality of death.

Regression Analysis

Regression analysis is indispensable as data analysis technique in many disciplines. In recent years, new techniques for regression analysis have been developed and many are being increasingly used in many branches of the social sciences including demography. In areas like fertility, mortality, migration, nuptiality, family planning, population projections and demographic models regression analysis has been found to be a very useful tool. Regression analysis is one of the multivariate techniques of analysis. There are other multivariate techniques as well as path analysis, factor analysis and multiple classification analysis, among others. Whereas regression analysis allows one to estimate the impact of the predictor variables, path analysis provides a means of analysing and representing relationships implicit or explicit in multiple regression analysis. The factor analysis helps in finding out relationships between sets of variables.

Multiple Regression Model

Model:

$$\underline{Y} = X\beta + \varepsilon \quad \quad (10.16)$$

Assumptions:
$\underline{E}(e) = 0$

$E(ee') = \sigma^2 I_n \quad \epsilon_i$ and ϵ_j are uncorrelated

$\text{Rank}(X) = \text{Rank}(X'X) = k < n$

Ordinary Least Squared (OLS) estimates:

$$\hat{\beta} = (X'X)^{-1}XY \quad\quad\quad (10.17)$$
$$E(\hat{\beta}) = \beta$$
$$E(\hat{\underline{\beta}} - \underline{\beta})(\hat{\underline{\beta}} - \underline{\beta})' = (X'X)^{-1}\sigma^2 \quad \text{(known } \sigma^2 \text{ constant variant).}$$

If σ^2 is unknown, its estimate is given by
$$S^2 = \hat{\sigma}^2 = \underline{e}'-\underline{e}/n-k^2$$

where
$$\underline{e} = \underline{Y} - \hat{\underline{Y}}$$
$$\underline{Y} = \hat{\underline{Y}} + \underline{e}$$

therefore
$$\underline{Y} - \overline{\underline{Y}} = \hat{\underline{Y}} - \overline{\underline{Y}} + \underline{e}$$
$$= \hat{\underline{Y}} - \overline{\underline{Y}} + \underline{e}$$
i.e.,
$$\underline{Y} = \hat{\underline{Y}} + \underline{e}$$
i.e.,
$$Y'Y = \hat{Y}'\hat{Y} + \underline{e}'\underline{e}$$
TSS = ESS + RSS; which is

Total Sum of Squares = Error of Sum of Squares plus Residual Sum of Squares

Therefore,

$$\frac{\hat{Y}'\hat{Y}}{Y'Y} = \frac{1-\underline{e}'\underline{e}}{Y'Y} = R^2 \quad\quad\quad (10.18)$$

which is called the multiple coefficient of determination.

Tests of Hypotheses

(1) Testing β_i's individually

i.e.
$\quad H_0 \quad : \quad \beta_i = 0$

against
$\quad H_1 \quad : \quad \beta_i \neq 0$ for each i.

Assuming that $\epsilon_i \sim N(0, \sigma^2)$, we use

$$t_i = \frac{\hat{\beta}_i - 0}{\sigma_i \sqrt{\left(\underline{e}'\underline{e}/(n-k)\sigma^2\right)}}$$

$$= \frac{\hat{\beta}_i - 0}{\sqrt{\left(\text{Var }\hat{\beta}_i\right)\left(\underline{e}'\underline{e}/(n-k)\sigma^2\right)}}$$

therefore

$$t_i = \frac{\hat{\beta}_i}{\sqrt{(X'X)_i^1(\underline{e}'\underline{e}/(n-k))}}$$

$$= \frac{\hat{\beta}_i}{\sqrt{C_{ii}(\underline{e}'\underline{e}/(n-k))}}$$

$$= \frac{\hat{\beta}_i}{\text{S.E.}(\hat{\beta}_i)} \qquad (10.19)$$

which is student t distribution with n-k degrees of freedom compare this calculated t with table t (n-k).

Note
$t_i^2 \sim F(1, n-k)$

In this case compare this calculated F (1, n-k) with the Table F (1, n-k).

(2)   Testing the combined effect

H_0: $\beta_1 = \beta_2 = \beta_3 = ... = \beta k = 0$
H_1: anything else

Alternatively

H_0 : $R^2 = 0$

against

H_1 : $R^2 \neq 0$

By using the ANOVA table or the covariance test

$$F = \frac{R^2/(k-1)}{(1-R^2)(n-k)} \sim F(k-1, n-k) \quad (10.20)$$

or

$$F = \frac{Y'Y/(k-1)}{e'e/(n-k)} \sim F(k-1, n-k)$$

Compare the calculated F-values with the corresponding table F-values to reject or not to reject the hypotheses under consideration.*

Basic Problems in OLS Method

(A) When the assumptions do not hold

- Multi-collinearity, i.e. Rank $(X'X) < k$ $[X'X] = 0$ or close to zero

    Solution: Remove all the dependent columns except one and then apply the OLS

- Unequal variance (heteroscedasticity)

- The random error $\in_i$ is assumed to be uncorrelated with any of the independent variables selected or $\in_i$ is and $\in_j$ are uncorrelated.

If they are, OLS cannot be applied immediately. This is particularly common in time-dependent data.

Solution: Generalised Least Square Method, Aitken's (GLS)

If

$$\underline{Y} = X\underline{\in} + \underline{\in} \quad (E(\underline{\in}\,\underline{\in}') = Vs^2 \quad \quad \quad (10.21)$$

where V is symmetrical, positive definite.

Then

$$\hat{\underline{\beta}} = (X'V^{-1}X)^{-1} X/V^{-1}Y \quad \quad \quad (10.22)$$

where V^{-1} is assumed to exist.

---

* The SPSS package was used for this regression analysis

(B) When the dependent variable is dichotomous

Suppose

$$y_i = \sum_j \beta_j X_{ij} + \epsilon_i$$

such that $y_i = 0$ or 1, or $0 \leq y_i \leq 1$

In the predictive model, i.e.,

$$y_i = \sum_j \beta_j X_{ij}$$

We cannot guarantee that the value of $\hat{y}_i$ will be in the 0-1 range for any value of X_{ij}. ϵ_i cannot be normal—it is discrete, etc.

One way of avoiding these difficulties is to apply a monotonic transformation to the probability in such a way that the resulting variable, has the range $(-\infty, \infty)$.

One such transformation is the logit transformation.

Define

$$P_i = \frac{e^{\Sigma \beta X}}{1+e^{\Sigma \beta X}} \quad \text{............(10.23)}$$

This implies that

$$1 - P_i = \frac{1}{1+e^{\Sigma \beta X}}$$

Therefore

$$\ln \frac{P_i}{1-P_i} = \ln e^{\Sigma \beta X} = \sum_j \beta_j X_{ij} \quad \text{............(10.24)}$$

which is the logit transformation.

Note that P_i can be expressed by

$$P_i = \frac{1}{1+e^{-\Sigma \beta X}}$$

For $X_{ij} = -\infty$, $P_i = 0$
For $X_{ij} = \infty$, $P_i = 1$
Thus, for $-\infty \leq X_{ij} \leq \infty$, $0 \leq P_i \leq 1$.

Case Studies

Demographic techniques have been used to identify determinants of mortality or other demographic processes. However, there are a number of questions that can be answered through the use of multivariate regression framework. If we wish to find the effect of the combined, the regression analysis approach is a better one.

Mott (1979) examined the trends and differentials in infant and neonatal mortality from a multivariate perspective using the KFS data. The regression variable listing as shown in the table below:

Table 10.9: Regression variables for infant and neonatal mortality in Coast, Nyanza and Western Provinces

Residence:	1 = lives there, 0 = lives elsewhere.
Age at birth:	Continuous variable of age of mother at birth of child (in years).
Primary school:	1 = completed some primary school, 0 = other.
Secondary school:	1 = completed some secondary school, 0 = other.
Male birth:	1 = birth was male, 0 = birth was female.
Urban residence:	1 = survey residence was urban, 0 = other.
Polygamy:	1 = mother in polygamous marriage at survey, 0 = other.
Parity one:	1 = first birth, 0 = other.
Parity 5-9:	1 = birth was parity 5 to 9, 0 = other.
Parity 10 plus:	1 = birth was parity 10 or above, 0 = other.
Birth 1 to 10 years:	1 = birth 13 to 119 months ago, 0 = other.
Birth 11 to 20 plus years:	1 = birth 120 to 239 months ago, 0 = other.
Birth 20 plus years:	1 = birth 240 or more months ago, 0 = other.
Infant mortality:	1 = infant died, 0 = infant survived.
Post-neonatal mortality:	1 = infant died months 2 to 12, 0 = infant died first month.

Source: Compiled by the authors.

When all these factors mentioned above are included in one regression where the dependent variable is whether or not an infant survives the first year of life, it may be seen that with the sole exception of current urban or rural residence, they are all statistically significant predictors given. It also shows that there is a strong temporal factor at play. Births during recent years are more likely to survive than births further in the past even after controlling for all the demographic factors.

Table 10.8 shows that the post-neonatal model parallels the overall infant mortality results. The post-neonatal prediction models are limited to the sample of infant deaths where the dependent variable is whether or not an infant death was post-neonatal (dying months 2 through 12) or neonatal (dying the first month of life).

Kibet (1981) used the multiple linear regression model to study the association of child mortality with some demographic, socio-economic, disease and health variables. By fitting the predictive model to the data of 41 districts of Kenya through the OLS method, he obtained the following regression coefficients with their corresponding t values as shown in Table 10.10.

Table 10.10: Regression coefficients

Regression variables		Regression coefficients	The t-values
X_2 =	% urban population	-0.9095	-0.936
X_3 =	Population per health facility	0.0001	0.193
X_4 =	Per capita high potential agricultural land	-4.8308	-1.740
X_5 =	Total Fertility Rate	4.3573	0.468
X_6 =	% malaria cases	2.7055	3.634
X_7 =	% female adults literate	-1.9485	-1.888
X_8 =	Hospital beds per 1000 persons	3.8569	+0.449
X_9 =	Kilometres of roads	-0.1620	-1.441
X_{10} =	Population density	0.1085	2.090

Source: Compiled by the authors
The constant = 98.465
The dependent variable is $X_1 = q(2)$ per 1000
$$R^2 = 0.594$$

The raw data can be found on pages of Kibet's thesis (1981:140-141). From the regression Table 10.10, it can be seen that women's education, malaria, agriculture and population density are the most significant variables that affect child mortality. Kibet found out that these four variables explained 52.5 per cent of the variation in the child mortality index as compared with 59.4 per cent explained by all variables put together.

The regression equation further suggests that when the percentage cases of malaria are reduced by 30 per cent, the effect would be to reduce child mortality level by 81 deaths per 1,000 births. If the percentage of women with five or more years of education is raised by 30 per cent and allowance is made for the other variables, the result would be a reduction of child mortality level by about 58 deaths per 1,000 births. These results suggest that malaria control would have a tremendous impact on child mortality.

If all the factors are taken together, the corresponding F test value of 5.039, applying formula 9.20, shows that the multiple coefficient of determination of the nine explanatory variables sufficiently exceeds zero.

Analysis of variance technique was applied in testing the significance of the sum of squares due to regression and it was found to be significant.

The analysis of variance for the multiple regression yielded the following summary results (Table 10.11).

Table 10.11: Analysis of variance to multiple regression

Source of variation	Degrees of freedom	Sum of squares	Mean SS	F value
Due to regression	9	53,724.76	5969.42	5.039
Deviation about regression	31	36,724.24	1184.65	-
	40	90,448.97		

Source: Kibet (1981)

While Kibet (1981)used multiple regression analysis at a macro (district) level, Ondimu(1988) used it at micro (individual) level to study factors that influence child mortality. He had two sets of variables with corresponding regression coefficients.

Table 10.12: Coefficients of variables

Variables	Coefficients
MCH1	0.820*
MARRD	0.129*
WRKHB	-0.039*
OTHERS	0.083*
AGE	-0.004*
HEDUC	-0.072*
PEDUC2	-0.072*
PEDUC1	-0.057*
URBN	-0.029*
WRKW2	-0.020**
EVUSE	-0.013***
POLYG	-0.015***
WRKHF	-0.009
WRKW1	0.007

Constant 0.193, $R^2 = 0.737$, N = 6581
***   Significant at 0.10 level with critical value = 2.71.
**   Significant at 0.05 level with critical value = 3.84.
*   Significant at 0.01 level with critical value = 6.64.
Source: Compiled by the author

The SPSS package was used for this regression analysis.

Table 10.13: Description of variables used

SURVE	is a binary dependent variable indicating whether or not a child survived to age 3 years.
PEDUC1	is a binary variable indicating whether or not 1-4 years of formal education was completed.
PEDUC2	is a binary variable indicating whether or not 5-8 years of formal education was completed.
URBN	= 1 if mother resides in an urban area.
	= 0 otherwise.
WRKW1	indicates whether or not mother is currently working or not; i.e. current work status
WRKW2	= 1 if mother worked in the past.
	= 0 otherwise.
EVUSE	is a binary variable indicating ever use of contraceptives (1 if ever used at least one method, 0 if otherwise).
MARRD	= 1 if married.
	= 0 if otherwise.
OTHERS	= 1 if divorced, separated, or widowed.
	= 0 if otherwise.
MCH1	is binary variable indicating whether a mother visited a maternity clinic or not.
WRKHF	is a variable indicating work status of father (1 if working in farms, 0 otherwise).
WRKHB	= 1 if father is working in business.
	= 0 otherwise.

Age of mother is the only continuous variable.
Source: Compiled by the uthor

The following categories were left out of their respective variables so as to act as reference categories:

- mothers who have no education;
- mothers who are currently resident in the rural area;
- mothers who have never worked; are single;
- mother who have never used contraceptive method;
- whose husbands are currently not working and;
- mothers who are in monogamous unions.

The mothers' attendance at maternity clinics seem to be the principal variable that determines childhood mortality up to the age 3 years. The factor accounts for 72 per cent of variations. At the level of individual contribution two variables are not significant: WRKHF and WRKW1. When all the variables are included in the equation their combined effect is statistically significant (Tables 10.13 and 10.14).

Table 10.14 Infant mortality rate for all cases and by education of mother—probability of dying at age 3 and 5

Region	All cases		NE	PE	SEC+
	1969	1979	1979	1979	1979
KENYA	113	96	114	86	50
NAIROBI	75	72	97	74	41
CENTRAL	76	61	72	53	35
Kiambu	74	65	76	60	35
Kirinyaga	95	76	84	58	39
Murang'a	85	67	77	54	43
Nyandarua	65	62	66	53	36
Nyeri	57	44	64	42	27
COAST	131	101	140	96	59
Kilifi	154	150	158	105	64
Kwale	132	140	141	111	65
Lamu	105	138	148	93	48
Mombasa	105	92	98	87	55
Taita-Taveta	142	99	101	88	57
Tana River	144	139	139	127	65
EASTERN	105	84	97	68	41
Embu	96	73	81	64	40
Isiolo	154	104	107	78	56
Kitui	124	118	120	94	60
Machakos	104	82	96	73	42
Marsabit	130	97	97	97	69
Meru	93	67	77	50	33
NORTH EASTERN	132	107	108	90	65
Garissa	116	104	108	81	40
Mandera	114	111	112	96	87
Wajir	156	106	105	102	85
NYANZA	165	133	136	118	66
Kisii	104	82	85	69	41
Siaya	185	157	153	140	94
South Nyanza	192	162	163	145	88
Kisumu	176	150	155	133	68
WESTERN	131	120	125	107	71
Kakamega	123	114	117	104	68
Bungoma	121	108	110	104	70
Busia	178	152	153	132	87
RIFT VALLEY	89	85	90	73	46
Baringo	115	129	138	91	49
Marakwet	79	102	112	82	79
Laikipia	82	65	71	59	34
Nakuru	103	79	81	74	43
Kericho	77	73	72	66	41
Nandi	88	86	84	78	54
Narok	79	73	72	68	53
Samburu	81	67	64	75	77
Trans Nzoia	124	87	88	81	51
Kajiado	72	66	56	59	43
West Pokot	113	145	151	89	45
Turkana	113	106	107	99	82
Uasin Gishu	77	74	75	68	46

Source: Compiled by the author.

Table 10.15 Kenya life expectancies at birth

Region	Anker 1969	Kichamu 1969	Kichamu 1979	Mudaki 1979
KENYA	47.82	48.08	51.82	51.29
NAIROBI	56.76	56.39	57.13	56.82
CENTRAL	60.11	56.26	59.82	57.10
Kiambu	60.14	56.60	58.75	57.71
Kirinyaga	52.79	52.06	56.14	55.59
Murang'a	53.94	54.13	58.33	58.67
Nyandarua	62.60	58.77	59.52	60.17
Nyeri	63.15	60.68	64.15	60.45
COAST	43.43	44.43	50.70	43.01
Kilifi	40.23	40.64	41.37	40.88
Kwale	42.60	44.38	43.14	45.42
Lamu	47.30	49.78	43.51	44.06
Mombasa	49.20	49.88	52.60	51.76
Taita-Taveta	44.10	42.77	51.09	51.69
Tana River	43.35	42.34	43.36	42.03
EASTERN	49.80	49.91	54.36	51.52
Embu	52.03	51.71	56.90	56.12
Isiolo	39.98	40.75	50.09	50.56
Kitui	45.13	46.08	47.10	48.12
Machakos	47.02	50.01	54.86	55.92
Marsabit	44.30	44.92	51.51	49.57
Meru	53.90	52.50	58.33	56.75
NORTH EASTERN	47.25	44.46	49.50	49.21
Garissa	47.82	47.62	49.95	47.76
Mandera	50.33	47.86	48.57	47.54
Wajir	45.05	40.39	49.72	48.43
NYANZA	38.56	38.98	44.28	43.85
Kisii	49.39	50.03	54.95	53.25
Siaya	34.21	35.95	40.28	41.45
South Nyanza	34.62	34.92	39.86	41.32
Kisumu	37.97	37.24	41.46	43.97
WESTERN	43.35	44.80	46.87	45.92
Kakamega	44.75	46.16	47.96	51.63
Bungoma	48.08	46.67	49.14	54.21
Busia	36.10	37.00	41.09	42.01
RIFT VALLEY	54.13	53.35	54.14	53.07
Baringo	48.33	47.68	45.10	45.36
Marakwet	57.73	55.45	51.51	51.71
Laikipia	56.08	54.79	58.30	56.21
Nakuru	51.18	50.20	55.59	55.36
Kericho	57.68	56.05	56.91	55.38
Nandi	55.75	51.87	54.03	53.92
Narok	55.27	55.65	56.89	55.96
Samburu	59.10	55.17	58.40	53.89
Trans Nzoia	50.23	46.09	53.68	54.71
Kajiado	59.73	57.17	58.55	54.91
West Pokot	47.02	45.95	42.26	43.41
Turkana	52.20	48.10	49.62	47.59
Uasin Gishu	59.30	56.07	56.68	57.56

Source: Compiled by the authors.

Conclusion

From the above discussion, it is clear that there is a lot of room for further work as far as conceptual framework, methodology of data collection and methodology in data analysis are concerned.

Conceptual framework

The Mosley-Chen conceptual framework which has been used at the PSRI, has basically five sets of variables, namely, marital factors, dietary intake, environmental factors, personal illness and accident or injury. No work seems to have been done on accidents or injury which is one of the boxes mentioned by Mosley and Chen.

Mosley and Chen do not seem to have directly mentioned cultural variables that affect infant and child mortality. An important cultural component that might affect mortality is household structure, such as type of marriage and family structure in general. Ethnicity and religion also play an important role. A number of alternative theoretical frameworks do exist such as those by Schultz's neo-classical economic framework, Jain's analytical framework for infant mortality, Gandotra's and Das's framework, Nag's framework, Mosley's alternative framework based on the reproductive lifecycle and Venkatacharya's framework. The most elaborate one is that by Mahedavan who considers over 200 variables.

Data collection

Census data is good for macro-analysis but not for micro-analysis. Thus, for micro-analysis and in-depth analysis, sampling is preferred. Most of the national demographic surveys have used the two-stage cluster sampling procedure while individual surveys differ from survey to survey. Most of the individual surveys, however, use random sampling and multi-stage sampling. The major problem with individual surveys is determining their accuracy. The problem with census and surveys is that the information is obtained through questionnaires which basically probe the 'when' and 'what' questions, the 'how' and 'why' questions are not considered. Thus, the anthropological approach is required in such situations.

It is also noticed that the questions asked are retrospective. Thus, epidemiological approaches are also required to take care of retrospective studies, cohort studies and case-control studies.

Data analysis

In demographic analysis, the Brass type models have been used exhaustively while in statistical analysis the regression model is predominant. No check-up is done in applying these techniques, however, to find out whether or not the necessary assumptions hold. If the assumptions do not hold, alternative methods should be sought such as the logit model, hazard model and other non-stable models. Corresponding to anthropological and epidemiological data proper statistical procedures should be applied. Before rigorous statistical analysis is done, exploratory data analysis is useful to have a quick look at the data.

Recommendation

Most of the studies in the area of infant and child mortality from a demographic point of view have concentrated on identification. Which areas are high, very high, medium or low mortality regions? What factors affect infant and child mortality?

The direction of research should now be operationally oriented to examine the impact of culture and technology on child survival at village level. Appropriate, simple technologies-such as food technology, textile technology, building technology, education, and training-should be introduced or enhanced in villages.

11 Anthropological Techniques for Demographic Field Studies: A Case Study

A. B. C. Ocholla-Ayayo

Introduction

It has been observed during many demographic surveys in Kenya that many socio-cultural problems have remained unsolved by the collection of data and as such analyses always end up with insufficient and incomplete interpretation of the data. This is because of the inadequate survey methods of data collection. In Kenya, like many African countries, almost all demographic data come from censuses, surveys and very few registration systems. Unfortunately, in all these sources of data, there are insufficient methods of collecting qualitative information. Many socio-cultural variables are not simply answered by a "yes" or "no" type of survey questionnaire. Furthermore, the enumerators are not sufficiently trained to probe and get the proper information required for socio-cultural variables. Traditionally, young enumerators cannot interview or ask certain questions which are considered private. Some questions cannot be asked by outsiders or young persons, particularly those from another clan, lineage or ethnic group.

Another problem with censuses is that the information is collected within 24 hours and there is no way of following up to check whether the information given is correct. Demographic surveys still have no method of checking age misreporting, age preferences, information on children ever born, dead children and household information. Yet this is information which is most central to demographic analysis.

Anthropological Approaches to Information Collection

Anthropological methods may broadly be divided into three categories. The first category deals with the techniques of collecting information about people's behaviour and material culture, technology of production, values and beliefs attached to these objects and how the value attached to these man-made objects affect the human social life. The second category is concerned with man and his non-physical world — norms, values and belief systems. The third category deals with the problems that necessitate relationships, interactions and the values that are invested in them.

Anthropological data can be collected by a variety of field techniques ranging from intimate social contact such as participant observation and focus group panels to transient contact in formal social surveys. In the past, social anthropologists preferred to work towards the participant observation end of the continuum. Anthropologists had to adopt "a long stay in technique" in which a researcher stays in a selected sample area and waits for the events and records them as he sees them happening.

Another variation of the "stay in and observe" in a selected sample area is participating in the people's activities as one records them. Here a researcher is not only sitting and observing, but is also taking part in events. The long stay of a researcher usually enables him to cross check the variations of events. Sometimes he is part of the change which he calls variations in the system in which he is staying.

Techniques of observation may be classified into three major categories: those that focus on acts (in this case demographic processes); those that focus on relationships of persons and those that focus on the object of group attitudes. The first two categories are further subdivided according to whether one or several acts or persons are to be observed, for example, activities of women in markets, hospitals, family planning clinics, burial and rituals ceremonies, among others. Similarly, if the focus is on persons, the observer may concentrate on an individual as a representative of a status or category or on a group, for example, observation on the groups behaviour towards a childless woman, towards a woman with single gender children and towards a woman without a male child. Six types of observation categories have been determined focusing on an activity; a category of acts; the object or person which is the centre of attention of a group; an individual as a representative of a status category; a dyad and a setting.

The number of observations of a given "type one" records depends on the generalisations one wishes to make. A good technique is to write a hypothetical statement and analyse its logical implications. It is important to note that being able to say that the Kano or Kakamega people have a high fertility rate is not enough. We may further demand to know the mechanisms which sustain these high rates. What are the types of social structures? What are the kinds of marriage patterns and the norms of procreation? How do they affect stages of the reproductive cycle of a woman? What is the procreation behaviour? If we wish to compare the behaviour of categories of women who have high fertility, we may contrast their marriage patterns, age at marriage, marriage place, original home place, level of educational attainment, regional links, status and socio-cultural ideologies of marriage. One does not need to sample even a quarter of the population in order to make generalisations out of mechanisms.

By using the Focus Group Panel (FGP), for instance, the problems of reliability often raised do not come about. The FGP method uses observation and recording by two or more individuals. It is a recommended procedure because it both controls for slippage in definition of focus, grasping the discussions, recording and reduces or eliminates altogether the observer bias. No survey technique is able to map out social structures like marriage pattern norms and beliefs which affect fertility, for instance, but FGP can and so can other anthropological methods of data collection. Some anthropologists have made distinctions in observation methods, namely the direct and indirect approaches. For example, in the case of funerals, an anthropologist may use the direct observation method. In both cases, anthropologists may use a prepared schedule for observations. These schedules may take form in terms of themes or topics.

Anthropologists have also merged the qualitative and quantitative methods Kluckhohn (1989) pointed out that: "If we deal with any problem such as that of the acquisition of culture by individuals in a way which is reducible to actual by behaviours, generalisations must be given a quantitative basic." Driver (1963) also stressed this point when he said that: "if we are going to use more mathematics, we must organise fieldwork with this in mind. We must retain more qualification of every kind wherever possible to do so…. If one of the goals of anthropology is to arrive at patterns, configurations, or structures of culture, these must be determined inductively from adequate numbers of actual facts if they are to satisfy the standards of science." However, Gluckman (1950), using statistical methods and anthropological methods, found that the frequency of divorce is correlated with the social structure of a particular type. Fortes (1949a) advanced the same idea from a different point of view that social facts viewed as cultural phenomena can only be dealt with by direct methods; therefore, statistical methods are inapplicable to them. It is clear from the above discussions that there is a need for both qualitative and quantitative approaches in demographic studies.

Anthropological Techniques

In this section, some observation techniques that have been used in the study of the impact of socio-cultural phenomena on development projects in Nyakach/Kano-plains in Kisumu District, Kenya, are described. Specifically, the study looks at the impact of socio-cultural norms and values on family planning programmes, settlement and irrigation schemes, health and nutrition programmes and other development programmes.

The objectives of the case study were expressed in the form of questions as follows: To what extent are a people's social actions and responses to innovations a reflection of their traditional beliefs, doctrinal convictions or ethical prescriptions? How much have the actions been a constraint to initiating development in a rural area (i.e. on the family planning programme) and how do the malignant traditions of the people reflect the ecological environment? How much have they been a problem in major developments such as pilot schemes, family planning programmes, rice schemes, water projects, sugar projects or any other development projects, including health care programmes or the proposed resettlement scheme from the construction of the Miriu Dam? Finally, can these malignant traditions be mapped to enhance rapid rural transformation in these same regions for a major irrigation project or reduction of population growth rate? To study the impact of socio-cultural values, norms, beliefs and practices on family planning programmes, both the case study and focus group panel techniques were used.

One of the cases reported on family planning programmes is about a man who broke his wife's arm for having used contraceptives. The man came home from work and his only son told him that his mother got small beautiful pills from a mobile clinic, but the mother had hidden them behind the box. The father, being curious, looked behind the box and found the contraceptive pills. A quarrel broke out and the wife was beaten up and her arm broken. The matter was taken to the council of elders. The husband explained that he wanted a child to name after their grandmother who had just died. The elders fully supported the man, but pitied the woman for the pain. The matter did not end there, for the woman's brother took the matter to the law courts which passed a three-month sentence. When the man completed his prison sentence, he filed a divorce case.

Anthropologists call these complex events social situations and they are used to describe the behaviour or actions of individuals and groups within the norms and beliefs as presented in a given situation. They exhibit the nature of the social structure, valued norms, conflict between customary and modern laws, attitudes and practices towards family planning.

A point worth noticing at this point is that whether a researcher used a survey method, focus group panel or other forms of judgement observation methods, the information that he wishes to obtain from the respondents is for the most part dependent on the powers of memory of the respondents. The question is: how do you know that the information you have collected is reliable? The search for reliability presents serious random errors, yet it is easier to correct these than other types of errors in the research process. Vansina (1969:27-30) outlined three modes of testifying as follows: "the group testimony, dispute testimony and interrogation of informants (respondents) testimony." The "group testimony" was modified to fit FGP and the "dispute testimony" was used to obtain cases such as those referred to above. Interrogation of respondents may have some variations. For very rigorous interviewing, it is the principal researcher who has to do the probing; otherwise, the ordinary enumerators can do the interviewing. When one is making observations by "dispute testimony" one has no way of interrogating the respondents. A researcher simply records the cases as they present themselves. In this case of the group testimony and "interrogation of informants," one is able to ask as many questions as possible and seek clarifications and confirmations.

A point which ought to be recalled at all times before interviewing your respondents is to distinguish two kinds of questions: those which do and those which do not indicate the kind of reply expected. If one looks at questionnaires from surveys, one finds questions such as, "Tell me what you know about family planning? What can you tell me about methods of family planning?" Clearly, the replies are oriented in a specific direction by questions of this kind and must be regarded as testimonies that spring from the narrator alone. The questioner merely creates the occasion for giving the testimony. The imminent question is, how should we frame the questionnaire?

For the study of fertility and mortality patterns, we used the Stratified Focus Group Panels and Multi-phase Focus Panel Methods. These methods of data collection are not a substitute for statistical sampling techniques, but can be used along with them.

Stratified Focus Group Panel

The Stratified Focus Group Panel (SFGP) technique is a relatively new, in-depth qualitative method of socio-cultural research. It may be simply defined as a FGP discussion in which a small number of selected discussants, under the direction of moderators (normally principal researchers themselves) sit to talk about topics believed to be of special importance for the investigation. The group must be made homogenous by some kind of stratification so as to allow free discussion. The technique is designed to answer questions of a discrete nature as well as the" why" and"how" questions, which normally are not sufficiently answered in random sampling schedules. It is also used in mapping systems and knowing how an established social structure actually functions. Knowing the behaviour patterns of a social structure means being in a position to predict what is likely to happen in that or a similar social structure in the future.

In the case study presented in this chapter, the Stratified Systematic Random Sampling Focus Group Panel technique has been used to select discussants for various focus groups. The population of both Nyakach and Kano were first divided into administrative boundaries, locations, sub-locations and clans. Each clan was further sub-divided into clusters known as *mjikumi*. (One *mjikumi* is composed of 100 heads of homestead.) In Stratified Systematic Random Sampling Focus Group Panels, the size and selection of sampling units and FGP discussant units coincided in most cases. For instance, for every 100 *mjikumi*, we selected 10 FGP discussant units.

The total of FGP discussant units of a sub-location was taken to form a sample size of the same sub-location for statistical interviewing. For qualification purposes, every known Kth respondent was picked through the same list starting with a randomly chosen number 1 and K inclusive. In this case, the total population size of a sub-location was, 3200 heads of household, but by the FGP selection procedure, 320 FGP discussant units were obtained (Table 11.1).

Table 11.1: Sample size in West Agoro Sub-location, Nyakach, Kisumu District

Clans	No. of	Total Mjikumi	FGP No. population	Discussants
Kanyigot	5	500	5	50
Kanyibana	8	800	8	80
Wareya	3	300	3	30
Kamiholo	5	500	5	50
Kasirunda	2	200	2	20
	32	3,200	32	320

Source : Compiled by the authors.

Let K = N/n = 3200/320 = 10. A number between 1 and 10 was selected at random and this determined the first sample number. This number was 5; the sample was, therefore, composed of number 5, 15, 25, 35, e.t.c. The list from where to draw these numbers was constructed from the FGP list which enabled the compilation of a full list for systematic random sampling. A total number of 120 sessions were covered in Nyakach alone with a total of 1,020 discussants from 59 clans.

In Kano, the statistical method was first used then followed immediately with the FGP discussions. In Nyakach it was the other way round, beginning with the FGP method and following with the random sampling.

Judgement selection is necessary in this Stratification Systematic Random Sampling Procedure for the FGP discussions. This is because some members of the population cannot be accepted in the sample size for instance, a group of discussants that includes a mother and her daughter or father and his son will not yield any positive results. If the stratification does not exclude such relationships in a session, it must be done by judgement selection.

The topics that were raised for the discussions were directed to the objectives of the research, designed to elicit as much information as possible in order to establish the impact of socio-cultural norms and practices affecting the reproductive cycle of married women in these regions.

One of the principal researchers observed and recorded sign language and impressions of inner feeling for discrete practices and convictions. The shorthand and dictation recorded during the discussions were later edited and grouped into code classification. The statements and responses, questions, objections, comments and affirmations recorded on every topic at each session were compiled and classified so that they could be transcribed and analysed for particular themes.

The records from non-verbal interactions of significant body movements stimulated by responses, statements and symbolic language used at certain points

Table 11.2: Retrospective marriage, fertility and mortality patterns from the FGP discussion

A. Kabodho East: Kandaria, Sigoti Area

Name, age of father interviewed	No. of Father's wives and children of interviewee	Discussant interviewees in FGP
1. Respondent, 61 years Komwono-Kojwang	5 wives, 44 children 1. Kadianga (Nyakach) 6-3 7-1 7 2. Nyakach Koguta 5-3 9-5 14-4 3. Konyango Kasipul 5 4-1 9-1 4. Agoro (Nyakach) 3 5-1 8-1 5. Nyakach Kogutu (-Nil)	1 wife from Nyakach Koguta, 13-4 15 children 8-1 15-1
2. 58 years Kandaria-Kotema	1 wife from Nyakach Agoro, 16 children 2-1 14-1 16-10	1 wife from Konyango (S. Nyanza) 9 children 6-1 3-1 9-4
3. 69 years Kandaria Karek	4 wives, 17 children Nyakach Koguta 1. Kabondo 2 3 5 (some died) 2. Nyakach Wasare 1 - 1 3. Kano Kabonyo 2 1-1 3-1 4. Kabongo 3-2 5-3 8-5	1 wife from 10 children 7-2 3 10-2

4.	50 years
	Kandaria Kojuando
		4 wives, 18 children
		1. Kano
		   5-1 4-1 9-1
		2. Kisumu
		   - 2-1 2-1
		3. Seme
		   3 3-2 6-2
			3 wives, 27 children
			1. Karachuonyo
			   6-1 8 14-1
			2. Kano
			   9 3 12
			3. Kano
			   - 1 1

5.	55 years
	Kandaria-Karek
	(plus G.M.P)
		6 wives, 41 children
		1. Kodera
		   2-1 7-1 9-5
		2. Kano Kabar
		   3 5 8
		3. Kano Kabonyo
		   5-2 3-2 8-4
		4. Kisa
		   2 1 3
		5. Karachuonyo Kakdhimu
		   3-1 3-2 6-3
		6. Kasaye
		   1 6-5 7-5
			1 wife from Nyakach
			Nyakach Ramogi
			5 9 14
			(14 Children)

6.	40 years
	Kojwang
	Kandaria-Kojwang
	(plus G.M.P)
		5 wives, 44 children
		1. Nyakach Kadianga
		   6-3 7-1 13-4
		2. Nyakach Koguta
		   5-3 9-5 14-8
		3. Kasipul Konyango
		   5 4-1 9-1
		4. Nyakach Koguta
		   - Nil -
		5. Nyakach Agoro
		   3 5-1 8-1
			1 wife from Saye
			6 children
			4 2 6

7.	61 years Kandaria-Kandenga	1 wife from Seme 7-2 2-1 9-3 (9 children)	1 wife from Kano Kogelo, (6 children) 3-2 3-2 6-4

8.	65 years
	Kandaria-Kogwang
	Plus. old sin 4
		2 wives, 10 children
		1. Karachuonyo
		   4-1 6-4 10-5
		2. Jimo
		   - Nil-
			2 wives, 14 children
			1. Kadianga
			   7-1 2 9-1
			2. Jimo
			   4 1 5

167

B. Sigoti Area Session

	Name, age of father interviewed	Father's wives and children of interviewee	Discussant's son interviewed in FGP

1.	73 years
	Kandaria-Kojuondo
	Kadenga
	* Plus S. in L.P.
		2 wives, 13 children
		1. Katieno (Mumbo)
		   3-1 3-1 6-2
		2. Seme
		   4-3 3 7-3
			1 wife from Kano
			Kolwa, 10 children
			5-2 5-2 10-4

2.	43 years
	Kandaria-Kojuondo
	
	*Assisted by
	Grand M.P
		3 wives, 7 children
		1. Kisumu
		   - Nil-
		2. Kabondo
		   5 2 7
		3. Kabondo
		   - Nil -
			1 wife from Kano
			10 children
			6-1 4-2 10-3

3.	Kasaye
	Kandaria-Kojuondo
	- Katieno
	
	*(Asst. by Sister in Law Present)
		5 wives, 23 children
		1. Kabar (Kano)
		   8-4 2 10-4
		2. Konyango (S. Nyanza)
		   2 2 4
		3. Ramogi (Nyakach)
		   2 21 3
		4. Agoro (Nyakach)
		   1 2 3
		5. Agoro (Nyakach)
		   1 2-1 3-1
			2 wives, 13 children
			1. Kabondo
			   6-1 4 10-1
			2. Kabondo
			   - 3

4.	70 years Kandaria-Katieno *(Asst. S. In L.)	5 wives, 23 children	1 wife from Nyakach Ramogi, 10 children 4-2 10-2
5.	Born in Nyakach Agoro *(Asst. by Grand M.P.)	1 wife from Seme 4-2 2-2 6-3 (6 children)	Husband had 5 wives She has three children 2-1 1 3-1

6.	67 years
	Nyakach Kamgan
	Kandaria-Karek
		2 wives
		1. 4-3 2-1
		= 6-3 5-1 3
			Husband had 4 wives
			37 children
			1. Karachuonyo
			   8-1
			2. Kakdhimu
			   3 6 9
			3. Kakdhimu
			   10-10 4-4 14-14
			4. Kamgan
			   5-2 6-2

Anthropological Techniques for Demographic Field Studies: A Case Study of Kenya

7. 61 years of in Uyoma
Katwenga in 1940
Kandaria Kojwang

   1 wife, 8 children

   Husband married
her alone
10 children
3-1 5-18-2
7 3 10

8. Born in Nyakach,
Age

   Kadianga
Kandaria-Kogwang

   *(Asst. by Grand mother)

   6 wives, 24 children

   1. Kabondo
   2-2
   2. Kobuya
   1 2 3
   3. Koguta (Nyakach)
   4 3 7
   4. Koguta Sango
   3-1 3-1
   5. Wasare
   3-1 2 5-1
   6. Jimo
   1 3 4

   They are two wives
She had 8 children

   6-1 2   8-1

9. 47 years
Born in Nyakach
Koguta
Kandaria-Kojwang
*(Ast. by Errand mother)

   2 wives, 7 children
   1. - Nil -
   2. 5-1 2 7-1

   Married alone
She has 15 children
7   8-1 15-1

Average:

1. Average polygamist
   3.2 wives
2. Average No. 2.
   children per woman
   = 6

Average polygamist
has 2 wives
Average No. of
children per woman
= 8 children

Total No. of women = 55
Total No. of children = 310
Average for the whole group:
Polygamist - 3 wives
No of children per woman = 6
- Women = 84, - children = 527

Total No. of women = 29
- Children = 217

169

C. Kabodho East—Manyatta Area Marriage Places and Children Born Dead

Place of marriage			Boys	Girls	Total	
Kano:	Kabonyo	1.	4	1	5	
		2.	5	2-1	7-1	
		3.	4-3	2-2	6-5	
			13-3	5-3	18-6	
	Kobura	1.	4-1	2	6-1	
		2.	-0	-	-	
		3.	-0	1	1	
		4.	4-1	2	1	
		5.	3-1	4	7-1	
		6.	-	1	1	
		7.	2-1	1	3-1	
		8.	5-5	5	10-5	
		9.	1	6-1	7-2	
		10.	3	7.4	10-4	
		-	22-9	29-6	51-15	
	Kolwa	1.	2-1	4-2	6-3	
		2.	-	-	-	
			2-1	4-2	6-3	
	Kochogo	1.	3-1	5	8-1	
		2.	1	4	5	
			-	4-1	9	13-1
	Kobar ± =		1.1	3-2	4-2	
			42-14			
	Kabondo	1.	2-2	2-2	4-4	
		2.	1	-	1	
		3.	3	2	5	
		4.	3-1	6	9-1	
		5.	1	-	1	
		6.	2-1	3	5-1	
		7.	3	4	7	
		8.	2	2-1	4-1	
		9.	2-1	2	4-1	
		10.	1	1-1	2-1	
		11.	-	3	2	
		12.	5	5-1	10-1	
			25-5	30-5	55-10	

Anthropological Techniques for Demographic Field Studies: A Case Study of Kenya

Nyakach:	Ramogi	1.	2	6	8
		2.	1	-	1
		3.	1	-	1
		4.	5	7-1	12-1
			9	13-1	22-1
	Kadianga	1.	1	1-1	2-1
		2.	1-1	1	2-1
		3.	1	2	3
		4.	4-2	1-1	5-3
		5.	1-1	-	1-1
			8-4	5-2	13-6
	Agoro	1.	2	2	4
		2.	1	1-1	2-1
		3.	4-1	5	9-1
		4.	2	2	4
		5.	5-1	6-1	11-2
		6.	1-1	1-1	2-2
			15-3	17-3	32-2
	Koguta	1.	3	4	7
		2.	5	2-1	7-1
		3.	2	2-1	4-1
			10	8-2	18-2
	Jimo	1.	1	4-1	5-1
	Rae	2.	-	-	-
	Karachuonyo	1.	4	4	8
		2.	1	5-3	6-3
		3.	1	1	2
		4.	1	1	2
		5.	3-1	-	3-1
		6.	2	1	3
		7.	1	1	2
		8.	4	2	6
		9.	4	2-1	6-1
		10.	1	6	7
		11.	1-1	1-1	2-2
		12.	3	5	8
			54.7	26-2	29-5
	Kisumo	1.	2	2	4
		2.	1	1-1	2-1
		3.	2-3	1-2	3-5
		4.	2	1	3
		5.	2	1	3
		6.	2	2-1	4-1
		7.	4-1	2	6-1
			25-8	15-4	10-4
	Seme	1.	3	-	3
		2.	3	5	8-1
		3.	3	-	3
		4.	2-2	2	4-2
			18-3	11-3	7
	Sakwa	1.	3	2	5

Source: Compiled by the author.

in these discussions were later integrated for each particular theme and topic. These were found to be very useful in the interpretation of the verbal messages from one group session to the other and when compiling a complete profile of the group practices, interaction patterns, feeling and attitudes towards such innovations as family planning programmes, use of contraceptives, resettlement plans, irrigation and introduction of new cash crops, among others, all of which may never be available in the random sampling data obtained through questionnaires. One of the discussion rules is that each participant must make a statement or respond to the topic for discussion. Each member of the panel is encouraged to ask questions, or comment on others' responses, or make statements regarding the topic for discussion. This enables individual records to be kept for quantification. There are individual biases towards the majority opinion, but the most important thing is that corrections of facts put forward are done on the spot.

The information obtained from stratified focus group sessions can help to estimate the levels of fertility, marriage patterns, mortality levels, types of marriages and a retrospective study of the region under consideration. This is a new social anthropological research method which sets to map out the existence of norms and beliefs.

Each discussant was given a code number to the investigators and research assistants. These numbers were associated and attached to their responses for the statistical 10. Seating arrangements were made according to clan and *mjikumi*, an arrangement that allows for seniority of the clan elders and seniority by age. It was noted that discussants were not always heads of the clans or lineages, but were mostly heads of the homesteads or households.

Multi-Phase Focus Panel Method

This method is a combination of a statistical sampling method and a socio-anthropological discussant observation of the focus group panel method. Multi-phase sampling is a type of sampling design in which some information is collected from the whole sample using the random sampling method and additional information is collected either at same time or later from sub-samples of the full sample by the FGP method. This method was meant to answer the "why" and "how" questions which are descriptive and, therefore qualitative in nature.

Multi-phase sampling with one sub-sample is called two-phase sampling or double sampling. Since the multi-phase method does not depart from the principle of random sampling, the first sub-sample of the study's sampling was by statistical method, but the more qualitative information was collected through the FGP discussions.

Conclusion

The chapter has clearly indicated that anthropological techniques can be used to supplement demographic studies by enriching the latter with qualitative information. With the help of anthropological techniques, demographic problems for obtaining qualitative information can be greatly reduced. Anthropological techniques, however, still need to be improved so as to provide statistical and qualitative information at the same time. Attempts made in recent studies show that statistical techniques can either be used simultaneously or one can be carried out immediately after the other. While planning to use the two techniques, it is important to identify items which could compromise the survey type of interviews and those which should be left for focus groups or other anthropological methods of observation. Items of these kinds ought to come from the same theme or topic.

An improved anthropological technique such as FGP can be applied expeditiously. The length of anthropological fieldwork has always been another problem; this can be considerably reduced. When both methods are used at the same time, they are accomplished almost simultaneously. The FGP, however, takes more time compared to surveys, but this is a shorter period when compared with the normal anthropological "stay and observe" and record method.

Anthropologists prefer to begin an investigation from kinship systems and marriage because of their multi-undimensionality and accumulativeness. Kinship systems may spell out who is permitted by custom to marry whom. It may also spell out who is not permitted to have sexual relations with whom. Kinship systems and marriage spell out whether a marriage is matrilineal or patrilineal and further whether a marriage is exogamous or endogamous. Patrilineal exogamous marriages, for example, have a negative effect on child nutrition because the wealth comes from one direction and women move in the other direction. Patrilineal exogamous marriage may also neutralise inheritable diseases.

Through kinship systems and marriage, the anthropologist can identify existing marriage patterns which are important for explaining fertility differentials and levels. The study enables us to understand how polygyny affects the TFR and how the rules and practices affecting the reproductive cycle of a woman can be studied.

Anthropologists argue that scientific samples are not needed for research in which the subject of inquiry is homogenous. Bernard (1988) said that there is no need for scientific sampling in phenomenological research in which the object is to understand the meaning of expressive behaviour or simply to understand how things work.

Anthropologists take expressive behaviours to be related and to form a system so that a theory of causation that accounts for variables being related to one another can be stated. This makes the anthropological approach easier in multivariate analysis. Anthropologists can work easily with the so-called "elaboration method" of multivariate analysis. This requires nothing more than careful construction and inspection of percentage tables and the use of bivariate statistics. This is useful in the field because it enables one to work with data as it is obtained. Anthropologists find the multiple regression models useful because they build more complex equations that tell how to weigh each of several independent variables such as marriage pattern and fertility rate. Ethnicity is a concept which when defined implies several socio-cultural ideologies which have positive or negative effect on fertility. Anthropologists have been exposed to covariant analysis when they collect information such as, most women say that they really wanted fewer pregnancies, but claimed that this was not possible so long as the men require them to produce at least two sons to work on the land, or name after their deceased ancestors.

The techniques used in the case study have helped to get information which would not have otherwise been obtained by the usual type of questionnaires. It has been possible to identify some patterns of marriage that existed in the distant past and assess their fertility and mortality patterns.

Nyakach people have both patrilineal exogamous and patrilocal exogamous marriage patterns while Kano people have only patrilineal exogamous marriage. Both Kano and Nyakach people have had a high percentage of polygynous unions over the past generations. Formerly, the fertility of polygyny was quite high, probably to compensate for the high mortality which was experienced at that time. Though a number of studies in Kenya show that first wives in polygynous unions have fewer children that those of junior wives, this study shows that fertility decreased from the senior down to the youngest wife. From this study, it is possible to associate high fertility, high child mortality and high specific sex of the children with the place of origin of women and the place of marriage.

12 The Health of Migrant Labourers in Colonial Kenya

Isaac Sindiga

Introduction

Although the role of population movements in the expansion of disease into new areas has long been recognised, most recent studies of migrant labour in Kenya focus on the economics of labour movement. Other studies deal with methods of labour recruitment and the development of an urban labour force respectively. Kanogo's (1987) study of squatters in Rift Valley Province is concerned with political economy. She discusses squatter welfare only in terms of access to land, wages and restriction of movement. The health and well-being of migrant workers has received little attention. Quite often problems of food supplies, health, disease, medical care, housing and sanitation have been given only passing attention.

Workers from different parts of the country had to adjust to new environmental-health circumstances, usually with serious danger to their lives. Other changes included diet, crowded insanitary settlements, and meagre medical attention. In all, some 375,000 African men (including 125,000 casual labourers mainly in agriculture) were registered as wage earners in Kenya by the late 1940s. This number, which formed some 13 per cent of the total population of the country at the time, is sufficiently large to warrant close study. Moreover, the contribution of migrant labour to Kenya's colonial political economy was quite substantial.

This chapter follows the fortunes of migrant labourers with regard to nutrition, health, disease and their general well-being at their places of work for the period 1900 to 1950. It is, however, not concerned with the occupational health of workers in various work stations such as in industry, mining, quarrying and transport since this has been pursued in other studies.

Background to Migrant Labour

The completion of the Kenya-Uganda Railway (KUR) from Mombasa to Kisumu in 1902 provided the immediate impetus to European colonisation of Kenya. The headquarters of the East Africa Protectorate (EAP) which had long been established in Mombasa was transferred to Nairobi in 1905. Citing the economic significance of the railway, Commissioner of the EAP, Charles Eliot, encouraged European settlement in the Kenya highlands. The story thereafter is well-known. Kenya attracted a sizeable European farming community which was granted large tracts of land. This historical background is important to understanding the beginning of migrant labour in the country.

The railway itself attracted many African labourers. Urban centres were developed all along the line. Manufacturing industries were subsequently established in some of the urban centres such as Mombasa, Nairobi, Nakuru, and Kisumu as well as mining areas such as Magadi, to name a few. All these required labour. In addition, African workers were needed on European plantations and ranches. The coming of the KUR and European settlement thus provided a need and reason for hired labour which had to be moved from various parts of the country.

Labour Recruitment

Most of the labourers came from Nyanza, Western and Central Provinces. Towards the end of the 1940s, 75 per cent of the migrant labourers in Kenya were Abasuba, Luo, Luhya, Kikuyu and Akamba. The Kikuyu were moved into Rift Valley Province and Coast Province to work both in urban areas and on European farms. As early as 1912 the majority of African wage workers in Coast Province were Kikuyu. Similarly, the Luo and Luhya moved into KUR urban centres and European farms in the highlands.

The task of recruiting potential labourers fell to the chiefs. The recruits had to be young able-bodied males. It was a statutory requirement under the Master and Servants Ordinances, 1910-1918, that "all labour recruited by a labour agent must be inspected as to its physical condition by a government medical officer before a contract may be registered" (Kenya, 1923:54).This provision was followed to the letter. Of the 13,618 recruits presented before a medical officer during 1922, for example, 4225 (or 24%) failed to pass the physical fitness test. This rigorous selection procedure, however, was not backed by the provision of basic amenities and health care at the places of work. Migrant labourers were exposed to extremely poor working conditions leading to high levels of morbidity.

On passing the medical examination, a recruit began his contract as a migrant labourer. Usually he would trek long distances under guard to get to the nearest station to obtain transportation. Many recruits walking along the Murang'a -Nairobi road contracted malaria due to camping at night near swamps.(East African Protectorate, 1913). Labourers whose work destinations were along the KUR travelled by train. Those from Nyanza province were transported in congested closed iron trucks which were attached to goods trains. These trucks had neither toilets nor water facilities. This caused high rates of morbidity for those who had to travel long distances (Kenya, 1923: 42). It may be noted that although the Native Labour Commission of 1912-1913 recommended unlocking KUR wagons and removing congestion from trains, the same conditions were reported more than a decade later and persisted long after that (Kenya, 1927).

Conditions at the Work Place

By far the greatest problem facing the African migrant worker was separation from home and family. It may be noted, parenthetically, that the postal services during the colonial period were poor. Correspondence could not be delivered regularly. Moreover, low literacy levels negated most attempts to use the post, as a medium of keeping in touch. Differentials in sex-ratio were quite high, especially in urban areas. This had adverse consequences on migrants' social life. In Nairobi, for example, the male-female ratio was 9:1 before 1945, and 3:1 in 1958 according to Werlin (1974:61). It appeared to have been thought by the colonial authorities that Africans did not want to live with their wives and other family members at their places of work. The problem was shamefully low wages and lack of family housing.

Throughout the colonial period, Africans received wages which would not provide for their families. In 1938, the minimum wage for Africans was Kshs. 38.00 increasing to Kshs. 56.50 in 1952. At the same time the cost of maize meal, the staple food, rose by 800 per cent between 1939 and 1953. Moreover, Africans were paid less for the same job than non-africans. Northcott (1949:14) found that, "the present arrangement for the pay of the several races are that African salary scales is assessed on the basis of 50 per cent of the salary scale for corresponding Asian posts".

Housing

In urban areas, African workers lived in poor and insanitary accommodation as attested by frequent outbreak of plague, pneumonia, influenza and malaria, among others. Under the Master and Servants' Ordinance employers were required to provide housing for their employees, but this provision was never observed. In fact, government and municipal employees in Nairobi were the worst housed. In many cases, their dwellings were completely unfit for human habitation.

There was no housing for Africans in early Nairobi; they put up mud-walled thatch-roofed structures in the eastern part of the urban centres. These were insanitary and lacked toilets and water. When some accommodation was given to Africans after World War I, it was in congested dormitories. Of the 40,000 plus Africans in Nairobi during the 1940s, some 5,000 of them had nowhere to sleep (Nairobi, 1948). The KUR authority, which had a reputation as the best employer, housed Africans in rooms which were 10 square feet. The situation in mid-1947 was that "2271 married men were provided with one room per family while 4,013 were treated as 'single' men." Most of the so-called single men were housed two per room while a large number lived three per room.

The African labourers on farms and plantations fared no better in terms of accommodation. They were required to put up grass-thatched huts on arrival. Most of these cottages were poorly ventilated, lacked water and were insanitary. As a result of these conditions in labour camps, dysentery was frequently reported among newly labourers recruited from African areas. Leys (1973: 300-301), asserted that the prevalence of dysentery among migrant labourers was related to lack of safe latrines and also because some latrines were quite shallow and formed an ideal habitat for flies.

Nutrition

Yet another problem which migrant labourers faced was undernutrition and malnutrition. In the early years of colonial rule, an African labourer's food ration comprised two pounds of maize meal (*posho*) per day, two pounds of beans per week and some salt (Kenya, 1926: 46). Meat and fruits were seldom supplied and green vegetables were generally unavailable. The problem of food quantity and quality was particularly severe in urban areas. Evidence given by Mr. E.W. Hickes, an engineer with the Magadi railway to the Native Labour Commission of 1912-13 showed that a total of 4,200 Africans were working on their projects. Their food comprised of maize meal and beans. No meat was provided; rice was given after three months, work. (East Africa Protectorate, 1913: 76). Other testimony to the Commission revealed great deterioration in the health of workers due to a lack of variety of food over a long period.

Ordinarily, Africans eat many types of food. John Ainsworth, Provincial Commissioner of Nyanza, noted that the Luo "normally had three meals a day and was constantly chewing something in between whiles (sic). Those living near the lake also ate a good deal of fresh fish". Similarly, Leys reported that "The Kikuyu in his own home eats at least eight different articles of food everyday and the absence of meat from his diet makes both large quantities and large variety of vegetable food necessary" (East Africa protectorate 1913: 273).

For the squatters, particularly the Kikuyu who moved into the Rift Valley to work on European farms, life at the beginning appeared normal. According to Kanogo (1987), some of them were large stock owners in Central Kenya and were looking for grazing land; others had already been rendered landless by the capitalistic tendencies of their own brethren. Yet others lost their lands to European settlement. Whatever the case, the early history of squatters points to the accumulation of wealth and relative food security. Squatters were allowed to live on European farms where they were required to provide labour and in return were allocated plots of land to cultivate and keep stock. They appear to have had plenty of food. However, this situation lasted only up to about 1923, after which Europeans demanded longer working hours and restricted the size of plots squatters could put to their own use. The story thereafter was one of progressive impoverishment of the squatters themselves. Inevitably, diets were affected.

If squatters could not supplement their food rations because of restrictive laws, the situation in urban centres and plantations was more problematic. Workers did not have another source of food. Consequently they became weak and liable to disease and infections. The labour commission of 1912-13 concluded that:

> Wastage of labour has undoubtedly occurred through insufficiency of medical attention given, resulting in deaths and loss of vitality, and the latter is often occasioned by malnutrition (East African Protectorate 1923:324). The problem, however, persisted throughout the colonial period in Kenya. Men "leave home too early for much breakfast. In the mid-day, they have too little time to return home and eat a meal. Their wage rates do not permit of the expenditure of 30 cents daily for lunch at the municipal canteen. The problem of food was closely interlinked with the wage levels paid to Africans.

Disease

In essence, the phenomenon of migrant labour thrust Kenyan Africans into new environmental health circumstances. In certain instances, people moved from cold to warmer areas. Such change often was a challenge to health. Perhaps the example of the Meru from the Mount Kenya region, who were taken to work on the Mombasa waterworks at Shimba Hills, is instructive. The morbidity rate among the 500 Meru people working at the coast was quite high. McGregor Ross, Director of Public Works, told the Native Labour Commission that the Meru "sickness-rate and death-rate were both excessive, the latter rising as high as 176 per thousand per annum". This forced the government to repatriate them after three months. Further recruitment from Meru was stopped.

The major causes of morbidity were malaria, dysentery and diarrhoea. The three diseases were essentially a result of poor environmental sanitation at the labour camps. In addition, the Meru, who came from a high altitude, malaria-free environment, could not withstand a malaria-endemic place without appropriate medical care and public health precautions. The Kikuyu who went to work at the Kenyan coast also faced the same problem. Leys noted that the mortality rate among the Kikuyu at the Kenya coast was six times higher than among the resident population. He attributed this to malaria and dysentery arising from poor sanitation, housing and food. In contrast, the Luo, who came from a warm malaria-endemic area, appeared to survive better at the Kenyan coast than those from colder areas. In the absence of superior sanitation and medical care for the Luo, this evidence points to their relative immunity to malaria attacks.

Workers originally from warmer areas who were taken to work in the colder highland areas faced new problems as well. A case study of the Uasin Gishu railway workers as reported by the Colonial Principal Medical Officer illustrates this (Kenya

1923). While these workers faced the same political and economic problems as their counterparts elsewhere, the geographical difference between their areas of origin and new places of work was significant to the type of illness they contracted.

The Uasin Gishu Railway Construction Workers

The Uasin Gishu railway began construction in Nakuru. In 1922 there were some 4,500 labourers who lived in 34 camps. Most of these workers were Luo and Luhya from the relatively warm areas around Lake Victoria. Like other workers elsewhere, their food comprised essentially two pounds of maize meal per day and one pound of beans per week. These workers suffered from severe pneumonia. The major cause of illness was lung disease due to the very low temperature. Uasin Gishu District, which is 6,000 feet in altitude is, on average, 2,000 feet higher than the native lands of the Luo and the Luhya. There were a total of 436 deaths in 1922 of which 205 were caused by respiratory disease, essentially pneumonia. There were some cases of dysentery, diarrhoea and influenza. Some 98 died of unknown causes.

In addition to the coldness of the Uasin Gishu plateau, there were heavy rains during 1922. Given the circumstances of poor and over-crowded shelter and lack of adequate blankets, pneumonia was always a threat. Further, the labour camps were insanitary. Medical attention was virtually absent; in fact there was no medical officer at the site (Kenya, 1923). One who was called in belatedly was stranded in Nakuru because of lack of transport.

This case demonstrates the appalling conditions which migrant labourers experienced. With little medical attention, disease was a constant feature in labour camps and worker output was considerably reduced. The Government of Kenya did not take responsibility for medical care for labourers in private employment. The law none the less required employers to provide medicines for labourers and "medical attendance in case of serious illness; if procurable. (Kenya, 1931: 26). Such attendance was virtually impossible to obtain for lack of facilities and the paucity of medical personnel.

The spatial movement of labourers also abetted disease expansion to new areas. Malaria parasites, for example, appear to have been carried from the traditionally malaria-endemic areas of Nyanza and the coast into the highland. Before World War I, malaria was not reported in areas lying beyond 1,600 metres. After that period malaria diffused into Trans Nzoia, Uasin Gishu, Nandi and Nairobi Districts.

The high rate of labour movements from Nyanza Province into these areas ensured a constant supply of malaria parasites. However, the presence of malaria parasites was only one condition for the mosquito vector, female Anopheles in particular. Under normal circumstances, mosquito breeding is inhibited in cold areas. In the highlands of Kenya, artificial mosquito habitats were created through building and road construction which left behind pits and quarries which could shelter mosquitoes. The mosquitoes themselves appear to have been carried by automobiles and trains to the highlands.

In subsequent years, malaria became an epidemic disease in highland Kenya.

European settled-districts were so affected that the country promulgated a malaria policy in 1926. In Uasin Gishu and Trans Nzoia Districts the malaria epidemic led to panic. "The value of land was said to have dropped as a result of health conditions" and "dislocation of business occurred." Consequently, special mosquito teams were sent to the areas to investigate conditions leading to malaria in rural areas establish medical units and hospitals in African areas and conduct investigations into malaria expansion (Kenya, 1929: 9). For the first time, malaria, which had always afflicted African workers, received attention.

Another health problem which afflicted the African male workers was venereal disease. Apparently, the long separation from their partners led men to become sexually mobile. The problem became serious, especially among returning labourers to Nyanza, that chiefs were unwilling to recruit more labour (East African Protectorate, 1915). Fearing that returning labourers were spreading venereal disease in their home areas. Indeed, the greatest incidence of treated cases of syphilis came from trading centres and urban centres from around the country, and from Central Kavirondo District and South Kavirondo District. These conditions continued beyond 1950.

Conclusion

The poor environmental conditions at work stations, insanitary housing, low wages and inadequate food were a bane in the migrant worker's life. The monotonous and unbalanced diet led to the physical deterioration of workers. Productivity plunged. Migrancy became a most hated colonial institution. Workers often ran away to their home areas. As early as 1912, a commission was set up to investigate the reasons for the labour shortage. Many of the colonial officers testifying before the commission noted that poor housing, inadequate food of poor quality, poor transportation, mistreatment by employers and lack of medical attention were the leading causes of desertion by labourers.

Medical examination of returning labourers after completion of contracts revealed "great loss of weight apart from sickness". In all, labour conditions had led to very high sickness and death rates and to widespread discontent. Continuing declining work output and desertion led to the appointment of another commission in the mid-1942. Dubbed the African Labour Efficiency Survey, the Northcott Commission found that average worker conditions had not changed since 1913. The Commission's findings revealed a large labour turnover in Nairobi among railway employees of the 5,979 employees; for whom records are available, 30 per cent had less than one year of service and 50 per cent had 3 years or less. Only 17 per cent had served at least 10 years.

By independence in 1963, medical attention to African migrant labourers was still inadequate, wages had not been improved and housing remained the insanitary dormitories. These conditions had a significant impact on the health and well-being of male migrant workers throughout the colonial period. They also played a role in spreading diseases, like malaria, which had previously been limited to specific environmental zones, into new areas to infect new populations. These aspects of Kenyan migrant labour have not received adequate attention from researchers despite their importance to understanding broad changes in health and the environment in Kenya during the first half of the 20th century. This chapter has emphasised these relatively unexplored factors in order to bring more attention to them in further studies.

13 Engineering Education and Related Development in Kenya

Francis J. Gichaga

Introduction

The late Prof. S. H. Ominde was an educationist par excellence who inspired many scholars not just in the humanities and social sciences, but also in the broader perspective of scholarship. This chapter is dedicated to this wider perspective in education which would encompass engineering education.

The phenomenal development that has taken place in Kenya in the last quarter of a century has largely been due to the heavy investment that has been made towards the development of basic infrastructure such as transport and communications, housing, energy, education and health services, among others. Among the key actors in the planning and development of the basic infrastructure have been engineers of various specialisations. Engineering education can, therefore, be considered as an important variable which has made significant contribution towards developing Kenya. The availability of the basic infrastructural services was very limited at the time of independence in 1963. However, following carefully planned development of the basic infrastructural services by the various professionals including engineers, most Kenyans today have access to those services within reasonable bounds of space and cost.

For engineers to have been able to make major contributions in the development projects, they would have to be properly educated academically, be exposed to engineering problems in the real world where they acquire practical experiences, and be provided with an enabling environment to make it possible for them to plan and develop projects which are affordable and relevant to the country's needs.

This chapter considers education and practical training that has been provided to engineers in the country. It also highlights developments which have resulted from the engineer's input and concludes by considering the future in terms of policy issues such as indigenisation of the industrial sector and technology choice and transfer for development.

Engineering Education

Historically, engineering education was introduced in Kenya following the establishment of the then Royal Technical College of East Africa in 1956 (later called the Royal College in 1961, University College, Nairobi in 1963 and University of Nairobi 1970). Until the early 1970s, the Engineering Faculty of the University of Nairobi provided engineering education for the East African countries — Kenya, Uganda and Tanzania. Following the creation of more public universities in Kenya, engineering education is now also available at Moi University, Jomo Kenyatta University of Agriculture and Technology and Egerton University.

Engineering education involves the systematic learning of the art and science of harnessing the resources of nature for the benefit and comfort of society. Thus, the student must acquire scientific and technological knowledge applicable in solving the relevant problems.

Engineering education consists of theory and practical (or industrial) training which a graduate engineer must undergo to be a professional engineer. The practical or industrial training (or internship) takes about three years after graduation; the intern is expected to present to a professional examination board documentation showing how he/she has achieved the required training. The graduate must also sit for both oral and written examinations conducted by the board.

University Engineering Education

Engineering students should have a strong scientific background, especially in subjects with a computational bias, including mathematics and physics. Engineering education emphasises understanding of safety of engineering works (such as buildings or bridges which can collapse if inadequately designed or dams which can be washed away during floods) and the economic viability of a project.

University engineering education provides three basic components of training namely, theoretical, experimental and industrial training. The theoretical component involves drilling the student through the various scientific and mathematical tools of engineering, thereby being made to conceptualise engineering problems in terms of basic discrete elements. Only after this, can the students apply engineering tools to solve problems. Conceptualising engineering problems is developed through experience which is acquired through exposure to a variety of problems.

The experimental work, which takes place in laboratories, in the workshops or in the field constitutes an important component of engineering education since it gives students practical experience. Experimental work also provides a useful scientific basis required to make decisions on the use of materials for work. Such decisions depend on the use of materials.

The industrial training component exposes students to the real conditions prevailing in the field. It also helps students to widen their scope of engineering problems in the real world.

Engineering education in Kenya is provided under four basic disciplines namely, agricultural engineering, civil engineering, electrical engineering and mechanical engineering. These generic disciplines are further subdivided. For instance, civil engineering is divided into structural engineering, water engineering, highway and traffic engineering, public health engineering, sanitary engineering and foundation engineering, among others, as areas of specialisation.

The agricultural engineering programme is biased towards the needs of agriculture and relates to such aspects as soil, water, farm animals, farm structures and agricultural machinery and implements. The agricultural engineer is expected to bring technology closer to the farmer and should be able to build shelter for human beings, animals and also food, to solve problems of sanitation and waste disposal and irrigation and to develop and provide energy for the farm at a reasonable cost. For example, the engineer should know how to build canals to irrigate the crops on the farm, conserve soil and ensure its maximum productivity and maintain farm machinery and plants.

Civil engineering deals mainly with construction materials like rocks, soils, bitumen, cement, steel, concrete, bricks, timber, oils, water and waste-water. Civil engineers thus learn planning, designing, constructing, operating and maintaining engineering works such as buildings, bridges, roads, railways, airports, harbours, dams, irrigation works, water treatment works, sewage treatment works, urban planning and general infrastructure planning. The civil engineer will normally be concerned with the structural behaviour of the solid materials and the flow characteristics of the fluid materials. Consequently he is trained in various analytical techniques necessary to understand the structural behaviour of materials used in construction works.

The electrical engineering programme is structured to produce graduates capable of planning and dealing with power generation including electrification and designing and producing telecommunications equipment like radios, telephones and televisions. Additionally, students train in electronic controls used in many machines and plants. There are examples from the medical field where electrical engineers have been involved in the development of machines like X-ray and scanners.

The mechanical engineering programme aims at equipping the graduate with knowledge to design and develop machines and equipment which make work easier for man. Mechanical engineers are involved in studying production processes, operations and maintenance devices as well as machines and engines in addition to refrigeration and air conditioning.

Practical or Industrial Training of Engineering Graduates

Graduates of engineering undergo a period of industrial in order to training learn how to apply their technological knowledge and principles in solving practical engineering problems that face the society. However, this programme is faced with many problems.

In the 1960s, many firms in Kenya used to send fresh graduate engineers overseas for industrial training. This was only possible because there were fewer graduate engineers, but as the numbers increased, it was no longer practical to do so and only those who could not obtain training opportunities in Kenya were sent overseas. By 1970 many organisations had developed their in-house training programmes, although they experienced problems of limitations with regard to facilities and qualified trainers. A study carried out by the author among trainers found that graduate engineers were suffering from apathy and laxity, lacked discipline, were not duty-conscious and were not tolerant where difficult field conditions were encountered (Gichaga, 1971).

In the same study the graduate engineers complained of lack of systematic guidance on the job, making it difficult for them to understand how academic knowledge could be applied to problem solving, undue emphasis on the period of training rather than on the quality of training; lack of appreciation of the academic achievement of the graduate engineer by the senior engineers and lack of machinery for monitoring and evaluating the implementation of training programmes. In a seminar involving the Engineering Registration Board (ERB), the problem of industrial training for graduate engineers was discussed and it was agreed that an executive body be set up which would develop, approve and monitor industrial training programmes. It is hoped that problems of industrial training for graduate engineers would be solved by the implementation body.

It is instructive to indicate that engineers need continuing education in order to keep abreast of the latest technologies in their areas of specialisation. This can be done through short courses, seminars and conferences. Academic postgraduate programmes in all the sub-disciplines of engineering are now available at the University of Nairobi, whereas in the 1960s and 1970s it was mostly the self-tailored postgraduate programmes that were available and those were very limited in scope.

Controls in the engineering profession

The engineering profession in Kenya is controlled by the ERB which was set up by an Act of Parliament in 1969. The body registers graduate engineers after being satisfied that they have acquired the necessary academic training from an approved university. The ERB also registers professional engineers who have gone through an approved industrial training scheme and have satisfied the requirements for registration.

The Institute of Engineers of Kenya (IEK) is registered under the Societies Act (Cap. 108) and caters for the welfare of the entire engineering fraternity. The IEK is represented on the ERB by three members out of a total of seven board members who are appointed by the Minister for Public Works.

Related Development Problems

Within the last quarter of this century, Kenya has experienced extensive developments, most of which have had substantial engineering input. It is vital to review some of them in the context of development problems that the country has faced since independence in 1963.

Developmental problems

At independence, Kenya faced three major problems, namely ignorance, disease and poverty. In crude terms, the engineer has a role to play in each of these problems in that he is involved in the planning and building of schools for eliminating ignorance. The engineer is also involved in the planning and building of hospitals for fighting disease. The works constructed by engineers provide employment opportunities thereby reducing poverty. However, the problems of development are more sophisticated than the above explanation. The issues of development are issues of planning infrastructural services so that the ordinary citizens can have access to basic services like health, education and social services. Due to historical reasons, certain parts of the country were provided with comprehensive social services at the expense of others, which had very sparse infrastructure including planning, designing, and constructing the infrastructures to help in providing services to the community. However, the country was faced with a shortage of qualified engineers to carry out planning and construction. Thus, major engineering works were contracted to foreign consultants and contractors who prescribed solutions based on their experience elsewhere, rather than on local conditions.

The country also faced the problem of developing an industrial base. Attractive infrastructure in the form of basic services was necessary to attract private investment. It is important to note that studies have shown that industrial growth in any country has a direct relationship to the number of engineers available. Thus, developed countries like the United States, the Russian Federation and Britain, which have many engineers and technicians per capita, enjoy a high standard of life in terms of per capita income as shown in Table 13.1.

Response to Problems of Development

The per capita income of most LDCs (which is a measure of the standard of living of the people) ranges between US$ 100 and 1500 (Kenya's per capita is below $500). It has been shown that GNP can be raised by increasing industrial sector output. An analysis of 83 countries was carried out to obtain a relationship between the output from the industrial and the agricultural sectors of the economy and the GNP. The results showed that growth in GNP requires increased production in the industrial sector which in turn requires increases in both the number of people employed in manufacturing industries and in their productivity. Productivity will basically depend on the capital invested per employee, energy consumed per employee and distribution of skills among employees.

Table 13.1: Commonwealth countries' engineers and technicians needed by 2000: Scenarios A,B and C

County	Engineers			Technicians		
	A	B	C	A	B	C
Bangladesh	10,575	19,035	31,725	42,300	76,140	126,900
Ghana	6,870	12,023	20,610	27,480	48,090	82,440
India	164,027	268,407	462,257	656,106	1,073,628	1,849,026
Jamaica	2,723	4,752	7,376	10,890	19,008	29,502
Kenya	10,800	18,900	32,400	43,200	75,600	129,600
Lesotho	621	1,070	1,863	184	4,278	7,452
Malawi	1,785	3,035	5,177	7,140	112,138	20,706
Malaysia	32,760	49,920	76,752	131,040	199,680	307,008
Nigeria	131,508	230,139	361647	526,032	920,556	1,446,588
Papua N.G.	2,565	4,523	7,560	10,260	18,0990	30,240
Sierra L.	1,116	1,860	3,009	4,464	7,440	12,276
Singapore	11,160	16,136	20,228	44,640	64,542	80,910
Sri Lanka	3,798	6,330	10,445	15,192	25,320	41,778
Tanzania	5,973	9,774	16,833	223,992	39,096	67,332
Trinidad	5,760	8,304	10,416	23,040	33,216	41,664
Uganda	4,374	7,655	12,393	17,496	30,618	49,572
Zambia	3,968	7,073	11,040	15,870	28,290	44,160
Zimbabwe	8,352	15,138	26,100	33,400	60,552	104,400

Source: Mordell and Coales (1983)

The study by Mordell and Coales (1983), involving 83 countries, developed a relationship between professional engineers and technical personnel per 1000 of population against GNP per capita (Figure13.1). Mordell and Coales further isolated 18 LDCs of the Commonwealth and carried out projections of required professional engineers and technical staff for three levels of growth in GNP up to the year 2000. In this exercise the breakdown of professional engineers and technical staff was done using the following three main classes:

    Engineers and high technologies   15%
    Technicians   60%
    Others   25%

The results of the projections are shown in Table 13.1. Scenario A presents zero growth while scenario B represents 2.5% per annum growth and scenario C represents 5.0% per annum growth in per capita income. It must be noted that increasing the number of skilled people alone does not guarantee an increase in GNP, unless these people are productively employed.

At the current rate at which engineers from local and overseas institutions are produced, it is estimated that by the year 2000, about 4,000 additional engineers will be available in the market in Kenya. This is a fairly optimistic estimate. During the same period and based on previous trends, it is estimated that an additional 31,000 technical staff will be available in the market, giving an additional total of professional engineers and technical staff of about 35,000. In 1989 there were about 5,000 engineers and about 25,000 technical staff, giving an existing total of about 30,000. Thus, by the year 2000 there will be approximately 65,000 professional engineers and technical staff in Kenya assuming minimal losses in the existing numbers (Kenya, 1989). Table 13.1 indicates that with that level of engineers and technicians, the country is unlikely to improve its GNP. It is also notable that there will be a need to train many more technicians to improve the situation.

With the 8-4-4 education system, which lays emphasis on development of technical skills, it is likely that slightly higher numbers of technical staff will be available than the above estimate, but such additional numbers are not likely to improve the situation significantly. This does not give much hope for growth in GNP by the year 2000.

In terms of relevance of training programmes for engineers, it is worth noting that the course structure for the various degree programmes at the University of Nairobi has undergone changes over the years since the early 1960s so that today the courses covered are more relevant to the needs of industrial growth of the country than was the case in the 1960s. There is need for continuous development of these courses to reflect the requirements of industry and the changing technological world.

Producing graduate engineers in large numbers is not a sufficient condition for meeting the challenges of development. They must be productively employed, to make their contributions. It is also important that these engineers are properly versed in the techniques of managing technology in order to enable them to make it possible for the Kenyan citizen to have access to the right type of technology and to affordable prices.

Whereas considerable achievement has been recorded in the public sector in terms of having local engineers manning key sectors, there is need to develop a strategy of enabling Kenyans to have more control over the private sector and more specifically the industrial sector. In other words, there is need for indigenisation of the private sector.

Fig 13.1: Productivity in the industrial and agricultural sector in relation to GNP

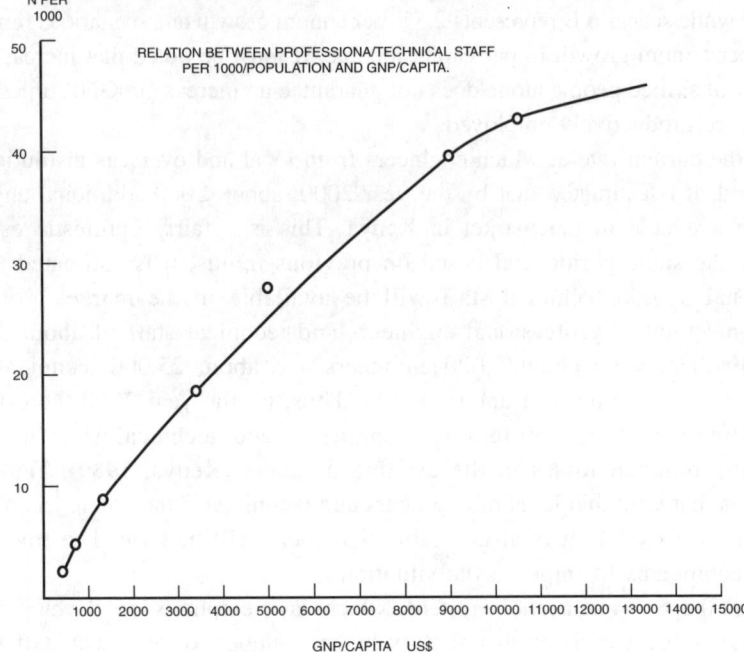

Fig 13.2: Relationship between professional/technical staff per 1000 population and GNP/capita

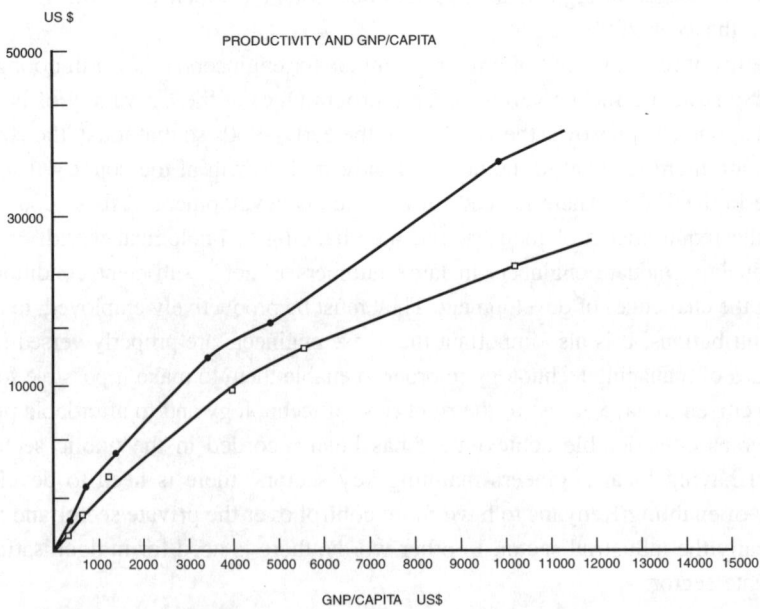

The Future

Example of Development Achievements

Kenyan engineers have recorded major achievements in the areas of irrigation, food storage silos and ploughing.

Civil engineers have also contributed greatly in the areas of transportation, buildings, dams and water supply works. The rural access roads programme has substantially enhanced economic development in the rural areas of the country. Requirements for low-cost sanitation and cheap housing for low-income people have received considerable attention by civil engineers among other professionals including architects.

The electrical engineers have also made an impact in the rural areas especially in terms of rural electrification, while the mechanical engineers have been involved in the development of various forms of energy like wind energy, solar energy and biomass.

Engineers have also built major works like airports, large dams and hydro-electric schemes, major highways, satellite stations, oil pipe-lines, postal and telecommunications structures and railways. Engineers have built major industrial works in many urban centres in the country.

However, it is quite clear that the most effective way to accelerate development will involve encouraging development strategies and utilisation of appropriate technology including rural development and agricultural machinery. There is need to develop small-scale industries which are capable of providing for the needs of rural development and the domestic market generally. Indeed there is an urgent need to check on the rural-urban migration by decentralising opportunities for work and basic social services, by locating more industries in the rural areas.

There is no doubt that more will be achieved by engineers in the future if the country adopts modern technology to be managed by them.

Policy issues

The policy issues raised in this chapter address problems that relate to the education and training of engineers, the development of the private sector in technology to ensure harmonisation of the direction of industrial development and the question of transfer of technology. These are subjects which could be extensive, but from the highlights in this chapter it is clear that engineering education should be carefully regulated to ensure that it is relevant to the requirements of the country. Additionally engineering education must relate to scientific discoveries, inventions and achievements in technology so that the graduate is comprehensively trained to be able to select an appropriate solution from a set of solutions. Such a selection should be guided by the prevailing situations and circumstances in the country as well as the likely conditions in the foreseeable future. The industrial training of engineers must receive due attention, to ensure that graduate engineers are not inefficiently utilised and that training is carried out in accordance with the governing professional norms.

The development of the country is dependent on the growth of its industrial sector. This, therefore, dictates that the industrial sector must be regulated by the government. More Kenyan entrepreneurs with ability and competence should be encouraged to enter the industrial sector as they will not face difficulties in harmonising their plans to fit within the national plans and priorities. Indeed, the indigenisation of the industrial sector must be seen as an objective which should be achieved sooner rather than later.

The issues of transfer of technology remain difficult because technology is generally available in the developed industrial countries whose priorities are different from those of the LDCs. Under current practices, the LDCs has little control on the nature of technology that is prescribed for it by the MDCs. If they have to develop and catch up with the MDCs, LDCs must develop strategies of acquiring independence in the selection and utilisation of technology. It is the level of independence in this respect that will determine how fast real development can take place in LDCs.

Part IV

AGRICULTURAL DEVELOPMENT

ISSUES IN RESOURCE MANAGEMENT AND DEVELOPMENT IN KENYA

Table 14.1: Raw and physiological population densities (persons/km^2), 1969–1979

Province/ district	Total population 1979	Land area (km^2)	Medium/high potential	Raw	Pop. Density
NAIROBI	827,775	684	n.a	1,210	n.a
Kiambu	686,290	2,448	1,320	280	520
Kirinyaga	291,431	1,437	1,080	202	270
Murang'a	648,333	1,476	1,570	261	413
Nyandarua	233,302	3,528	2,650	66	88
Nyeri	486,477	3,284	1,600	148	304
CENTRAL	2,345,833	13,173	8,220	178	285
Kilifi	430,986	12,414	3,510	34	123
Kwale	288,363	8,257	2,880	34	100
Lamu	42,299	6,506	3,260	6	13
Mombasa	341,148	210	210	1,624	1,624
Taita Taveta	147,597	16,959	529	8	284
Tana River	92,410	38,694	1,310	2	70
COAST	1,342,794	83,040	11,690	16	115
Embu	253,173	2,714	2,560	96	103
Isiolo	43478	25,605	–	1	n.a
Kitui	464,283	29,388	12,040	15	3.9
Machakos	1,022,522	14,198	8,960	72	144
Marsabit	96,216	73,952	40	1	2,405
Meru	830,179	9,22	3,360	83	247
EASTERN	2,719,851	155,759	26,960	17	101
Garissa	128,867	64,931	–	2	n.a
Mandera	105,601	26,470	–	3	n.a
Wajir	139,319	56,501	–	2	n.a
EASTERN	373,787	126,902	–	2	n.a
Kisii	869,512	2,196	2,196	395	395
Kisumu	482,327	2,093	4,610	230	230
Siaya	474,516	2,522	–	188	–
South Nyanza	817,601	5,714	5,714	143	143
NYANZA	2,643,956	12,526	12,520	211	211
Baringo	203,792	9,885	2,500	20	81
Elgeyo Marakwet	148,868	2,279	1,040	65	143
Kajiado	149,000	19,605	220	7	677
Kericho	633,348	3,931	3,800	161	167
Laikipia	134,534	9,718	1,300	13	102
Nakuru	522,709	5,769	3,300	90	158
Nandi	299,319	2,745	2,340	109	128
Narok	210,306	16,115	9,080	13	23
Samburu	76,908	17,521	1,400	4	55
Trans-Nzoia	259,503	2,078	2,078	124	124
Turkana	142,702	61,768	120	2	1,189
Uasin Gishu	300,766	3,378	3,270	89	92
West Pokot	158,652	9,090	1,030	17	154
RIFT VALLEY	8,240,402	168,884	31,480	19	188
Bungoma	503,935	3,074	2,530	163	199
Busia	297,841	1,074	1,626	183	183
Kakamega	1,030,887	3,495	3,250	294	317
Western	1,832,663	8,196	7,406	223	247
Total (national)	15,327,061	364,162	99,420	27	154

Note: * Three categories are defined as follows:
    High potential: annual rainfall of 857.5mm or more (over 980mm in Coast Province).
    Medium potential: annual rainfall of 735mm – 857.5mm (735mm – 980mm in Coast Province and 612.5mm – 857.5mm in Eastern Province).
    Low potential: annual rainfall of 612.5 mm or less
    n.a: not applicable

Source: Compiled by the author

ecological, historical and economic. Geo-ecological factors include altitude and related rainfall pattern, distribution of fertile volcanic soils and location of main tsetse fly areas. Generally, there is a strong positive relationship between altitude and rainfall, on the one hand, and between these aspects of ecological potential and population distribution/density, on the other. The fertile volcanic soils of the Kenya highlands are of great agricultural importance to the country. Finally, the colonisation of certain areas by the tsetse fly such as the Lake Victoria Basin, has inhibited their occupancy and although this menace has been eliminated, the population is still reluctant to settle in these areas.

Historical factors are mainly connected with the history of settlement both by the indigenous population before colonialism and by European settlers during the colonial period. The indigenous groups, after carving their own ethnic territory have tended to hold firmly to their acquisitions, only temporarily circulating between these and places of employment. European settlers alienated most of the fertile arable land, thereby exacerbating population density in this so-called African Trust Land. Reversal of these historical factors has influenced spatial population distribution and redistribution within well-defined boundaries following the resettlement programme implemented during the post-independence era.

Economic factors relate to where socio-economic development has concentrated. The *core-periphery* model of development adopted by the colonial administration meant that development was concentrated within the core and the "mobilised periphery," leaving virtually underdeveloped the rest of the periphery. A few economic islands thus jutted out of the ocean of poverty, subsequently influencing population redistribution, including internal migration and the distribution of core urban areas which further underdeveloped or marginalised rural areas (Obudho, 1997: 239 - 252). The present configuration of Kenya's population and socio-economic development is, therefore, a function of the interplay of these and other closely related factors. As these factors are pertinent in Kenya's spatial demography, they will recur in subsequent sections of the chapter.

Spatial Population Distribution and Movements

This section focuses on two indices of spatial population distribution (distribution and density) and the two related phenomena of internal migration and urbanisation. Kenya's population is clustered in three areas of the country, namely, the Lake Victoria Basin (comprising Nyanza and Western Provinces, and the outlying districts in Rift Valley Province); northern and southern mobilised peripheries focused on Nairobi and covering districts in its proximity in both Central and Eastern Provinces; and, the coastal belt around Mombasa, Kenya's second largest urban centre. A striking feature is the concentration of the large urban centres in these three population clusters.

Urban-rural distribution has changed only slightly during the period 1948-79 while by 1948, only 5.1 per cent of the population was urban, the "proportion urban" changed to 7.8 per cent in 1962, 9.9 per cent in 1969, and 14.6 per cent in 1979. This implies the paradox of urbanisation: although more than 80 per cent of Kenya's population is rural, urbanisation is increasing rapidly (Obudho, 1983/1994 and Obudho and Waller, 1976). Population is clustered in relatively much smaller administrative districts in these three areas. The much larger, ecologically hostile districts are sparsely populated. The current state of technology polarises Kenya's districts as overpopulated and underpopulated, contrasting features which probably influence the Kenya government's attitude towards the country's rapid population growth. This polarisation may be better understood by considering population density which explains area-specific population concentration.

Population Density

In spatial demography, population density is a meaningful indicator of the land-population relationship because, among other things, it indicates the land carrying capacity and population pressure. Geographers have cautioned on the application of crude densities which mask considerable disparities and which hardly explain the realities. This *arithmetic density* is a crude quantitative measure of population without description of its qualities. The two indices relevant to this chapter are *physiological density,* which indicates the ratio of people to arable land and *agricultural density,* referring to persons per cultivated area. Population density within Kenya's medium and high potential agricultural land better explains land-population relationships (see Table 14.1). Particular emphasis is placed on hectares per person, raw as well as physiological densities.

With a population of 15.3 (adjusted to 16.1) million returned in the 1979 census, Kenya recorded an average crude density of 27 persons/square kilometre. Nyanza Province recorded a density of 211 persons/ square kilometre and Western Province recorded 223 persons/square kilometre while the equally densely settled Central Province had a density of 178 persons/square kilometre (Table 14:1). These pockets of dense population are also important subsistence and cash cropping areas where land acreages, due to population growth and land fragmentation, are fast increasing, resulting in land units too small to be economically viable.

Comparison of raw and physiological densities provides interesting insights. The national physiological density is six times greater than the raw density. In all the districts there is a startling difference between these two indices of population density. Central, Western and Nyanza Provinces, in descending order, are by far the most densely settled parts. Even those districts of Rift Valley Province which have low densities exhibit high physiological densities. However, the true picture is

14 Demographic Implications of Population Distribution, Density and Movements Within Kenya's Arable Lands

John O. Oucho

Introduction

Kenya has such limited arable land that the country's rapid population growth poses serious challenges to national development efforts. Specifically, there are important demographic implications for the spatial dynamics of population. Of Kenya's area of about 582,646 km^2, only about 17.4 per cent is arable. But even this small area is occupied by 16,178 km^2 forest plantations and forest reserves thus reducing the amount of agricultural land to only 83,000 km^2 (14.4%). The arable land has influenced not only population distribution and density, but also internal migration and urbanisation in Kenya. There exists in the country a *laissez faire* policy whereby the government's concern centres around the coexistence of high rates of growth in some areas with sparsely settled areas, which has far-reaching demographic implications. Yet the country's population policy underwent fine-tuning following the establishment of the National Council for Population and Development (NCPD) in 1983.

This chapter considers the demographic implications of these phenomena within Kenya's overworked arable land in which development projects have been concentrated. The geo-ecological, historical and economic factors which are responsible for the present configuration and movements of population within the arable land are explained while the pattern of these three phenomena during the census era 1949-79 is described. Population movements are analysed in terms of population redistribution affecting settlement patterns as well as contemporary internal migration and urbanisation (Obudho, 1983 and Obudho and Waller, 1976). The demographic implications of Kenya's spatial demography are rapid population growth, more acute population pressure and further population redistribution. Kenya's spatial demography can be better understood if placed in the context of its arable land and future research should address the problem of the relationship between demographic trends and agro-industrial developments within the small amount of arable land before technological advances permit the exploitation of Arid and Semi-Arid Lands (ASAL).

Factors Influencing Spatial Population Distribution and Redistribution

Three categories of factors influencing spatial population distribution and redistribution (attributed to migration and urbanisation) can be identified as geo-

concealed by an important post-independence development, namely, both large-scale and small-scale farms developed during resettlement of the former scheduled areas which were formerly reserved for non-African settlers. For example, Turkana and Kajiado Districts stand out as the two districts with the highest physiological density, significantly contrary to raw density. The same situation occurs in both Marsabit and Meru Districts in Eastern Province and Taita–Taveta District in Coast Province; the three districts are also covered by national parks which mask the true physiological densities. Although there are wide differences between the two density types in Central Province, the differences are negligible in the equally densely settled Nyanza and Western Provinces. This is due to the fact that the latter two provinces in western Kenya have arable land, much of which is populated rather than occupied by forest plantations and forest reserves (as is the case with most districts in Central Province).

The picture painted by the foregoing analysis explains why rapid population growth poses serious challenges to Kenya's development which is currently confined to arable land. It is not surprising that population pressure is reflected in the increasing dominance of the small-scale farmer, that excessive land fragmentation signifies land shortage and that indigenous spontaneous settlers have occupied even marginal lands thereby disturbing ecological balance. Yet raw densities do not explain these subtle but significant characteristics. The so called "sponge effect" of rural areas to absorb more population through intensive agriculture is greatly threatened in the densely settled areas.

Internal Migration

The establishment of a dual economy (comprising traditional and modern sectors) in Kenya during the colonial period precipitated four patterns of internal migration—rural-urban, rural-rural, urban-urban and urban-rural which remain significant in the post-independence period. The relative magnitude of the four patterns has been recognised by the World Bank as follows: rural-rural accounts for 40 per cent: rural-urban, one-third; urban-urban, 4 per cent and; urban-rural one-quarter (World Bank, 1980: 34). Thus the first two are by far the most important in the country.

Both rural-urban and rural-rural migration patterns have been studied at both macro and micro levels. Macro studies by Rempel (1978) and Oucho (1974) have been based on census data while micro studies have derived from several surveys within specific modern sector settings. Migration has been found to be highly selective of some ethnic groups, especially the three largest – the Kikuyu, Luo and Luhya – relative to others; of the youngest, best-educated, and most development conscious members of the society, and of males relative to females. Both positive and negative effects of rural-urban migration have been recognised. A major shortcoming, however, is that migrants have been surveyed at their destinations (generally the urban end of migration) thus biasing the results. Surveys of return migrants have the potential to shed light on retrospective views of former rural-urban and rural-rural migrants, who have already re-entered ancestral home areas or acquired new homes elsewhere.

Kenya's eight provinces are nearly equally divided in their net migration patterns. The net in-migration provinces are Nairobi, Rift Valley and Coast. The five net out-migration areas are provinces of Nyanza, Western, Central, Eastern and North Eastern Province (Table 14.2).

Table 14.2: Internal migration one year before the 1979 census by province

Province	Enumerated in province	Resident in province	In-migration	Out migration	Net migration
Nairobi	799,397	719,989	180,265	95,857	+84,408
Central	2,258,836	2,263,709	83,301	88,374	- 5,073
Coast	1,295,499	1,273,599	62,325	40,425	+21,900
Eastern	2,620,937	2,644,352	56,253	79,669	- 23,416
North Eastern	365,319	369,435	8,251	12,367	- 4,116
Nyanza	2,549,178	2,595,000	65,557	111,379	- 45,822
Rift Valley	3,115,630	2,055,473	138,203	78,046	+60,157
Western	1,759,315	1,792,271	58,359	91,315	- 32,956
Total (national)	14,764,111	14,713,828*	652,514	597,432	+55,082

Note: * Excluding children under 10 years of age.

Source: Kenya (1981: 6).

Migration patterns are attributed, to inter alia, economic disequilibrium in the country an urban employment-oriented system of education and urban-rural imbalance. These problems are being solved by re-orienting spatial physical planning to benefit the rural areas which were neglected in the colonial period and; overhauling the education system to inculcate technical skills and eliminating urban-rural imbalance by various strategies including the DFRD strategy since 1984. The densely settled provinces, predominantly arable land with rapidly increasing populations, experience large-scale, incessant population movements.

Urbanisation is an interesting paradox in Kenya. Although there are only a few large urban centres, the urban population is increasing rapidly, especially within the arable land. Of all the 91 urban centres recorded in the 1979 census, the majority are within arable land. For example of the 27 urban centres with population above 10,000, a total of 24 (88.9% of all urban areas in that category) are within that land potential category. This suggests urban centres with population below 10,000 have increased exceptionally rapidly, though in some cases they encompass basically "rural" populations in the process of boundary extensions (Table 14.3), (Obudho, 1983, 1990: 214 - 234 and 1994: 198 - 212).

Table 14.3 : Growth of urban centres by size and number 1948, 1962, 1969 and 1979

Size	Number			
	1948	1962	1969	1979
100,000	1	2	2	3
20,000 - 99,999	1	2	2	13
10,000 - 19,999	2	3	7	11
5,000 - 9,999	3	11	11	22
2,000 - 4,999	10	16	26	41
	17	34	48	91
Total urban population	276,240	670,950	1,082,437	2,238,800
Per cent urban	5.1	7.8	9.9	14.6

Source: Obudho (1983b:117).

Conclusion

The demographic implications of Kenya's spatial distribution of population must be considered against current population trends which are likely to persist for the rest of the present century. They must be considered within the framework of the country's phenomenal population growth rate and spatial redistribution which are likely to aggravate the situation of population-land relationship. That Kenya's population growth rate has consistently increased from 2.3 per cent per annum in 1948 to 3 per cent in 1962, 3.3 per cent in 1969 and 3.9 per cent in 1979 raises important implications for the land-population relationship. In spite of large capital outlays for family planning (aimed at regulating fertility and, therefore, reducing rapid population growth), Kenya's rate of population growth has never abated. The decline of Total Fertility Rate from 7.7 in 1984 to 6.7 in 1989 may be a temporary feature and a bias inherent in the survey methodology which has characterised national surveys in the country (Kenya, 1989). Higher or even stabilising population growth rates spell doom for the country's arable parts whose carrying capacity could be adversely affected. With increased population densities, populations (especially in the most densely settled parts) could become more responsive to family planning. If family planning becomes more effective, Kenya's policy aimed specifically at reducing the population growth rate may pay greater dividends than it has ever done before.

Population pressure on the land and other resources is expected to exacerbate in the future. Instead of the few population pressure melting pots in the country, there will be many more of them coalescing to form population pressure belts. Already, there is evidence of population pressure in such melting pots overspilling into the marginal lands, thereby threatening these relatively fragile environment. Imagine a situation by the year 2000 when Kenya's population will have doubled and when, therefore, physiological densities will have more than doubled. Unless development of marginal lands is realised the situation will be intolerable. The government, anticipating this future scenario, created a ministry in 1989 to deal specifically with rehabilitation and subsequent exploitation of ASAL, including wastelands. The new ministry is the custodian of hope for every Kenyan and it is hoped that resources will be injected to enhance its activities.

As Kenya's economy advances steadily, increased internal migration and urbanisation will result in greater redistribution of population. This will trigger urban-urban migration from smaller to larger urban centres and the reverse. Urban migrants may also become more established, preferring permanent urban residence to return migration to their rural origins. Given the weak urban infrastructure and rampant urban unemployment in the country, there will be increased urban problems which civic and national efforts alone may not solve satisfactorily. Migrant selectivity also has implications for sex ratio patterns and fertility behaviour. Sex imbalance resulting from migrant selectivity affects household and community gender roles to which populations do not readily respond. Migrant selectivity also affects fertility behaviour, especially in cases where married couples live separately for long periods; such conditionally "separated" couples are likely to have different fertility performance from that of couples cohabiting most of the time. The process certainly has implications for population structure by various socio-demographic attributes.

It is evident from the foregoing analysis that Kenya's spatial demography is better understood by considering the situation within the arable land. Physiological density paints the dismal picture that even areas of low raw densities suffer from population pressure which is already affecting agricultural pursuits. As population redistribution continues, concern must be raised about whether the co-existence of rapid population growth and sparsely settled areas is a necessary or challenging situation in Kenya. There is need to explore possibilities of redistributing population from congested rural areas to land surplus rural areas, although strictly speaking, surplus land is a misnomer. Although population redistribution was implicitly a primary objective of Kenya's land settlement programme at independence, it was eroded by stronger waves of land settlement objectives. The demographic challenges analysed above necessitate a more comprehensive population policy covering population distribution, redistribution, migration and urbanisation alongside the current one focusing on fertility regulation and mortality reduction. Kenya's academic institutions and political climate are now more favourable to the formulation, adoption and implementation of a comprehensive population policy than prior to the 1980s.

15 Food Production and Population Growth in Kenya

D. A. Obara

Introduction

Kenya's rapid population growth calls for a rapid increase in food production. This chapter examines the problems of food production and growth in the country, a phenomenon which has characterised the economies of Kenya's rural areas since the 1960s where there has been the gradual decline in the total output of food relative to national population growth. Within agriculture, there is a disproportionate growth of the production of cash crops relative to food crops. These imbalances have been accompanied by regional disparities. In comparison to conditions in urban centres, the situation in the rural areas has continued to deteriorate. This has contributed to massive rural-urban migration. The situation has further been exacerbated by the fact that population growth in both rural and urban areas is increasingly threatening to outstrip agricultural production and the availability of food. The situation is all the more serious since agriculture is the backbone of the Kenya's economy.

Although Kenya has gone ahead to develop agriculture and processing industries, the former constitutes the dominant sector of its economy. About 80 per cent of the population is presently engaged in agricultural production both for subsistence and export earnings. All post-independence development plans have focused sharply on the development of rural areas to tap the potential of the farmers and activate the underutilised agricultural areas. The implementation of agricultural policies is expected to alleviate rural poverty through increased food production to generate rural employment and to improve the standards of living (Kenya, 1981c).

Between 1976 and 1995, the rural and urban labour force was estimated to have increased from 4.0 to 11.2 million and 1.0 to 3.1 million, respectively. These numbers will be slightly higher by the year 2000. However, most of the new entrants into the labour force will probably have to be absorbed into primary and processing activities within agriculture. Moreover, Kenya is faced with the pressing problem of income distribution and growth. Not only is there a wide disparity between average incomes in rural and urban areas, but also the distribution is highly skewed within these two areas. Equity and growth of income in the initial stages will depend largely on the development of the agricultural sector. Agriculture contributed 37.1 per cent of the Gross Domestic Product (GDP) in 1977 but its share has since then slightly declined because the agricultural sector accounted for 34.4 per cent in 1979, 33.4 per cent in 1982 and less than 30 per cent in 1995. Nevertheless, it is still the most important activity as the other major sectors such as service, manufacturing and commerce only contributed about 14.7, 13.3, and 9.7 per cent, respectively, to GDP in 1982. Consequently, the agricultural sector accounted for more than 65.0 per cent of foreign exchange earnings.

Kenya is primarily a nation of small-holding agriculturalists who maintain themselves through their own farms and a variety of off-farm activities. Despite Kenya's limited endowment of arable land, the majority of the population live in rural areas and derive some or all of their livelihood from crop cultivation and livestock raising. Because of the shortage of arable land, most of the rural population are small-holding agriculturalists, accounting for about three fourths of the total agricultural output, two thirds of the land is devoted to arable agriculture which constitute 85 per cent of total agricultural employment and about 70 per cent of total employment in the economy. Despite the agricultural basis of the economy, good quality land is a scarce resource and this implies that adequate food production by the year 2000 is a real challenge under the current population growthrate of about 3.4 per cent per annum.

Food and Agricultural Production

Agricultural production is the core of African economics. However, it is well-known that there is a continent-wide food crisis, whose features include widespread food shortages, chronic malnutrition and sporadic famines that have been widely publicised in the last two decades. Despite enormous advances in the technical aspects of food production, it seems that the ability of Africans to feed themselves is seriously at risk. Most observers predict that this situation will continue to deteriorate in the near future.

In the face of increasing demand for food by the growing Kenya population, the government has developed policies to address the problem of self-sufficiency in food production. Food security and self-sufficiency in food production are goals to which the government has committed itself and a great deal of work has been done within the country to formulate plans of action by which these goals are to be achieved (Kenya, 1981 and 1986). If the goals and targets laid down are to be achieved, agriculture will have to provide food security for a population of almost 35 million in the year 2000; generate farm family incomes that grow by at least 5 per cent per annum; absorb new farm workers at the rate of over 3 per cent per annum with rising productivity; supply export crops sufficient for a 150 per cent increase in agriculture export earnings by the year 2000 and stimulate the growth of productive off-farm activities in the rural areas.

Kenya's food security remains a major objective of the government. The country will continue to be self-sufficient in maize, beans, potatoes, vegetables, milk, beef and meat products, among other. However, the greatest challenges will be to intensify maize and milk production so that output can keep pace with rapid population growth with a large increase in land devoted to these commodities. The country has been importing wheat, vegetable oil and rice to meet its needs, but the government has strived to alleviate the import gap in these three commodities (Kenya, 1986). Meanwhile the rapid growth of export crops is part of a strategy for food self-sufficiency. Agriculture will continue to be a major contributor to foreign exchange earnings to enable food imports when shortfalls occur and to cover intermediate needs until increased production can reduce chronic deficits. Another element in self-sufficiency is growing rural cash incomes which may ensure that rural households can purchase food when necessary.

It is clear that if Kenya is to reach the year 2000 with a healthy balance between rural and urban economic activities, agricultural prosperity is essential. There have been few effective policy tools to stimulate the growth of non-farm rural activities and employment. In the end, the only way to achieve such growth is to ensure a prosperous agriculture that will in turn stimulate off-farm activities. It will become apparent later that to attain goals for agricultural production the widespread localised marketing of inputs, especially fertiliser, will be crucial. If output does expand, localised processing and marketing of grains, milk and other commodities will also have to expand. Farmers will need maintenance and repair, transportation, financial and other services available in nearby small and intermediate centres. And rising farm incomes will generate demand for many consumer goods some of which can be made in local workshops and for the construction of housing, water supplies and other farm improvements. These developments, which are essential to agriculture will also provide opportunities for rural entrepreneurs and the self-employed to create new jobs at satisfactory incomes. Although this chapter will focus more on the means to promote agricultural growth than on the essential linkages, the latter are as important for Kenya's long-term welfare as agriculture itself.

In order to achieve these goals for agriculture, three broad strategies are necessary. All three must operate within the constraint of available high-potential land because no easy means exist to bring large amounts of new land into productive use. Firstly, within existing crop patterns, farmers have to be encouraged to adopt more productive practices, especially the wider use of improved seed varieties, fertiliser and disease and pest control. Pricing policies, marketing policies and institutions and extension service are the main instruments for obtaining much higher yields through known techniques. Secondly, research into new varieties especially of maize and other grains should be redesigned and accelerated to generate the new, high-yielding varieties that are essential to keep pace with consumption. Thirdly, to a limited extent, the production pattern should be diversified in favour of crops such as tea, coffee and vegetables that produce much higher incomes and generate considerably more employment per hectare than other crops and livestock activities. Small shifts in land use can yield relatively large gains in income, employment and export revenue when these crops are involved.

Scarce productive land is the central fact of Kenya's agriculture. Of Kenya's 44.6 million hectares of land, only about 8.6 million hectares are medium-to-high potential agricultural land. Of this, about 60 per cent or (5.2 million hectares) are devoted to crop and milk production. Much of the rest is used for extensive grazing for beef and stock production or is taken up by national parks and forest reserves. Although 500,000 hectares of land could be brought into production under irrigation, drainage or flood control by and large Kenya's agriculture will have to provide for both food security and export growth on its existing crop and dairy land.

Milk production *per se* accounts for almost 47 per cent of the 5.2 million hectares devoted to farming, while maize plus beans accounts for almost another 23 per cent, as shown in Table 15.1. These three plus root crops of sorghum and millet account for 84 per cent of the farmed land. Yet these same six basic foods produce only 43 per cent of the total value of the commodities. In contrast, three crops—coffee, tea and vegetables—produce 37 per cent of the total value using only 5 per cent of the land

Two strategic implications can be derived from the Table. One, rapid growth of rural incomes and GDP would be served by greater output of coffee, tea and vegetables through both intensification and acreage expansion. Two, given the importance of maintaining self-sufficiency in maize, milk and most other basic foods, it is essential to intensify production on existing lands without encroaching on land devoted to higher income-earning crops.

Until recently, coffee, tea, vegetables and pyrethrum employed between 1.4 and 2.0 person–years per hectare, compared to only 0.3 to 0.6 for milk, maize and beans, root crops, sorghum and millet. Thus employment goals also dictate an expansion of the high-value crops. Similarly, the need to expand agriculture's net export earnings also requires an expansion of tea, coffee and horticultural production and other cash crops such as coffee, tea and vegetables which earn five to ten times the foreign exchange per hectare than can be saved by import substitution for food grains.

Seven commodities are central to achieving the development goals established for agriculture. Coffee and tea expansion is the foundation for growth of both agricultural incomes and exports. Maize, wheat, milk and meat production must be adequate for food security. Other commodities – especially sorghum, millet, rice, root crops, sugar and oil crops – will remain important to farmers and to the government.

Table 15.1: Estimated areas, values and value per hectare for selected commodities, 1983 and 1987

Commodity	Area		Value (d)		Value per ha	
	% of total	(Rank)	% of total	(Rank)	K£/ha	(Rank)
Milk	46.6	(1)	16.3	(3)	70	(16)
Maize and beans (a)	22.6	(2)	16.6	(2)	153	(12)
Root crops (b)	7.9	(3)	8.1	(5)	205	(9)
Sorghum and millet	6.7	(4)	1.5	(11)	48	(17)
Coffee	2.9	(5)	21.6	(1)	1,489	(1)
Wheat	2.2	(6)	2.1	(10)	191	(10)
Cotton	2.1	(7)	0.4	(18)	32	(18)
Fruits	2.1	(8)	3.1	(9)	296	(7)
Sugar	1.7	(9)	-3.6	(21)	-432	(19)
Tea	1.6	(10)	11.9	(4)	1,325	(2)
Sisal	1.1	(11)	1.1	(12)	137	(14)
Vegetables	0.7	(12)	3.4	(8)	913	(3)
Cashewnuts	0.5	(13)	0.4	(15)	162	(11)
Ground-nuts	0.4	(14)	0.2	(20)	84	(15)
Barley	0.3	(15)	0.4	(17)	249	(8)
Sunflower	0.2	(16)	0.2	(19)	141	(13)
Pyrethrum	0.2	(17)	0.4	(16)	419	(6)
Rice	0.2	(18)	0.5	(13)	519	(5)
Tobacco	0.1	(19)	0.5	(13)	885	(4)
Beef	(c)	–	6.8	(6)	(c)	–
Sheep and goats	(c)	–	4.9	(7)	(c)	–
Others	(c)	–	3.1	–	(c)	–
Total	100.0(e)		100.0(e)		170(f)	

NB: (a) Because beans are typically interplanted with maize, the two crops are considered together; maize alone accounts for 13.3 per cent of total value.
(b) Includes potatoes, which account for 5.3 per cent of total value.
(c) No estimates available.
(d) Value at farm gate.
(e) The total area is 5.17 million hectares and total value is K£ 1,035 million.
(f) Excludes beef, sheep, goats and "others".

Source: Kenya (1986: 6).

Political System of Food Production

The political system of food production operates in a physical and cultural environment that has two important aspects, namely, the ecological aspect and demographic aspect. The ecological aspect consists of the different physical attributes of the land available for agriculture (size, location, topography, climate, rainfall and soil) and the cultural arrangements that regulate how the land is held and distributed. The demographic aspect involves the physical and cultural attributes of the farming population that is relevant to their farming performance (age and sex ratios, skill levels, modes of labour organisation, food preferences and taboos). These are the two aspects of the food production process in the sense that they place absolute limits on what is possible given the level of contemporary technology.

The role of the state in the food production process has been the manipulation of either or both the ecological and demographic factors so as to achieve the desired yields. This manipulation has been achieved by the organisation of production units; investment in inputs and implementation of pricing and extension programmes in order to make producers more productive. Prior to the economic liberalisation of the 1990s, the state typically bought all the cash crop and resold it on the international market at a price well in excess of the producer price. The GOK has had more difficulty in monopolising the purchase of the food crop, which has been distributed mostly through private merchants. The state controlled the price of food by decreed prices and policy enforcement. In choosing the policies to be used in their intervention in the food production process, African states are continuously reacting to both internal and external political and economic interests. Through information and implemention of food policy, the GOK meets the demands of influential constituencies while at the same time striving to generate and maintain the support of as wide a range of groups as they feel they need and can tolerate.

The extent of state intervention in the food production processes depends on four factors, namely, the state's knowledge of production units that would be most appropriate; type of inputs that are needed and the types of incentives that should be made available to farmers; availability of the resources for implementing the government's programme and political will to engage in the programmes. The problem of knowledge is the easiest one to solve. There is now available a mounting mass of technical information about appropriate methods to increase food production. However, the creation of new knowledge through research and development is affected by both limitations on resources as well as political considerations. Frequently, the direction and objectives of agricultural research are chosen on the basis of career considerations and political factors.

The government, in essence, has to invest in agriculture if Kenyans are to see improvements in food production. However, there are so many demands on the meagre resources that the food production sector frequently loses the competition to other claimants. Ironically, one of the expenditures that has superseded investment in agricultural development is food imports due to the acute nature of the political problems that food shortages can cause.

Sources of finance on which all African countries have had to rely are bilateral and multilateral development and food assistance. However, international aid has had only a marginal effect on increasing food production. In many cases, bilateral aid has been given in order to meet the needs of the donor such as trade expansion to stimulate donor imports, project aid given to encourage the purchase of equipment and materials from donor countries and aid given for ideological reasons. Another problem is that most of this aid is given as short-term solutions to crises and they tend to be used for food import, salaries of civil servants or payments for security equipment. In such cases, the aid may inhibit rather than encourage greater local food production.

Population Growth, Arable Land and Food Production

A broad and variable interpretation of the arable land is possible on the basis of experience and viewpoint. Strictly speaking, arable land is land which is suitable for cultivation. However, this definition does not indicate whether such land may be cultivated profitably at any set price level for agricultural products or whether the land will deteriorate rapidly under cultivation unless special agronomic and agricultural systems are used. There is a tendency to apply the term from a purely physical viewpoint and to consider any cleared land as arable provided it is not too stony, rugged topographically or too badly in need of drainage to grow crops. Furthermore, such arable land should not be located in an area of very limited precipitation during the growth season.

The official definition of arable land in Kenya is based primarily on the amount of rainfall received per annum (Kenya, 1982: 96). Annual rainfall in areas classified as high potential is 857 mm or more, medium potential 735.0-857.5mm and in areas classified as low potential is 612.5 mm or less. According to this classification, only 12 per cent of Kenya's land is of high agricultural potential and an additional 5.5 per cent has medium potential. High potential land is suitable for coffee, tea, pyrethrum, intensive livestock production, food crops and cotton. Medium potential land can sustain groundnut, cotton, tree crops, wheat, oil seeds and food crops. Other areas of lower potential are production rangelands where some crops are grown but drought is a continuing problem. About 80 per cent of Kenya's surface is arid and semi-arid and is largely used for nomadic pastoralism. Even the dry grassland plains receiving 250 to 500 mm of rainfall per annum have a relatively low stock-carrying capacity.

The distribution of Kenya's population has always reflected and continues to reflect the availability of arable land. Most of the high and medium potential land in the Lake Victoria basin, highlands and coastal region is densely populated. But population densities decline progressively in marginal, arid and semi-arid areas because of their low population-carrying capacity in terms of arable land. Specifically, 80 per cent of the population resides in the 17 per cent of the land which is suitable for agriculture, but increasing population pressure on the agricultural land is forcing people to move to marginal, arid and semi-arid environments where production of adequate food to sustain increasing population growth is impossible. The high rainfall areas along the coast and in the highlands and Lake Victoria basin are currently intensively cultivated for the production of food and cash crops. Because food staples and cash crops compete for declining arable land, it is questionable whether there will be enough good land in Kenya to produce the foods needed to supply an adequate diet for all the people by the year 2000. To obtain an answer to this question, we must know how many hectares of arable land are currently available and will be available by the end of this century.

During the greater part of human history in Kenya, the number of people was exceedingly small, particularly during the period when people in the country earned their livelihood through fishing and hunting and the gathering of wild fruits. Not until regular cropping was introduced did it become possible to accommodate more people and support greater population densities. Kenya had a population of about 5,408,000 in 1948; 8,636,000 in 1962; 10,943,000 in 1969; 15,237,000 in 1979; and currently it is estimated over 30 million. Since 1948 Kenya's population has quadrupled and is expected to reach about 35 million by the year 2000. The accelerating trend justifies the term "explosion" and is the most portentous feature of population development. Because of reduced infant mortality, Kenya's population is growing at a rapid and accelerating rate. The official growth rate figures for 1962, 1969, 1979 and 1989 census are 3.3, 3.4, 3.8 and 3.4 per cent per annum, respectively. As a result of the widening gap between fertility and mortality, the country has a large proportion of young people who have to be fed.

The high population in Kenya is usually attributed to medical advances and sanitary improvements. This explanation is insufficient and overlooks the crucial fact that people survive because they are well-fed. The current high population growth and low food production may lead to serious overpopulation and extensive famines, respectively, by the year 2000.

Until recently, Kenya has not fully enjoyed the advantages of science and technology in agricultural production. These advances have so far benefited the country only to a limited degree in terms of food production. In view of this difficulty, population control is the best option to avert massive hunger and misery by the year 2000. There is likely to be far too little food for far too many people and the issue is to adjust people to land resources and vice versa. To handle food apart from population will only exacerbate failure. A supreme effort must be made in the area of family planning to control population growth while simultenously stepping up food production through the application of modern science and technology.

A knowledge of the hectarage of agricultural land in different provinces and districts alike in relation to population leads to a better understanding of some of the food problems which will be confronting these administrative regions by the year 2000. In 1979, the average population density for the entire country was 27 persons per km^2; that is, a 40 per cent increase over the 1969 figure of 19 persons per km^2. Both averages hide marked and long-standing variations at the sub-national level which reflect the limited distribution of arable land. At the one end of the

Table 15.2: Population and high/medium potential land by province and district

Province/district	Total area	Population ('000)	% Agriculture land/high/medium potential	% of potential land	Population	Density person/km^2	Availability ha./person
NAIROBI	684	828	30**	0.2	5.4	–	
CENTRAL	13,173	2,346	96**	9.3	15.3	245	0.39
Kiambu/Murang'a	4,924	1,335	94**	3.9	8.7	341	0.29
Kirinyaga	1,437	291	100**	1.1	1.9	269	0.37
Nyandarua	3,528	233	98**	2.7	1.5	88	1.14
Nyeri	3,284	487	93**	1.6	3.2	304	0.33
COAST	83,040	1,342	17*	11.8	8.8	115	0.87
Kwale	8,257	288	36*	2.9	1.9	100	1.00
Kilifi	12,414	431	29*	3.5	2.8	123	0.81
Lamu	6,506	42	50*	3.3	0.3	13	7.69
Mombasa	210	341	100**	0.2	2.2	1,624	0.06
Taita-Taveta	16,959	148	8**	0.5	1.0	285	0.35
Tana River	38,694	92	4**	1.3	0.6	70	1.43
EASTERN	155,759	2,720	19*	27.0	17.7	101	0.99
Embu	2,714	263	100*	2.5	1.7	104	0.96
Isiolo	25,605	44	– –	0.3	–	–	
Kitui	29,388	464	53*	12.1	3.0	39	2.56
Machakos	14,178	1,023	66*	9.0	6.7	114	0.88
Marsabit	73,952	96	– –	0.6	–	–	
Meru	9,922	830	52**	3.4	5.4	247	0.40
NORTH EASTERN	126,902	374	– –	2.4	–	–	
Garissa	43,931	129	– –	0.8	–	–	
Mandera	26,470	106	– –	0.7	–	–	
Wajir	56,501	139	– –	0.9	–	–	
NYANZA	12,526	2,644	100**	12.6	17.3	211	0.47
Kisii	2,196	869	100**	2.2	5.7	395	0.25
Kisumu/Siaya	4,615	957	100**	4.6	6.2	208	0.48
S. Nyanza	5,714	818	100**	5.7	5.3	143	0.70
RIFT VALLEY	163,844	3,240	21**	31.6	21.1	103	0.97
Baringo	9,885	204	25**	2.6	1.3	82	1.22
Elgeyo Marakwet	2,279	149	53**	1.1	1.0	143	0.70
Kajiado	19,605	149	– –	1.0	–	–	
Kericho	3,931	633	100**	3.8	4.1	167	0.60
Laikipia	9,718	134	15**	1.3	0.9	103	0.97
Nakuru	5,769	523	59**	3.3	3.4	159	0.63
Nandi	2,745	299	100**	2.4	2.0	128	0.78
Narok	16,115	210	56**	9.2	1.4	223	4.35
Samburu	17,521	77	8**	1.5	0.5	55	1.82
Trans-Nzoia	2,078	259	100**	2.1	1.7	125	0.80
Turkana	61,768	143	– –	0.9	–	–	
Uasin Gishu	3,378	301	100**	3.3	2.0	92	1.09
West Pokot	9,090	159	22**	1.1	1.1	154	0.65
WESTERN	8,196	1,833	100**	7.5	12.0	247	0.40
Bungoma	3,074	504	100**	2.6	3.3	199	0.50
Busia	1,626	298	100**	1.7	1.9	183	0.55
Kakamega	3,495	1,031	100**	3.3	6.7	317	0.32
Total (national)	564,162	15,327	19**	100	100	154	0.65

* All or primarily medium potential.

** All or primarily high potential.

Source: Kenya (1982: 13 and 96)

scale are high rural population densities found within Central, Nyanza, Western and Coast Provinces, whilst at the other end of the scale are less densely populated Provinces of Rift Valley, Eastern, and North Eastern. There is a wide gap in population concentrations in the former provinces and the latter ones.

If population densities in 1979 are calculated on the basis of land classified as high and medium potential instead of total area (564, 164 km^2), the average density rises from 27 to 154 people per km^2, a dramatic indication of the limited availability of Kenya's main economic resource. This situation will get worse when the population doubles. Table 15.2 shows the total land area, population, the proportion of agricultural land within the province of high/medium agriculture potential and the province's share of; Kenya's total high/medium agricultural potential land and Kenya's total population and density per km^2 of high/medium potential land. The arable land available per person varies from less than one to about one hectare and from less than 0.10 to less than 8.0 hectares at the provincial and district levels, respectively. Population pressure on good quality agricultural land, although severe, is not homogeneous. There are provinces and districts whose share of the total population exceeds their share of the nation's good quality land. However, there are provinces and districts whose share of the total population is less than 20 per cent of their total agricultural land of high/medium potential.

District breakdowns pinpoint with greater precision the areas with extensive over- and under-utilisation of fertile land. Districts with extreme population pressure on good agricultural land are Kiambu, Kirinyaga, Murang'a, Nyeri, Taita-Taveta, Meru, Kisii and Kakamega while those with the least population pressure on good agricultural land include Nyandarua, Lamu, Tana River, Kitui, Baringo, Narok and Samburu.

From the foregoing analysis, it appears that the most common fallacy in appraising agricultural capabilities is to overlook the degree to which districts are supported by hectarages beyond sight, neglecting to take into account hectarages taken by aquatic harvest and the livestock.

Conclusion

A dramatic growth in Kenya's population has influenced the distribution of its people in two ways. First, it has intensified pressure on land in the traditional high density areas. Second, it has stimulated migration into less densely populated rural areas and into urban centres. Each of these trends has implications for food production by the year 2000. These implications provide a basis for concern about the possibilities for increasing food production to provide Kenya's people with an adequate diet.

It is agreed that the country's food production needs to be doubled to eliminate the present food shortages and provide every person with a minimal diet. If we further take into account that the number of people in the country will nearly double by the year 2000, the actual requirement is a four-fold increase in food production. Theoretically, all the districts may not have problems in meeting their food needs for the remainder of the present century, but practically it appears that inevitably there will be increasingly less food per person for millions of people because populations are increasing most rapidly in districts where increased food production strategies are most difficult to achieve.

Rapid population growth in Kenya makes urgent the need for a rapid increase in food production, but there is a scarcity of arable land for the extension of food production. Lack of suitable agricultural land for an increasing population and food production is one reason why farmers are expanding cultivation into rangelands and ecologically fragile environments. However, increasing farm fragmentation and competition of cash crops with food crops for limited arable land have also contributed to low food production in the country. These difficulties highlight the need for the government to implement the policy which it has formulated for food self-sufficiency and self-reliance.

16 Small-holder Food Production: A Geographical Investigation of Maize in South Nyanza District *

S. O. Akech

Introduction

Kenya's food policy emphasises domestic self-sufficiency in major food crops. These include maize, beans, rice, wheat, meat and dairy products. With regard to maize, it has been the intention of the GOK to maintain a strategic reserve of four million tonnes (Kenya, 1981c). The main strategy for achieving this objective is intensification of production, mainly in the small-scale farming sector because of its predominance in the country's economy. The country's food policy was formulated in recognition of the periodic marketing crises which have been witnessed within the maize industry and the rapidly increasing population. The consequent rise in food demand must be met from domestic resources given the foreign exchange constraints in the country. The main bottlenecks to increased domestic food production include the limited arable land which will continue to support the production of cash crops to boost foreign exchange earnings.

This chapter investigates the main constraints to increased food production in the small farm sector based on fieldwork conducted in Rongo Division, Migori District. Maize is the staple food and its demand is certain to increase in future given its largely untapped potential as livestock feed and as an industrial raw material. The country's capacity to be self-sufficient in alternative cereals such as wheat and rice is limited. As a result, these crops will continue to be imported. Although these cereals are increasingly consumed, especially by the urban population and the higher income groups, it is certain that maize will continue to be the major staple food for the majority of Kenyans.

For adequate domestic supplies of maize to be achieved, greater maize output has to come from the small-scale farming areas. To ensure this, maize hectarages have to be expanded where idle land exists. However, in the long run, increased maize production will be possible only through enhanced yields based on the development and acceptance of relevant technological innovations. This chapter therefore, examines the major factors influencing variations in maize yield, hectarage and the adoption of innovations in the study area. All three variables are closely

---

* South Nyanza District now comprises Homa Bay, Kuria, Migori, Rachuonya and Suba Districts

linked with the nature of crop competition within a given farm as well as environmental, socio-economic and agronomic factors. The main objectives of this study, therefore, are to establish the major crop combinations in the study area and explore their implications on maize production; analyse the influence of environmental, agronomic and socio-economic factors on spatial variations in maize yields and hectarages and examine the major factors influencing the adoption of hybrid seed and manure/fertiliser in maize production.

Literature Review and Conceptual Framework

No other food crop has attracted as much attention as maize in Kenya. Several studies have been undertaken on diverse aspects of its production. Theoretical studies on maize have tended to emphasise its introduction in Africa and its general growth requirements. Most empirical studies on maize have concentrated mainly on the influence of agronomic and physical environmental factors on maize yield. The role of socio-economic factors and the adoption of new innovations in maize production have also attracted attention, although to a lesser extent.

Maize is not an indigenous cereal in Kenya but rose to become an import staple food following its introduction by the early Portuguese explorers and Arab slave traders at the Kenyan coast in the 16th century. Maize was then of the Caribbean flint type which had low yields. This was later intercrossed with other varieties from South Africa in the beginning of 20th century to produce the Kenya flat white maize. There was a consequent spread of maize which by 1903 occupied only 20 per cent of the total food crop acreage in Kenya.

Other important food crops introduced into the country during this period included rice, wheat and beans. Acland (1971) attributed the early spread of maize in East Africa to the numerous advantages it had over traditional cereals such as sorghum. These advantages included higher yield potential, lower vulnerability to pests and diseases, less labour requirements, easier storage and better palatability.

The present maize varieties in the country are mainly the result of maize breeding and agronomic research programmes which began intermittently in the 1930s. The major breakthrough in these programmes was the production of hybrid seed varieties in the 1960s. By 1967, these varieties had been completely adopted within the large-scale farming sector. The adoption of hybrid seed and related technological innovations has been one of the major reasons for increased maize production in the country from the 1960s.

In Nyanza Province, maize served mainly as a cash crop due to the restrictive colonial agricultural policies predating the Swynnerton Plan. It was only in the 1950s that maize was accepted in most of the people's diet. Miracle (1966) observed that the importance of maize as a food crop in Africa was likely to increase in the shortrun but would be overshadowed in the long run by non-traditional cereals such as wheat and rice.

In an analysis of the demand and supply of the major cereals in Africa, Shah et al (1984) noted that there was a growing demand for non-traditional cereals like wheat and rice in the African continent. Although maize has become a major food item in the diets of the majority of Kenyans, there is a renewed interest in the promotion of alternative food items such as livestock products, horticultural crops, cassava, beans, ground-nuts and traditional cereals such as sorghum and millet to achieve self-sufficiency in food.

Theoretical studies on maize have pointed out that the crop can be grown under diverse physical conditions but has a strict pattern of water requirements during its life cycle. The crop is particularly sensitive to waterlogging during the early stages of its growth. The crop also requires well-drained fertile soils and has a particularly great demand for nitrogen. Acland (1971) estimated that maize yield in Kenya ranged between 110 and 1,350 kilogrammes per hectare and attributed low maize yields to poor husbandry methods especially insufficient weeding and low plant populations. He asserted that fertiliser use in maize cultivation could only be beneficial if accompanied by good husbandry.

Allan (1971) investigated the influence of agronomic factors on maize yields in western Kenya. He singled out time of planting as the most critical factor influencing maize yields. In addition, poor husbandry methods, especially low plant populations, poor weeding and the planting of inferior seeds, were found responsible for low maize yields. He cited ignorance, carelessness, inefficiency, inadequate supervision, labour shortage and excessive rainfall as the major reasons for low standards of crop husbandry.

The influence of both rainfall and solar radiation on maize yields in Kenya was investigated by Simango (1976). Through the application of the multiple regression model, he established a significant relationship between rainfall and maize yields especially between the periods of sowing and flowering. These findings point to the extreme sensitivity of maize to excessive rainfall and soil moisture deficit. Simango's (1976) study further established that variations in maize yields in Kenya could not be ascribed to solar radiation and temperature variations in the country which were found to be fairly uniform. Minimum and maximum temperatures of 18 and 30 degrees centigrade, respectively, are considered optimal for maize cultivation. The influence of previous crops on maize yields has been studied by Jones (1974) in Nigeria. He observed that maize fields previously planted with cotton and sorghum had lower yields than those previously planted with ground-nuts. He explained this finding to be the possible result of the nitrogen fixing capacity of ground-nuts.

While suitable environmental attributes and agronomic practices are needed for improved maize yields, socio-economic factors have a bearing on the actual management of maize farms. Poor farm management practices, especially late planting, have been found to be responsible for low crop yields in South Nyanza. The most affected crops in this respect are maize and subsistence food crops like sorghum, beans, cassava, finger millet and cash crops such as cotton.

In an analysis of maize yields in Vihiga District, Moock (1973) found that farms managed by women had lower maize yields. Furthermore, farmers who belonged to the Friends African Mission, a church denomination, were found to be more competent maize farmers. Formal education was also significantly and positively related to the overall managerial ability of the farmers. He concluded that maize output could be improved through an efficient allocation of available farm resources. Matovu (1979) investigated the efficiency of resource utilisation in small-scale farming of maize and cotton in Machakos and Meru Districts. He concluded that small-scale farmers were relatively efficient in the utilisation and allocation of resources between maize and cotton enterprises. He, therefore, suggested that expanded adoption of already existing technological innovations through appropriate economic incentives could lead to increased crop output.

Although much effort has been put into agronomic and breeding research in maize, comparatively little has been put into the actual adoption of innovations recommended from these investigations. The introduction of hybrid maize in South Nyanza District was empirically investigated by Johnson (1970). He found that very few adopters of hybrid seed were able to sustain its use. The main reasons for this phenomenon included lack of land, money and farm machinery coupled with farmers' desire to experiment with second generation hybrid. He stressed that the poor husbandry practices prevalent in the area contributed to low maize yields hence the subsequent rejection of hybrid seeds.

Gerhart (1974) used probit analysis to investigate the diffusion of hybrid seed in western Kenya. He found that the income accruing from cashcrops and farm size were both positively related to earliness of adoption. The most important explanatory variable in the adoption of hybrid seed was found to be agro-ecological zone. Farms located in high altitude and high rainfall areas in western Kenya had a greater likelihood of adopting hybrid seed. The presence of drought-resistant crops and off-farm work experience were, however, found to be negatively related to hybrid maize adoption. Other factors related to hybrid maize adoption included formal education, extension visits, knowledge of existing credit facilities and attendance of courses at the local farmers training centres. He, therefore, called for improved agricultural services including input supply and extension services to farmers.

Rundquist (1984) investigated the temporal and spatial diffusion patterns of hybrid maize in South Nyanza District and Central Province. The main finding of his study was that the adoption of innovation packages was closely associated with the existing socio-economic stratification of the rural areas. The farmers higher in the hierarchy were found to be better adopters of innovations and tended to adopt more complex innovation packages. He called for a welfare approach to the dissemination of innovations whose main factors are production, seed distribution points, credit and extension services. He concluded that without these measures, further dissemination of innovations would magnify the already existing socio-economic stratification.

Figure 16.1 is a conceptual framework for the investigation of maize production based mainly on previous works. It stresses the intricate interrelationships between the several facets of maize production.

Fig 16.1: A conceptual framework for a geographical analysis of small-holder maize production

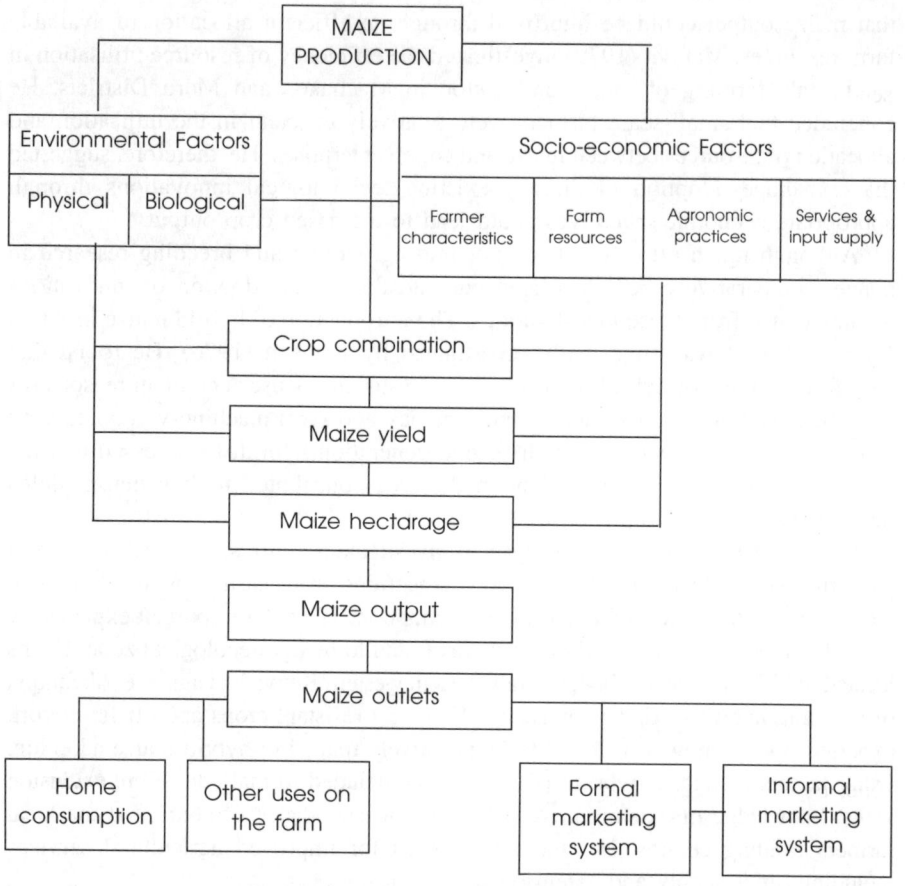

Source: Akech (1990: 27)

In general, maize production level is a function of the interaction between environmental and socio-economic factors. In the first place, the level of maize production on the farm depends on the degree of crop competition as represented by the number and types of crops grown together with maize. This crop mix is in turn related to the environmental factors which define the potential of a given piece of land to support the different crops. In addition, the crop mix is related to the diverse needs of the farmers. Maize yields and maize hectarage are both related. They are in turn influenced by environmental characteristics and socio-economic and agronomic factors including the adoption of innovations. The latter is a function of the individual attributes of farmers, as well as environmental and socio-economic factors.

The main hypotheses tested in this study were that spatial variations in maize yields and hectarages are not influenced, including the adoption of innovations (crucial in enhanced yields) which is a function of the individual attributes of farmers, environmental and socio-economic factors and that the adoption of hybrid seed and farmyard manure and/or commercial fertiliser are not influenced by environmental and socio-economic factors.

Study Objectives and Methods of Analysis

This study was undertaken in Rongo Division, one of the nine administrative units of South Nyanza District. The division occupies a total land area of approximately 483 square kilometres, which is about 8.5 per cent of the total land area of the then South Nyanza District (Figure 16.2)

The whole of Rongo is classified as high potential agricultural land. Lying between an altitudinal range of 4,300 and 5,300 feet, the area exhibits environmental conditions generally conducive to maize production. The area has a rugged landscape in the northern half marking the beginning of the Kisii highlands and a flat landscape in the south which is part of the Lake Victoria plateau. This flat surface attracted mechanised agriculture represented by the nucleus sugar plantation run by the South Nyanza Sugar Company (SONY).

Soils in Rongo Division are predominantly clays, with fertility ranging from low to high. The average rainfall ranges from 1,400-1,800 mm experienced mainly between March and July (long rains) and from September to December (short rains). Four major agro-ecological zones exist in the study area, namely, sugar-cane zone, the marginal sugar-cane zone, coffee zone and coffee-tea zone.

According to the 1979 census, Rongo had a total population of 85,647, giving a population density of 117 persons per square kilometre. Most of the farmers in the study area are middle-aged. Female-headed households are few and each household has an average of 5 persons. The majority of the farmers either have no formal educational background or never went beyond primary school. The main source of living is agriculture with many households receiving remittances from relatives working in the urban areas.

The main mode of land ownership is inheritance, although purchase, tenancy and lease are also increasingly becoming important. This system of land inheritance has led to land fragmentation and diminishing farm units. Farms are generally small averaging 4.9 hectares. The number of individual farm units per farmer

Fig 16.2: Location of Rongo Division, South Nyanza District, Kenya.

Source: Akech (1990: 30)

ranged from 1 to 4. The dominant land use in the area is agriculture characterised by small-scale crop cultivation and livestock raising. Large-scale farming is represented by the nucleus sugar plantation run by the SONY sugar company at Awendo. The main cash crops grown are sugar-cane, coffee and tobacco while the main food crops are maize, sorghum, beans, ground-nuts and bananas. Beans, bananas and ground-nuts double as cashcrops of local significance.

Maize is the leading crop grown in the area. In 1984 a total of 3,449 hectares were planted with maize in the study area. This figure increased to 3,549 hectares in 1986. The amount of land devoted to maize production is expected to decline in future as more farmers take on cash crop farming. All the farming operations in maize involve the use of simple hand or ox-drawn implements with motorised mechanisation playing a very limited role. The planting of maize during the long rains occurs between February and March. The use of improved hybrid maize varieties is not yet universal in the study area while the use of commercial fertiliser is mainly restricted to sugar-cane and coffee cultivation. Intercropping is a widespread practice in the area indicating increasing land scarcity and the need to lessen the risks of total crop failure. Most of the required inputs are distributed through local stockists in the study area. Most families depended on family labour for all their farm operations. Credit utilisation is quite low while extension services still do not reach the majority of the farmers.

The bulk of the data utilised in this study was derived from primary sources and supplemented by secondary data. Primary data collection involved informal interviews of government officials and community leaders. Formal interviews were conducted for a total of two hundred farmers in the study area using a structured standard questionnaire. The sampling design used in selecting the farmers was stratified random sampling. The whole area was stratified into its 20 sub-locations. Ten farmers were randomly selected from each sub-location using the District Land Registry as the sampling frame.

Two analytical models were employed in subsequent data analysis on a micro computer. These were the multiple regression model and factors analysis available in the GLIM and SPSS statistical software, respectively. The factor analysis model is useful in data reduction such that new and fewer variables are derived from the original data set. This model has been used in geographical studies by numerous scholars and was used in this study to establish crop combinations. The factor loadings were subsequently used in the mapping of the established crop combinations. Two specific models from the general linear regression model have been used in the analysis of yield variations, hectare variations and the adoption of innovations. The multiple regression model was used in the first two. The logit regression model was used in analysing the adoption of innovations.

The multiple regression model can be expressed as follows in its general form:

$$Y_i = B_0 + B_1X_1 + B_2X_2 + B_3X_3 + \ldots + B_1X_1 + E_1$$

where:

Y_i = the ith observation of the dependent variable Y
$B_{i\,'a}$ = regression coefficients
$X_{i\,'a}$ = the independent (explanatory) variables
E_1 = the random error term

The modelling procedure used involved fitting the first model with the highest number of variables then successively deleting insignificant ones until the model with the fewest number of explanatory variables was obtained. This simple backward elimination procedure was mainly meant to reduce the effects of multicollinearity and leave out insignificant variables which do not contribute significantly to the explanatory power of the model. The t-statistic was used to screen out insignificant explanatory variables. Two regression diagnostics were employed in the model fitting process. These were the analysis of residuals and leverage values. The latter are highly influential data points in the model fitting process. Identified outlier as well as highly influential data points were subsequently weighted out in each mode.

Logit regression is a special case of the general linear regression model used in dealing with dichotomous dependent variables. In its general form the model can be expressed as:

$$\text{LOG}(p/1-P) = B_{10} + B_1X_1 + B_2X_2 + \ldots + B_iX_i + E_i$$

where:

P = the probability of making a choice between two alternatives
B_{10} = the intercept value
$B_{i\,'a}$ = regression coefficients
$X_{i\,'a}$ = independent variables
E_i = random error term

The model was used to examine the major factors influencing the adoption of hybrid seed and farm yard manure and/or commercial fertiliser. The application of logit models in geographical studies is not yet widespread but considerable developments are taking place.

Crop Combinations

The most frequently occurring crops are also the major users of land. These are maize, sugar and beans. Only crops with at least 5 per cent occurrence were included in the data analysis. The correlation matrix for the crops showed that most of the crops have very low correlations. This indicates that the occurrence of any given crop is in most circumstances not related to the occurrence of the rest. Exceptions to this are maize and beans and maize and ground-nuts. These crops are frequently intercropped. Using principal axis factoring, the original twelve crops were collapsed to form four factors. The four factors extracted and their corresponding loadings after quartimax rotation are displayed in Table 6.1. The factors describe the existing crop combinations each of which can be identified through the factor loadings.

Table 16.1: Rotated factor matrix (quartimax rotation)

Factors	1	2	3	4
Beans	.96	-.05	-.20	-.07
Maize	.62	.22	.17	.08
Ground-nuts	.46	-.11	.27	.14
Potatoes	.46	-.05	.06	-.13
Sorghum	.41	.00005	.0003	-.17
Bananas	.40	.05	-.17	.25
Sugar	.40	-.05	.07	.38
Horticultural crops	.05	.71	.13	-.01
Fruits	-.03	.34	-.08	-.03
Cassava	.25	.01	.52	-.20
Coffee	.07	-.08	.001	.33
Finger millet	-.02	-.10	.17	-.28

Source: Akech (1990; 107)

Factor 1 is the most typical crop combination consisting of maize, beans, ground-nuts and potatoes. This may be called the food-crop combination. Sugar-cane loads highly on this factor underlining the fact that cash cropping is becoming important. The "subsistence" crop combination (Factor 2) comprises vegetables, fruits and maize with an apparent lack of cash crops. This crop combination should be typical of very poor farmers who have not yet gone into cash cropping and whose primary interest is to meet subsistence food needs. The "marginal food crop" combination (Factor 3) consists of drought-resistant crops which can be grown under inferior soil conditions. These include cassava, ground-nuts and finger millet. The fourth crop combination of the study area is the cashcrop combination (Factor 4) including sugar-cane, coffee and bananas. This crop combination should be characteristic of the richer farmers in the region.

The four crop combinations display distinct spatial patterns in the study area underlining the fact that crops do not occur haphazardly within the small-scale farming areas. Each crop combination has several implications for increased maize production. The areas where the "cash-crop" combination predominates also tend to have lower maize yields. This suggests that cash-crop production has a negative influence on maize production. The possible relationship between maize production and the growing of the principal cash crops, namely, sugar-cane and coffee are analysed further below.

Maize Yield Variations

In general, maize yields tend to decline from east to west in the study area. The highest yields average 31 bags per hectare while the lowest yields are between 10 and 20 bags per hectare. From an original list of thirty independent variables, a multiple linear regression procedure was used to explain the observed yield variations. Overall, eleven models were fitted. The final model adopted to explain yield variations in the study area consisted of 10 significant explanatory variables explaining 61 per cent of the total variation in the response variable from 179 observations. An examination of the excluded data points revealed that most were either very low or very high, suggesting yields measurement error perhaps due to misreporting during the data collection process.

Table 16.2: Regression results maize yelds analysis

Parameter	Parameter estimate	Standard error	t-value
1 (intercept)	33.710	4.929	6.84*
Maize hectarage	-7.678	2.039	-3.75*
Low rainfall	-9.341	2.886	-3.24*
Agro-ecological zone 2	10.210	2.735	3.73*
Agro-ecological zone 3	-0.016	3.007	-0.01*
Moderately fertile soils	13.510	4.241	-3.19*
Poor soils	-1.101	2.375	-0.46
Number of maize fields	2.941	1.052	2.80*
First weeding date-March	-2.506	2.616	-0.96
First weeding date-April	-11.850	4.016	-2.95*
Intercropping	-6.659	2.259	-2.95*
Sugar hectarage	3.429	1.634	2.10*
Farm-size	0.857	0.436	1.96*
Local seed	-0.283	2.400	-0.12
Second generation hybrid	-13.900	6.326	-2.20*
All seed types	8.890	3.548	2.50*

*Significant at p=0.05
Source: Akech (1990: 124)

The parameter estimates and the corresponding t-values for the final model are shown in Table 16.2. It should be noted that the value of each parameter estimate of a categorical variable is interpreted as representing differences in the dependent variable between the levels of the explanatory variable. The interpretation of the parameter estimates of dummy and continuous variables is the same as in conventional multiple linear regression analysis.It is evident that the agro-ecological zone within which a particular farm is located has a significant influence on maize yield. The parameter estimates reveal that farms located in the marginal sugar-cane zone (agro-ecological zone 2) have the highest average maize yields (44 bags/ha). Farms situated in the sugar-cane zone (agro-ecological zone 1) register a significantly lower average maize yield of approximately 33 bags per hectare. Areas designated as coffee and coffee-tea zone (agro-ecological zone 3) have the least average yields (33.7 bags/ha), although negligibly and insignificantly different from those in the sugar zone (agro-ecological zone 1). These results indicate that areas less suitable to the two principal cash crops (sugar-cane and coffee) have the highest potential for maize. A possible explanation for this observation is the fact that when maize is in competition with the two crops, farmers tend to allocate available farm resources in favour of the latter. This suggests the adverse effect cash crop production is likely to have on food crop production in the study area and other areas where the same conditions exist.

Areas having low average annual rainfall (1400-1600 mm) have significantly lower yields than those experiencing high average annual rainfall (1600-1800 mm) by approximately 9 bags. The parameter estimates further indicate that farms located in areas of medium soil fertility have significantly lower yields (22.2 bags/ha) than farms situated in areas of high soil fertility (33.71 bags/ha). Areas of low soil fertility also show the same trend although the difference is not significant (32.6 bags/ha). Farms found in this category constituted only 6 per cent of the total sample. These results confirm the important influence environmental factors have on yield variations in low technology areas. There is, therefore, a need to plant different maize varieties even within a small area. The results also call for the adoption of adequate soil management practices to maintain and improve inherent soil fertility.

The only significant agronomic variables found significant in explaining maize yield variations were date of first weeding, intercropping and type of seed planted. Farms weeded in the month of February or earlier had higher yields than those weeded in March by about three bags. However, this difference is not statistically significant. Maize fields weeded in April or later, however, have significantly lower maize yields (by about 12 bags) than those weeded in February or earlier. These results indicate the high responsiveness of maize to care. Early weeding prevents the competition of soil nutrients associated with weed growth. (Weeding is the most demanding of all the field operations in maize production in terms of the size of labour and amount of time required.) Late weeding is, therefore, a reflection of the difficulties faced by farmers in raising the necessary labour for the task. It may also be the cumulative effect of both delayed land preparation and late planting. It, is, therefore appropriate to increase farmers' financial resources to enable them to hire casual labour at reasonable rates of remuneration during the weeding period. The formation of farming groups whereby communal labour can be utilised is also a supplementary measure which can alleviate the problem of late weeding.

Intercropping affects maize yields through its reduction of the plant population (maize) per unit area. In other instances, crops interplanted with maize may have identical nutritional requirements as maize thereby leading to competition which can reduce yields. The parameter estimates indicate that pure stands of maize have significantly higher yields than intercropped maize by about 7 bags. Due to various reasons, intercropping is widely practised in the study area (70.5% of the total farmers sampled intercropped) and is likely to continue. However, the practice should be accompanied by a careful choice of the crops intercropped with maize. In this respect, preference should be given to legumes such as beans and ground-nuts which have complimentary nutritional requirements with maize. Such crops as sugar-cane, finger millet and sorghum currently intercropped with maize should be discouraged.

The development of suitable hybrid maize varieties and their consequent adoption by farmers has been a major strategy in boosting maize production in Kenya. The parameter estimates show that hybrid seed has a higher average yield than local seed by about 0.2 bags although this difference is not significant. Second-generation hybrid has a significantly lower yield than hybrid seed by about 13 bags. The highest yields are recorded in cases where combinations of all the three major seed types are planted. Although hybrid seed has higher yield potential than local seed varieties, this is only true when adequate management practices are carried out. Under circumstances of improper care, local seed can give higher yields than hybrid seed given the former's adaptability to the local environment. The negligible difference in yield levels between the two seed types predicted by the regression model should, therefore, be interpreted as indicative of the poor agronomic practices prevalent in the small-scale farming sector. Second generation hybrid seed poses the greatest problem to the achievement of high maize yields. These seeds are normally planted by farmers on the belief that they have identical genetic properties to the original hybrid seed they are derived from. On the contrary, hybrid seed loses most of its genetic qualities after the first harvest due to contamination from other seed varieties. It is because of this that the purchase of fresh hybrid seed is recommended for each season. The continued planting of second generation hybrid seed is a symptom of inadequacies in the extension and seed distribution services. Financial difficulties faced by farmers can also inhibit them from purchasing fresh seed every season.

The total hectarage under maize, number of separate maize fields, farm size and the total hectarage under sugar are the four socio-economic variables significantly related to maize yield. For every unit increase in maize hectarage, there is a corresponding drop in maize yield by approximately 8 bags. Given all the agricultural resources at the farmers' disposal, smaller hectarages of maize are likely to be worked more intensively than larger ones. Faced with scarce agricultural inputs, farmers tend to maximise their maize output by expanding the total land devoted to the crop with a corresponding investment in the purchase of seeds and fertiliser, early planting and proper weeding.

The maintenance of more than one separate piece of maize field is a strategy adopted by farmers against the possibilities of crop failure. The practice also represents farmers' efforts to utilise only the most favourable environmental attributes which are never uniform even at the single farm level. In addition, the practice enables farmers to space their field operations thus increasing the chances of better management practices. The parameter estimates indicate a significant and positive relationship between the number of separate maize fields and yields.

Farm size has a positive and significant relationship with maize yield. Larger farm sizes have several advantages over smaller ones which could account for this result. In the first instance, the observation of long fallow periods and its positive effect on soil fertility is more possible in larger farms. In addition, the practice of intercropping on account of land scarcity is less likely in such farms. Given the negligible application of fertiliser within the small-scale farming sector, farmers will continue to rely on the natural fertility of the soils. The observed relationship between farm size and maize yield points to the danger that increasing population pressure on available land poses to maize production in the study area.

The regression model predicts a decrease of 3 bags in maize yield with every unit increase in sugar-cane hectarage planted by a farmer. Moreover, it not only competes with maize for fertile land, but also with the major agricultural inputs necessary for proper crop husbandry. The allocation of these resources between the competing enterprise depends largely on the expected returns from each. Sugar-cane not only fetches higher prices than maize, but it also has a ready market provided by the sugar factory in the study area. The crop also enjoys better supportive services including land preparation and transport services as well as an efficient extension service. Due to all these advantages farmers are expected to allocate more of their resources towards sugar-cane than maize thus depressing maize yields. With decreasing farm sizes and the expansion of the SONY sugar factory at Awendo, maize production is expected to be on the decline unless special measures are taken to reverse the expected trend. Supportive services specific to maize should be initiated, mainly dealing with a more intensified extension and credit support. The credit should be made available for specific field operations and preferably be in kind rather than cash to avoid possible misuse. These services should be targeted mainly at sugar-cane farmers who are the most likely to neglect the production of maize.

Maize hectarage covered in this study, shows that land devoted to maize varied from minimum of 0.1 hectares to a maximum of 7 hectares. The average size of maize fields was 1.1 hectares with a standard deviation of 0.8 hectares. From these figures, it is evident that maize fields are generally small although wide variations exist. The highest hectarages are registered in the east and west of the study area while the lowest (0.5-1.0) are found in the north and south.

From an original set of 29 selected variables utilising all the regression diagnostics already mentioned, a total of eighteen regression models were fitted to explain maize hectarage variations. The parameter estimates of the final model adopted are shown in Table 16.3. The model explained 78 per cent of the total variation in maize hectarage. This was a considerable improvement over the first model fitted which had a total of 29 independent variables but explained only 54 per cent the total variation in the response variable.

Table 16:3 Results of regression modelling of maize hectarage

Meter	Parameter estimate	Standard error	t- value
1	0.7074	0.1996	3.5447
Number of maize fields	0.2840	0.0376	7.5496
Land preparation (January/February)	-0.4414	0.1287	-3.4301
Land preparation date (March/later)	-0.5448	0.1372	-3.9698
Soil fertility (moderate to high)	-0.6931	0.1757	3.9455
Soil fertility (moderate to low)	-0.1177	0.0808	-1.4570
Farm size	0.0457	0.0126	3.6377
Agro-ecological zone 2 (marginal sugar-cane zone)	-0.2749	0.0766	-3.5857
Agro-ecological zone 3 (coffee-tea zone)	-0.1577	0.0921	-1.7120
seed type (local)	-0.2047	0.0763	-2.6832
seed type (second generation hybrid)	-0.1704	0.1516	-1.1246
seed type (combination of all types)	-0.0110	0.1028	-0.1074
Ox-plough team ownership	0.3903	0.1479	2.6382
Coffee hectarage	-0.9058	0.3696	-2.4505
Presence of farm owner	0.1991	0.0821	2.4234
Length of fallow	-0.0759	0.0384	-1.9757
Number of fields and ox-plough ownership	-0.0898	0.0499	-1.7968

Source: Akech 1990:140.

The parameter estimates indicate that farms located in agro-ecological zone 1 (sugar-cane zone) tend to have more land devoted to maize than the two other zones. This suggests that the areas most appropriate to the main cash crops (agro-ecological zone) sugar-cane, coffee and tea tend to have higher maize hectarages.

It is interesting to note that agro-ecological zone 2 (marginal sugar-cane zone), which has the least land under maize was also identified above as having the greatest yield levels. These results conform with the inverse relationship already established between maize hectarage and yields. They also suggest that sugar-cane competes with maize mainly for the inputs required for the enhancement of yields rather than land. The parameter estimates for the soil fertility variable indicate that less land is devoted to maize as soil fertility decreases. This unexpected relationship could be representative of situations whereby crops such as cassava, millet and sorghum take precedence over maize on increasingly marginal land. These crops can tolerate inferior environmental conditions. It is interesting to note that within the "marginal food-crop" combination region identified earlier soil fertility ranged from medium to low.

The results further indicate that late land preparation has a negative impact on maize hectarage. As the land preparation date delays, less land is planted with maize. This is a logical result in small-scale farming areas where shortages of equipment and labour are common. Farmers in such circumstances are usually forced to prepare less land for timely planting to occur. There is, therefore, a need to help farmers acquire the necessary equipment and labour if maize hectarages are to be expanded, especially in areas where idle land exists.

Hybrid maize seed tends to be planted in larger areas than other seed varieties. This could be interpreted as a reflection of the possibility that richer farmers who can afford to plant hybrid seed also tend to cultivate more maize due to their ability to raise the necessary resources. The regression results show that as the length of the fallow period increases less land is devoted to maize. Long fallow periods increase the inherent fertility of the soil necessary for high maize yields. This possibly removes the pressure from the farmer to expand hectarage under maize. In the absence of adequate soil fertility enhancing practices, shorter fallow periods are likely to be reflected in lower yields. The observance of long fallow periods could also reflect the difficulties faced during the land preparation period resulting in less land planted with maize.

The total amount of land under maize increases with farm size. The amount of land available to each farmer for maize cultivation is expected to continue diminishing as a result of population pressure on land and competition from the newly-introduced cash crops. This trend calls for intensified maize production so that more output is realised from increasingly less land. This will involve the application of greater amounts of inputs to preserve soil fertility and improve maize yields.

Ox-ploughing is the most widespread means of land preparation in the study area. The ownership of an ox-plough team, consisting of at least two oxen and an ox-plough, has a significant influence on maize hectarages. The parameter estimates indicate that owners of ox-plough teams have larger amounts of land under maize than those who do not have them by about 0.3 hectares. There is, therefore a need to improve farmers' accessibility to the means of land preparation. This could be done through the provision of tractor-hire services at reasonable rates for timely preparation of land in larger units. On the other hand, communal groupings could be encouraged so that land preparation implements are pooled together. The results also suggest that livestock development should be consciously pursued as part and parcel of agricultural crop development. Better ox-ploughs should be designed to expand the range of field operations to which the implement could be applied. In addition, sugar-cane farmers should be required to devote an adequate minimum hectarage to maize every season.

Coffee hectarage is negatively and significantly related to maize hectarage. A unit increase in coffee hectarage lowers maize hectarage by about one hectare. This indicates that coffee is a better competitor for land than maize. Given the fact that coffee cultivation is not yet widespread in the study area, there is a possible future decline in maize hectarages as more farmers take up coffee cultivation. The expected decline in output can only be avoided by enhancing maize yields. The need to increase farmers' accessibility to important farm inputs for intensified maize production, therefore, cannot be over-emphasised.

The problem of absentee landowners has long been suspected for its undesirable influence on agricultural development. Results from this study indicate that less land is devoted to maize in farms whose owners are absent. Cases of absentee farm owners constituted 20 per cent of the total sample covered in this study. In 82 per cent of these cases, farm management was left in the hands of the female spouse. It is reasonable to suggest that women need greater control in farming decisions especially when they are left behind to manage farms. For families whose male heads are in employment in urban areas, the pressure to meet food needs from the farm is rather less because transfer payments received can be utilised in the purchase of grain and other foodstuffs. This could explain the negative effect absentee landowners exert on maize hectarages.

The interaction effect between the number of maize fields and ownership of an ox-plough team is also significant. This shows that for those with ox-plough teams, an increasing number of separate maize fields has the effect of reducing their average sizes. This reflects the possibility that those without ox-ploughs tend to cultivate maize in larger but fewer land units. This group of farmers normally relies on hired tractors and ox-plough teams, making it convenient to cultivate fewer but larger maize fields.

Adoption of Innovations

Table 16.4 shows the pattern of adoption of innovations for a sample of farmers in the study area. The largest number of farmers adopted manure and/or fertiliser followed by the adoption of hybrid seed. A logit regression procedure was used to find out the most important variables affecting the likelihood that a farmer adopts a given innovation. For the adoption of hybrid seed, a total of four models were fitted. The final model used to explain the adoption of hybrid seed is also shown. Parameter 1 is the anchor category against other parameters which are judged to lead to an increase or decrease in the likelihood of adoption and to represent the lowest level of all the categorical and dummy variables in the model. The p-values are the computed probabilities associated with each parameter in the model.

Educational level and age of the farmer are important variables explaining the likelihood of adoption of hybrid seed. The effect of having primary education in comparison to no formal education at all is to increase the likelihood of adoption to 0.43. The possession of secondary level of education also increases the likelihood of adoption but to a lower level. The higher the age group of the farmer, the lower the likelihood of adoption of hybrid seed. Only 13 per cent and 7 per cent of the farmers in middle and old age respectively are predicted to adopt hybrid seed. Universal basic education is, therefore, an important variable for the adoption of hybrid seed within the small-scale farming sector. The rural exodus of young men to urban areas poses a great danger to agricultural productivity since it reduces the proportion of young farmers who are predicted to be better adopters of hybrid seed in comparison to their older counterparts.

Table 16.4 The distribution of adopters for different categories of innovation

Category	Total no. of adopters	% total farmers
1. Hybrid seed in combination with local seed varieties	132	66.0
2. Manure and/or fertiliser	152	76.0
3. Hybrid seed only	107	53.5
4. Fertiliser only	25	12.5
5. Hybrid seed 7 local seeds/ fertiliserand/or manure	111	55.5
6. Hybrid seed fertiliser combination	19	9.50

Source: Akech (1990:160)

A unit increase in maize hectarage increases the likelihood of adoption of hybrid seed to 0.34. This reflects the fact that the richer farmers who can afford to have greater hectarages under maize are also better adopters of the innovation.

Table 16.5 Regression results Model 3: Partial adoption of hybrid seed

Parameter	Estimate	St. Error	t-value	p-value
1 (intercept)	-1.59	0.65	-2.43	0.17
Primary education	1.31	0.37	3.54*	0.43
Secondary education	0.74	0.75	0.99	0.30
Maize hectarage	0.92	0.32	2.91*	0.34
Distance to the nearest road	-0.49	0.19	-2.59*	0.11
Use of casual labour	0.90	0.36	2.51*	0.33
Distance to the nearest market	0.26	0.13	1.95*	0.21
Extension contact	0.90	0.44	2.03*	0.33
Age (45-64 years)	-0.31	0.40	-0.77	0.13
Age (= or > 65 years)	-0.97	0.55	-1.78*	0.07
Wheel barrow ownership	0.55	0.31	1.74*	0.26

* Significant variables (p = 0.05)
** p-value is probability value
St. Error is standard error.

Source: Akech (1990: 167)

The use of casual labour in farming operations and the ownership of a wheelbarrow are both positively related to the likelihood of adoption. The former increases the likelihood to 0.33 while the latter increases it to 0.26. The two variables can be considered as surrogates for the economic status of the farmer. These results, therefore, suggest that the farmers who are better off economically are also better adopters of hybrid seed. The further a farm is from the nearest road, the less likely that the adoption of hybrid seed would take place, showing that the adoption process is related to accessibility of farms to other places. Increasing distance from farms to markets increases the adoption of innovation. This is an unexpected result, but might suggest that the major market centres in the study area are not major information centres for innovation. As is expected, extension services increase the likelihood of adoption to 0.33.

The model explaining the adoption of farmyard manure and/or fertiliser was the fifth in the modelling process. The results of the model are displayed in Table 16.6.

Table 16.6: Regression results Model 4: Adoption of farmyard manure and/or commercial fertiliser

Parameter**	Estimate	Standard error	t-value	Probability
1	0.22	0.32	0.56	0.55
Primary education	0.94	0.39	2.40*	0.76
Secondary education	0.26	0.79	0.32	0.62
Soil fertility (medium)	1.25	1.06	1.17	0.81
Soil fertility (poor)	-1.15	0.39	2.92*	0.28
Number of farm animals	0.03	0.01	2.58*	0.56
Sugar hectarage	-0.51	0.27	-1.92*	0.43
Wheel barrow ownership	0.55	0.31	1.74*	0.68

* Significant variables (p=0.05)
Source: Akech (1990: 177)

The educational level of the farmer was isolated as a significant variable in explaining the likelihood of adoption of farmyard manure and/or commercial fertiliser. The possession of primary education increases the likelihood of adoption to 0.76 in contrast to the secondary level of education which increases the likelihood to 0.62. This signifies that at least basic education is necessary for the adoption of the innovation. It should be noted that those with a secondary education have less likelihood of adopting the innovation, perhaps reflecting their greater interest in non-farming occupations. Female-headed households have a greater likelihood of adoption of the innovation than their male counterparts. This calls for a restructuring of the agricultural support programmes to increase female participation and control in agricultural production.

Farmers located in the least fertile soils also have the least likelihood of adoption of the innovation suggesting that the infertile soil, therefore, requires greater extension effort to encourage farmers to adopt fertility-enhancing innovation.

The number of farm animals kept by a farmer increases the availability of farmyard manure while wheelbarrow ownership helps greatly in transferring the manure to the maize fields. Both variables are positively related to the adoption of the innovation. Livestock production should, therefore, be encouraged as an intrinsic part of efforts aimed at boosting maize production in the study area.

The negative impact of sugar-cane cultivation on maize production is shown by the parameter estimates. A unit increase in the hectarage of sugar cultivated reduces the likelihood of adoption of farmyard manure and/or commercial fertiliser. This shows that when sugar is in competition with maize, farmers respond by diverting more inputs towards cane cultivation.

Conclusion

The major findings of this study are that crops do not occur haphazardly within the small-scale farming sector, but in distinct crop combinations forming recognisable spatial patterns. The findings further suggest that cash crop production in the study area (especially sugar-cane) adversely affects maize production. In addition, maize yield variation in the study area is attributable mainly to variations in the physical environmental conditions and crop husbandry practices found in the area. Intercropping, late weeding and the planting of second-generation hybrid are responsible for reduced maize yields. Sugar-cane cultivation is also negatively related to maize yields. Maize hectares, on the other hand, are limited due to lack of land and land preparation implements. The adoption of hybrid seed and manure/fertiliser are related mainly to the individual attributes of the farmers as well as exposure to extension services. Sugar-cane cultivation particularly reduces the likelihood of applying manure and/or commercial fertiliser in maize fields.

In the light of the main findings of this study, it is concluded that maize production in the study area is certain to decline following the introduction of lucrative cash crops such as sugar-cane and coffee. In addition, future increases in the production of maize will have to rely on increased adoption of technological innovation to increase yields per unit area. In this respect, the agricultural support programmes must be intensified specifically for farmers lower in the socio-economic hierarchy. In terms of food policy for the country, the apparent conflict between cash and food crop production must be resolved.

17 Change, Persistence and Development in Small-holder Livestock Production System in Western Kenya

Collette A. Suda

Introduction

Empirical studies are rarely theoretical. In cases where the approach is not eclectic, a research agenda is usually set and formulated within a specific theoretical perspective considered most appropriate for understanding the problem under investigation and the substantive issues that are to be dealt with. Rural social scientists involved in research activities concerning livestock production always have to deal with both substantive and theoretical issues. This is because empirical work should be theoretically well-grounded.

This chapter analyses the various assumptions and assertions of the modernisation theory or the developmentalist perspective and examines its potential or adequacy in conceptualising farmers' attitudes towards and practices in livestock production in a small-holder community. Small-holder agricultural production in western Kenya is a mixed enterprise in which the production mixture is largely determined by the general consumption pattern and needs of the producers.

The empirical component of this chapter focuses on the differences between livestock and non-livestock farms and the varying perceptions about cattle and small ruminant production. Overall, the chapter deals with the issues of how the modernisation perspective looks at and interprets the farmers' perceptions of and preferences for cattle or small ruminants and how this bears upon their production management priorities in the context of changing social and economic systems. Since this is a micro-level study which takes the household as its unit of analysis, the relative appropriateness of the modernisation approach for such a study will be evaluated on the basis of its premises and level of analysis.

Theoretical Overview

Most studies on the organisation of production and changes in small-holder agriculture in Kenya are largely informed by the modernisation perspective. Discussions of the modernisation perspective which follow will attempt to delineate the specific arguments the approach makes on issues like the household division of labour and the production strategy in small-holder agriculture.

Modernisation theory came out of the evolutionary approach to structural-functionalism exemplified in the writings of such scholars as Parsons (1951) and Durkheim (1964). Theoretical discussions within the modernisation perspective tend to reflect this functionalist influence. The general orientation of the modernisation perspective is marked by a dichotomous conception of reality derived from the five-pattern variable scheme developed by Parsons (1951) but which has been used by Hoselitz (1960) to apply to the study of economic development and cultural change.

Different approaches were subsumed under the modernisation perspective. They include Rostow's (1978) stage analysis of economic and industrial development, McLelland's (1961) social-psychological analysis of individual motivation, the diffusion model, and Levy's (1966) discussions of modernisation processes and structural changes. All of these approaches are developmentalist in orientation and they use a simple traditional-modern dichotomy as their basic frame of reference. Change and development are thus conceptualised as evolutionary processes which necessarily require that the attributes of relatively modernised societies be transposed onto relatively non-modernised societies. Thus, the processes of social change and economic development can be viewed as representing a transition.

This model posits that roles in developed societies are typically universalistic, based on achievement and functionally specific whereas those in underdeveloped societies are particularistic, based on ascription and functionally diffuse. Development thus consists of replacing the ideal typical features of a traditional economy with those of modern economy; it involves transition from traditional forms of organisation to modern methods of production:

> ...economic development would be a process whereby a society undergoes a basic structural transformation from a type characterised by diffuseness/particularism/ascription to one characterised by specificity/universalism/achievement. (Stockwell and Laidlaw, 1981: 152)

According to this conceptualisation, as development takes place there is a concomitant increase in the division of labour and a growing structural complexity, differentiation and integration among the various units of the social system. Increased role specialisation and the complexity of the division of labour are seen as the key to great integration and high productivity that have come to be associated with economic growth and development in industrial societies. In the case of small-holder agriculture in western Kenya, development would consist of a highly specialised commodity production based on a well-defined division of labour. Based upon five pairs of pattern variables, the modernisation perspective assumes that "traditional" societies are characterised by a division of labour in which roles are functionally diffuse rather than specific and the level of farm production, for example, is generally low because of lack of modern technology, resource limitation and unsustainable use of productive resources.

Scholars who operate within the modernisation framework frequently argue that labour in small farm households in so-called traditional societies is allocated on the basis of ascriptive rather than achievement criteria. An analysis of gender-based division of labour in small farm households in western Kenya would support this argument. Household labour is differentiated by age, gender and the type of commodity produced.

The basis for role differentiation is thus predominantly ascriptive but it is also culturally defined. But since achievement is usually not a major consideration in the assignment of agricultural tasks, modernisation theorists contend that substantial human potential may be left untapped, particularly in matters regarding livestock management and farm decision-making which are traditionally reserved for male heads of household. In households where the male is away from home and the woman is left behind to function as *de facto* household head, she has to defer most of the major decisions to the head; there is a possibility that this could impede rather than facilitate productivity of certain enterprises and eventually reduce overall farm production. Although the modernisation perspective provides some useful insights into the organisation of household labour, the cultural, social and economic basis of these production practices are somewhat taken as given and rarely made problematic.

Developing economies are also characterised as particularistic rather than universalistic in terms of the ways in which productive tasks are assigned. This implies, for example, that small farmers in western Kenya are more likely to exchange labour with their relatives and friends than to engage the services of someone else from outside who may be more competent to do the job. This tendency to engage in exchange with members of one's family or kinship group derives from a strong sense of kinship obligation. The cultural beliefs and value systems are viewed by the diffusion model as presenting an obstacle to change in LDCs. The assumption is that change cannot be generated unless these traditional obstacles are overcome and significant changes made in the farmers' predispositions, attitudes, values and beliefs, which underlie and influence their behavioural patterns. The monetisation of the economy has weakened the traditional social support network and undermined the kin-based social obligations exemplified in reciprocal labour exchange. According to this orientation, change and development are considered possible through piecemeal reforms and do not necessarily involve a radical transformation of the system which produces and reproduces poverty.

Despite the diversity within the modernisation perspective, there is, however, a common thread that is woven through all the different strands of thought which comprise this perspective. Although modernisation theory tends to focus on the values, beliefs, attitudes and demographic characteristics of a given individual or groups, these are usually aggregated and used to characterise the nation-state as an appropriate unit of analysis:

The level of analysis which is of crucial theoretical significance is that of society and culture – the nation state is normally the focus of interest (ultimately, even individual modernity is of interest because of its implications for the societal level. (Lauer, 1977:304)

According to this assumption, development and underdevelopment tend to be explained in terms of the unique conditions and attributes internal to a particular nation-state. These units are often studied and analysed as discrete entities, largely in terms of their internal dynamics and processes, almost as if they have an existence independent of the world economic system of which they are a part. Usually, the major task of modernisation analysis is to identify the obstacles and design projects and programmes to help remove them in order to promote development or at least to initiate some changes. Once the obstacles are removed, then development, normally measured in terms of aggregate economic growth and improvements in the quality of life can proceed, regardless of how the benefits of such growth are distributed. Changes in individual perceptions, cultural patterns and economic systems are seen as processes that can be brought about or facilitated through exposure to outside influence. These processes enhance sustainable development in whatever enterprise the producer engages in. For example, modernisation theory would argue that infrastructurral development, improved livestock production and more social and economic opportunities for small-holder producers in western Kenya are indicators of development. But the specific consequences of some of these change processes have received relatively little attention. The emphasis of internal factors as the major explanatory variables in the analysis of underdevelopment is premised on the assumption that economic growth is largely an endogenous process, even if an LDCs receives aid from the MDCs. The social, cultural, economic and political institutions in the recipient country must create an overall atmosphere conductive to change. This line of argument suggests that even though modernisation takes place within the context of globalisation and increased interdependence between nations, it is largely seen as a function of internal structure because, ultimately, this is where the intervention occurs.

Because of its focus and emphasis on endogenous processes within each country, the modernisation perspective has been criticised for being ahistorical. The approach has had little to say about the colonial history and experience of LDCs which have structured the relationship between the North and the South or between the urban and rural areas in some fundamental ways. Development and underdevelopment can be seen as parts of the same historical process, namely, colonialism and the emergence and expansion of a stratified capitalist world economy characterised by an international division of labour which is locally reproduced at the household and community levels.

The majority of the farmers in western Kenya are engaged in small-holder agricultural and livestock production carried out largely on a subsistence basis. From the standpoint of the diffusion model, the small-holder sector is "backward" and an increase in livestock or food production calls for a change in attitudes as well as traditional farming practices. This, it is argued, could be achieved through greater exposure to new values, transfer of modern technology, capital and skills from the core to the periphery. When these various forms of innovations are diffused from outside and adopted by the indigenous target groups, this process is seen as basic to increased productivity. The value and efficacy of indigenous knowledge becomes epiphenomenal. Technology is thus held as a prepotent factor in agricultural and livestock development and its inexorable force and compelling effects sometimes tend to be overstated even though a particular technology might be hopelessly inappropriate for the resource conditions of the setting to which it has been transferred. Very often, the consequences of certain technological innovations tend not to be anticipated *a priori*. As Goss (1979) has pointed out, some technical changes in farm production could create and exacerbate inequalities. Moreover, many technical modifications and inputs, notably in farm mechanisation and improved animal husbandry techniques in Kenya and other parts of Africa, have consistently been shown to be hopelessly inappropriate. Such technical modifications are seldom incorporated into existing practices nor informed by indigenous knowledge held by the producers. And since the modern inputs are not usually combined with adequate fostering of creative skills and knowledge on the part of small producers, they seldom lead to a meaningful change in small-scale livestock enterprises.

The diffusion model, which is a variant of the modernisation perspective, tends to explain lack of development in small-holder agriculture in terms of what Roxborough (1979:20) has referred to as "some missing factor which was absent in these sectors and would account for their failure to alleviate economic growth." Included among the missing factors are capital, modern technology, achievement motivation, enterpreneurship, institutions, skills and sometimes labour although this is generally considered abundant on account of high population growth rates. Within this framework, improvement in livestock production would be conceptualised as a process involving the transfer or acquisition of the "missing factors".

From the foregoing discussion of the modernisation perspective and examination of some of its underlying assumptions, it seems clear that its conceptualisation of small-holder livestock production and how this relates to the broader question of agricultural development calls attention to the pre-existing forms of cultural, social and economic systems which need to be restructured or otherwise modified through contact with the larger market economy as a necessary part of the development process. Many of the insights from modernisation theory bear directly on the characteristics of the small farm households in western Kenya, specifically in terms of resource allocation, cultural attributes and economic opportunities. From the point of view of modernisation theory, therefore, understanding these cultural and structural conditions is essential to an understanding of the development potential of the livestock sector. What follows is an empirical analysis of livestock production and the role of livestock in the socio- cultural and economic fabric of the small-holder producers in Siaya and Kakamega Districts.

Mixed Crop/Livestock Production System

The prevailing farming system in western Kenya involves an intensive use of labour and land resources to produce crops and livestock on a small-scale subsistence basis. Farmers in these communities are not specialised commodity producers. Rather, they are small-holders whose principal goal is to meet basic household consumption needs. Although they occasionally exchange part of what they produce on the market, production is primarily for home consumption and not oriented towards profit maximisation through sales. The high ratio of production for domestic consumption to production for exchange demonstrates a partial integration of the small-holder agriculture into the market system. The household members consume much of what they produce and produce much of what they consume through essentially labour-intensive cultivation practices.

The overall production strategy is to achieve an efficient use of resources in a production process which is informed and guided by a different set of logic from those that underlie capitalist agriculture. Since all the major productive resources are in limited supply, one of the strategies employed as a labour- and land-saving device is growing sorghum and sweet potatoes which are among the basic staple foods in the area. Land and labour shortages also require that cattle, sheep and goats are all grazed together on the same pasture to take advantage of economies of scale. Based on a low level of technology, this system of production may not be highly efficient and cost-effective in the capitalist sense of rationality but it has been fine-tuned and well-adjusted to the total environment of the small farmers in Siaya and Kakamega Districts.

Traditionally, crop production has been the major enterprise within the small farm system. Livestock production, though important, is generally considered a secondary component. The predominance of crop over livestock production provides a social and economic basis for understanding the differential pattern of resource allocation and management between crop and livestock production systems.

Livestock production is not a dominant enterprise and this is evidenced by the fact that livestock ownership was limited to only some of the farms. Some 37.8 per cent of the surveyed farms had no livestock and there was no evidence or indication

Table 17:1 Number of households with livestock by district (Siaya and Kakamega)

Presence of livestock in the household	Siaya		Kakamega		Both districts	
	n	%	n	%	n	%
With livestock	22	61.1	24	63.2	46	62.2
Without livestock	14	38.9	14	36.8	28	37.8
Total	36	100.0	38	100.0	74	100.0

Source: compiled by the author.

that these households had engaged in any form of livestock production before. There was no significant difference in patterns of livestock ownership between the two districts under survey as shown in Table 17.1.

The study showed that livestock farms were generally larger than non-livestock ones. More livestock were found in households with more land and labour (Table 17.2). These households were also more likely to hire outside labour than their counterparts without livestock. In Kakamega District, for example, the number of livestock per household significantly increased with farm size and availability of family labour, which indicates that livestock production or specifically cattle ownership in this area is an important measure of wealth. In these communities, a large family is important. For many people, it represents a major form of security, particularly during times of need. It requires a large family with a substantial amount of labour and adequate land to be able to raise large herds of cattle. But, by the same token, a large family also needs a large number of livestock to be able to plough and plant its fields on time, sell them and obtain enough cash to cover the high and rising cost of household reproduction. Because cattle ownership is associated with status enhancement, livestock farms had greater access to other resources, including hired labour. Households which kept livestock were observed to be relatively well-off.

Table 17.2: Relationships between the number of livestock and the family and farm size by district

Variables	Siaya (N=36)	Kakamega (N=36)	Both districts (N=74)
Family size	0.03	0.27*	0.07
Farm size	0.07	0.36**	0.18

* $p < 0.10$
** $p < 0.05$

Source: Author.

Livestock Production and Use of Hired Labour

Despite the limited use of hired labour throughout the farms surveyed, it was observed that the use of hired labour varied significantly between livestock and non-livestock farms. Those households which had a large number of livestock were the ones that hired labour (Table 17.3). The majority of the households that hired no labour had very few animals or none at all. They relied entirely on non-wage family labour. This significant relationship between the use of hired labour and the number of livestock per household suggests that households with livestock were

Table 17.3: Use of hired labour by district

	Difference in mean number of livestock								
	Siaya			Kakamega			Both districts		
Hired labour	N	X	S.D	N	X	S.D	N	X	S.D
Yes	10	5.4	4.7	8	6.0	5.6	18	5.7	4.9
No	26	3.5	6.3	30	2.1	2.6	56	2.7	4.7

Summary of one-way analyses of variance

	Siaya (N=36)			Kakamega (N=38)			Both districts (N=74)		
Source	DF	MS	F	DF	MS	F	DF	MS	F
Hired labour	1	27.1	0.8	1	97.7	8.5	1	118.7	5.3**
Error	34	34.6		36	11.4		72	22.5	

NB: **$p < 0.05$ ***$p < 0.01$
* Table 17.3 reports the results of three separate analyses of variance, one for each district and another for both districts combined.

Source: Compiled by the author.

relatively better off than those which had none. It was also noted that households which hired labour also tended to have larger farms than those which hired no labour. These findings point to the differences and inequalities between livestock and non-livestock farms in the survey area.

Such inequalities have largely been engendered by the agricultural research and extension system. The system, like many other institutional organisations, was inherited from colonialism and seems to widen the gap between small and what are commonly referred to as progressive farmers. Traditionally, the extension system in Kenya has been attentive, responsive and attuned to the needs and interests of a small group of relatively wealthy farmers. In some parts of the country, this orientation has either become more pronounced or basically remained the same.

Role and Status of Animals

The study showed that the average number of livestock per household was 4.0 in Siaya and 2.9 in Kakamega. Many of these were cattle, but sheep and goats were also common. Cattle are kept for a wide variety of purposes ranging from consumption to cash. They are a major source of meat and the only source of milk. There are several cultural beliefs associated with the consumption of goat meat and milk. There is a strong belief in Siaya, for example, that goat meat causes stomach disorders and other related health problems. It is also customary in this part of Kenya for the elderly women to abstain from milk and poultry consumption. On the other hand, goat milk is generally believed to be medicinal.

Cattle are generally perceived as a traditional form of wealth, a status symbol, a source of prestige and a convenient source of liquid capital. Because of their role in the social and economic fabrics of the two communities, cattle are considered a higher status animal than sheep and goats (Table 17.4). They are used to provide draft power (which is a critical factor in timely ploughing); pay bridewealth at the time of marriage; provide cash for household expenses and purchase other goods and services essential for the reproduction of the household. Fomerly, farmers with large herds of cattle were also the ones who could afford multiple wives because polygyny required a substantial number of cattle to make bridewealth payments possible. According to Kayongo-Male and Onyango (1984: 64-65), "... the African man did not only value variety but he also needed many people to work on his land and marrying many wives satisfied this need".

Table 17.4: Status of livestock by district

Status of livestock	Percentage of households					
	Siaya		Kakamega		Both districts	
	n	%	n	%	n	%
Cattle (N=68)						
Most important	34	100.0	34	100.0	68	100.0
Least important	-	-	-	-	-	-
Goats (N=56)						
Most important	2	7.4	12	41.4	14	25.0
Least important	25	92.6	17	58.6	42	75.0
Sheep (N=57)						
Most important	19	70.4	14	46.7	33	57.9
Least important	8	29.6	16	53.3	24	42.1

Source: Compiled by the author.

These data indicate that there is a strong preference for cattle over sheep and goats in these communities. Although 75 per cent of the entire sample reported that goats are the lowest status animals after sheep and cattle, people in Kakamega District had basically the same preference for sheep and goats. But in Siaya District, only 7.4 per cent of the households considered goats as most important. This seeming anti-goat sentiment in Siaya is linked to the cultural taboos.associated with the consumption of goat products. However, it is evident from the data that, compared to small ruminants, cattle are the most preferred and highly valued animals in both communities.

Goats are kept for household consumption as well as for exchange. But they are essentially regarded as ritual animals that are used for ceremonial purposes during births, weddings and funerals. As with other types of livestock, goat production is traditionally a male responsibility. Women tend to be relatively less involved in all the major goat management and production decisions.

Despite the role of goats in the socio-cultural life of the people in the two Districts, goat production is generally considered a low priority enterprise. This disposition is due in part to the strong perception in Siaya District that goats, in particular, and small ruminants in general, are lower status animals which neither generate wealth nor provide real security nor status to the producer. Another major consideration is the feeling that goats are relatively more difficult to manage than cattle in terms of their labour requirements. Unlike cattle and sheep, goats were widely (90.9%) perceived by farmers in Siaya District as being hard to care for (Table 17.5). This strong negative perception about goats reflects a deep-seated belief system in Siaya District of the social role of goats and the consumption of goat meat and milk. No one in Siaya reported that goats are easy to tend. On the contrary, farmers in Kakamega District did not seem to care one way or the other whether they

Table 17.5: Livestock that are easy to tend by district.

Type of livestock that are easy to care for	Percentage of households					
	Siaya		Kakamega		Both district	
	n	%	n	%	n	%
Cattle (N=67)						
Easy	21	61.8	19	57.6	40	59.7
Average	13	38.2	4	12.1	17	25.4
Hard	-	-	10	30.3	10	14.9
Goats (N=65)						
Easy	-	-	6	18.7	6	9.2
Average	3	9.1	20	62.6	23	35.4
Hard	30	90.9	6	18.7	36	55.4
Sheep (N=67)						
Easy	20	60.6	8	23.5	28	41.8
Average	13	39.4	22	64.7	35	52.2
Hard	-	-	4	11.8	4	6.0

Source: Compiled by the author.

had sheep or goats, although many of them also indicated a strong preference for cattle and felt that they are easier to raise and the most valued animals.

These perceptions, predispositions and preferences are built into the overall production strategies employed by the small-scale farmers in the surveyed households. In Siaya District more than in Kakamega District, these preferences are also reflected in the farmers' management priorities. Given the social and economic role of cattle in the community's social structure, a major livestock management priority in the districts is to improve cattle production. Many of the farmers interviewed in Siaya District indicated some willingness to keep a few extra goats on their holdings, but not if it involved further resource investments and major changes in the current production system. This, they felt, they could not afford given their poor resource conditions. Some of them may be willing to make changes in the current production system but most of them are simply unable to do so. Farmers in Kakamega District were not particularly enthusiastic about a new goat enterprise either, but most of them were quite ambivalent about the idea. This ambivalence should be understood against the background of farmers' resource endowment and constraints. A significant majority of the small-holders in the area are faced with severe land and labour shortages.

It was pointed out earlier that livestock production is a secondary enterprise in a crop-dominated production system. It may be noted further that goat production is also a secondary undertaking within the livestock enterprise. There is an inherent crop bias in the entire system and a definite cattle bias within the livestock sector.

Conclusion

These findings have important implications for development efforts aimed at incorporating new livestock enterprises, such as high grade cattle and dual - purpose goats for meat and milk production into the existing production system in western Kenya. If livestock production is considered a low priority enterprise, as is evidenced by the distribution of livestock farms in the survey area and if small ruminants — notably goats — are even more secondary, then it is conceivable that significant increases in overall production can only be warranted by adequate access to labour, land, capital and other resources. Households that are relatively deprived in terms of major productive resources may be unable to increase or improve production by incorporating new enterprises in the future.

Part V

URBAN INFRASTRUCTURE

18 The State and Low-income Urban Housing Production and Consumption in Kenya

Gervase C. Macoloo

Introduction

State intervention in the affairs of its citizens is so widespread that it tends to be taken for granted, especially in the LDCs. It results in observable socio-spatial patterns and, therefore, directly affects development. Because state decisions influence people's lives, it is imperative to understand why state intervention occurs in the first place. However, a clear distinction must be made between what appear to be the reasons and the actual underlying motives for this form of intervention.

The evidence of state intervention in the LDCs may be found in all spheres of economic activity as well as in the regulation of local authorities in terms of policy formulation and the provision of funds for development projects. In the field of housing, the main form of state intervention in the past has been the construction of public sector housing, which has had only limited benefits because of the small quantity able to be constructed. In addition, state housing is often unaffordable to low-income people despite being principally intended for them. The limited quantity available is due to the fact that most governments regard housing provision as an unproductive use of scare resources and, therefore, invest little in that sector. The urban poor have at best been left to fend for themselves and, at worst, are frustrated in their efforts to provide themselves with forms of shelter that does not conform with government regulations. It is only within the last decade that a shift in government intervention in low-income settlements has become evident.

Before discussing the effects of state intervention in the production of low-income housing in Kenya during both the colonial and post-colonial periods, a brief theoretical exposition will be given of the relationship between the state and the economy as well as the determinants of the state intervention.

The Nature of State Intervention in Public Policy

The provision of housing for the urban poor has invariably been associated with the state. The form and role of the state must, therefore, be understood if low-income housing policies are to be explained correctly. The pervasiveness of state action in spatial organisation can be seen clearly in the urban mosaic in former European colonies where state action may be compared over time in changing political circumstances. Since the state forms part of the social formation in which it acts, it is important to account for the specific forms of state action because this action often seems contradictory. The contradictions may exist between the state and the economy, and even between sections of the state. For reasons of space, state theories will not be delved into, but the key arguments about the nature of state intervention will be summarised before illustrating how state action has shaped low-income housing production and consumption in Kenya from the colonial period to the present.

There is little, if any, consensus on the nature of state functions because the responsibilities that the state is assumed to bear are clearly linked with forms of state intervention, which in turn reflect these assumed responsibilities. The major theoretical perspectives of the state are pluralism, instrumentalism and structuralism. According to pluralism (or representational democracy), the state consists of a set of institutions which are highly autonomous from other societal institutions. The state is, therefore, portrayed as neutral in its functions and so cannot represent specific class interests. This position of superiority and externality, it is argued, is what enables the state to mediate in all conflicts in society in the general interest. The crucial implication of such an omnipotent view of the state is that the state apparatus is contested by sections of society that desire dominance. This perspective cannot adequately explain why irregular settlements have been bulldozed by the state though they meet basic needs of the poor majority.

The instrumentalist perspective maintains that the state is an instrument through which one group dominate the others. It is able to do this because state functionaries such as the administration, judiciary, police and military are largely drawn from backgrounds similar to those of the elite groups. Whether the domination is by the elites over the masses or the ruling class over the proletariat, it is the interest of capital which becomes paramount without relegating the state to a secondary position vis-a-vis economic interests. This view of the state cannot provide a theoretical base for analysing the welfare state in which the interests of the less powerful are also incorporated into the decision-making process.

According to structuralism, associated mainly with the work of Poulantzas (1973 and 1975), the state is viewed as responding to conditions of class conflict with the major aim of maintaining and reproducing the capitalist structure. Structuralism takes classes rather than individuals as the *scientific categories of political analysis* and sees the state as neither an instrument of class domination nor a centre of power independent from classes. The state is a representation of the balance of class forces and is relatively autonomous from any one class. However, it necessarily functions in the long-term interests of monopoly capital.

None of these perspectives by themselves can explain the complex nature of the intervention in low-income housing in the LDCs. A satisfactory explanation of such interventions must embrace different elements from these various perspectives.

The reasons as to why the state responds in specific ways to certain housing problems and the nature of state-economy relationship lie at the heart of evaluating aided self-help housing policies. The different strands of the Marxist theory of the state take issue with the liberal view that the state represents the general interest of the whole society and does not merely serve the interest of a single faction within a particular social formation.

The ambivalent or contradictory nature of the state becomes clear when we break down the role of the state into its specific categories. The functions of the state may include the provision of goods for collective consumption, regulation of class relations and sustaining conditions for capital accumulation. Lojkine (1976), for example, related state existence to the provision of infrastructure, a role which he admitted was antagonistic to capitalist profitability, while at the same time necessary for the overall reproduction of the capitalist social formation. This automatically puts the state in a contradictory position as it attempts to juggle its social character (obligation) and the need to sustain capital accumulation. Therefore, the categorisation of state roles focuses primarily on appearance (after they have occurred), what the state appears to be doing rather than the internal logic that propels it to do what it does at a specific point in time. For these reasons, Marxist approaches reject both the neutrality of the state and pluralism as merely ideological because they do not reflect the fact that there are crucial structural factors which circumscribe the freedom of the state. The identification of these structural limitations which curtail state action involves an examination of the state-economy relations.

Within the Marxist tradition, there are various strands in the conceptualisation of the relationship between the state and the economy. One such strand emphasises the influence of the economy on state action and borders on economic determinism and, therefore, fails to explain different forms of state action, some of which are clearly in opposition to the interest of the capitalist class. A reaction to this *mechanistic economism* resulted in the formulation of the concept of the *relative autonomy* of the state. This concept emphasises the political activities of the state but without fully incorporating the limitations that may be imposed on the state by the way in which it is related to the process of capital accumulation.

Day (1978) argued that the relationship between the state and the economy must be viewed in a dialectical way as the state is simultaneously in unity and conflict with the economy and it is the interpenetration of the state and the economy which produces changes in both instances and informs policy formulation. In such a relationship, the state may intervene to alter the operations of the economy by, for example, raising or lowering the taxes on profits. Similarly, the economy may also interfere with the operations of the state, as is normally the case during economic crises when the state tends to cut down its expenditure on goods of collective consumption due to the fall in its revenue from the various sectors of the economy.

This power of the economy to shape state policy has also been discussed by

Harms (1982) who illustrated how the capitalist state has historically tended to propound a housing policy of self-help during crises in capitalist accumulation. In such situations, the state may act to mitigate the dysfunctions of capitalism and it is perhaps for this reason that Saunders (1980) concluded that state intervention is reformist and, given that it does not change capitalist ownership relations, it works in the long-term interests of monopoly capital. Peattie (1979: 1019) has also come to a similar conclusion concerning state intervention in low-income housing in the LDCs and rejected the simplistic view of the state as an instrument for exploiting the poor when she posed the question:

> Why do the governments of capitalist developing countries characteristically behave, in the housing field, in a way which seems to counter the interests of the capitalist employing class?

She posited that the answer to the question may lie in examining the multiplicity and institutional importance of organised interests relating to housing. She concluded that an overall class analysis of state intervention in low-income housing may not be very helpful. Analyses should be based on the specific constellations of power and conflict in temporally and spatially specific contexts. The zig-zag nature of low-income housing policies in the LDCs may, therefore, be understood within the dialectical relationship between the state and the economy. This line of argument is maintained in the analysis of state intervention in the production of low-income housing in Kenya and evaluating who the winners and the losers are.

Colonial State Intervention in Urban Housing

Outside the coastal region the majority of the present urban centres in Kenya were colonial creations (Obudho, 1983 and 1994: 198 - 212). Kenya was a settler colony and so the imprint of colonial state intervention in all economic spheres is still visible today due to the intensity and ruthlessness with which it was conducted. It involved the extraction of labour from pre-capitalist social formations through the proletarianisation of peasants and other strategies which accelerated rural-to-urban migration to the foreign-owned commercial farms.

Urban wages were low and this began to manifest itself in housing problems thus necessitating state intervention. This was thought to be necessary as illegal settlements had began to appear on the periphery of urban centres and, if continued unchecked, it was feared that the colonial government would lose control of the African labour force and law and order would break down. The need for social control grew even stronger with the emergence of urban-based African nationalist politics and the illegal settlements were viewed as the breeding ground for this discontent. The nature of this early intervention was to create racially-segregated residential neighbourhoods which the colonial state argued were necessary as they would ensure that different races were able to practise their cultures without interference from other groups.

The consequence of this intervention was the creation of what were known as the African locations which in effect were one-roomed dormitories rented out to Africans at sub-economic rents since they could not afford to purchase them. Up to the 1920s, there were no official African housing programmes underway, and the colonial government was extremely intolerant of the so-called temporary houses constructed by Africans themselves.

In Mombasa, the situation was radically different. By 1927, the village layout schemes, consisting of temporary Swahili houses, had been officially accepted as the only way to provide accommodation for low-income people in that urban centre. The restrictive policies implemented in other urban centres did not deter the urban poor from providing accommodation for themselves. Hence, the construction of illegal settlements continued. One of the biggest of these settlements was the Pangani Native Village.

The area around modern-day Pumwani in Nairobi, was chosen as the first native location and in 1919, some 400 acres of land was acquired in the area for that purpose. Communal ablutions and latrines were built and the 343 plots of 1500 square feet each were laid out. This may be considered as Kenya's first site-and-service scheme. The allotees were to pay a monthly charge to the Nairobi City Council (NCC) for ground rent, services and rates. From 1922, buildings began going up. A house was to occupy less than 50 per cent of the plot area and each house was expected to house a maximum of 15 people. However, as time went on, the house owners added extra rooms to cover the entire plot and even extended the houses on to the spaces left between them. These developments made Pumwani take on the appearance of an unplanned settlement where houses were rented by the room, and it attracted prostitutes. At independence Pumwani had become an eye-sore whose future hung on a tight rope.

In colonial Kenya, therefore, low-income housing become synonymous with African housing (Obudho and Oduwo, 1989:17 - 30). Such segregation, tempered with the need for social and political control, has been reported in other former settler colonies such as Zimbabwe (Wekwete, 1988). The first public housing in Nairobi for Africans was constructed next to Pumwani and called the Pumwani Extension. This housing estate, known as Shauri Moyo by the Africans, was constructed between 1937 and 1938 to receive the inhabitants of Pangani when the settlement was demolished. The scheme consisted of 175 four- and six-roomed houses costing Kshs 2,400 and Kshs 3,300, respectively, and consisted of a central corridor giving access to individual rooms. The total cost was K£46,000.00. Pangani was subsequently destroyed in 1938 and the house owners paid an average of

Kshs 1,000.00. They were expected to relocate to Shauri Moyo. But they could not afford to buy the new houses so the houses were rented out by the NCC on a room-by room-basis.

The imperative to provide housing for the African population as a vehicle for social control soon began to contradict the economic interests of the European urban dwellers who were the rate payers. It was argued that because rental units had to be heavily subsidised they could not be reproduced on a large scale. This meant taxing the non-African property owners to provide houses for Africans. This was an unwelcome burden to most non-Africans, making it a source of conflict between the state and the landed class.

Apart from the need for control, another reason that shaped the nature of later colonial state intervention in low-income housing could be described as *utilitarian*. Studies in other parts of the colonies illustrated the need for a stabilised work-force if capital accumulation was to proceed at an efficient rate. The need for a stable work-force implied improved salaries and housing conditions for the African population. It also implied the removal of all barriers to urban property ownership by Africans. Although these went against the grain of the colonial policy about African urbanisation, some rudimentary policy of self-help was devised in which serviced plots were provided in the African locations in Nairobi. In Thika, some tenant-purchase houses were constructed in 1951 hoping the African work-force would purchase them. An evaluation study of this scheme in 1972 revealed three discouraging tendencies. First, the houses had been rented out on a room-by-room basis thus tripling the number of people in the scheme. Secondly, about 84 per cent of the houses were not occupied by the original purchasers. Finally, many landlords were collecting total rents which were approximately double the amount of their monthly repayments to the council.

The *East African Royal Commission Report* (published in 1956) also strongly argued for the removal of any barriers which could prevent Africans from full participation in urban life and called for the lowering of building restrictions on Africans. In Mombasa, this idea of Africans providing housing for themselves was tried in the Changamwe Repooling Scheme in the late 1950s. The scheme covered 200 acres of freehold land on the West Mainland acquired by the GOK. The irregularly-shaped individual plots were grouped together and subdivided into 1,000 plots, each measuring one-eighth of an acre. These plots were allocated to people with tenure security and they were urged to construct high-quality Swahili houses. By 1974, more than one-third of the plots were still undeveloped as people preferred to build cheaper houses in the unplanned adjacent areas such as Majengo where they could recoup their investment faster.

It has been illustrated how the reasons for colonial state intervention changed with circumstances and created a residential mosaic in urban Kenya, particularly in

Nairobi, where people were segregated by race. In the next section, we examine how the two main reasons for colonial state intervention—the need for social and political control and economic expediency—differ from the motives of post-colonial state intervention.

Post-colonial State Intervention

Most scholars who have studied post-colonial urban development policies in Africa have reported greater continuity than change from colonial times. In Kenya, as in other former settler colonies, urbanisation restrictions imposed on the local population were lifted at independence. This accelerated the rate of urban growth. At independence in 1963, Kenya was already suffering from an urban housing crisis, for example, 70 per cent of African urban households occupied single rooms, 53 per cent of which were occupied by three or more persons, thereby contributing to overcrowding.

The newly independent GOK had to intervene in the urban housing scene for reasons that bear an uncanny resemblance to their colonial predecessors. Similar to the way in which the colonial government had imposed the restrictions to announce the dawn of its authority and the breakdown of traditional authority, the new GOK had to show that there was a break with the past. As part of this legitimisation process, the provision of low-income housing for Africans became the main urban housing policy (Kenya, 1966). This was intended to show that the new GOK was committed to its electoral promises.

The construction of single-room dormitories (initiated by the colonial government) was now replaced by two-bedroomed flats coupled with the demolition of irregular settlements and attempts to resettle their inhabitants elsewhere. Two major phases of the post-colonial state intervention in low-income housing have been isolated. The first was the short-lived provision of conventional housing, which was immediately followed by demolition or slums settlements, aimed at controlling and containing urban growth (Obudho and Oduwo, 1989:17 - 30). The second phase consisted of an apparent change of heart, with the government espousing supportive policies to provide an enabling environment for aided self-help.

As a continuation of the colonial hangover, the squatter settlements were viewed for the whole of the 1960s by the independent GOK as an eye-sore and a hide-out for criminals. Nobody ever considered them as the product of rational decision-making on the part of the urban poor to provide their own housing in the face of a hostile, elitist and uninformed planners and decision-makers.

Earlier academic works were also hostile and apathetic to squatters, portraying them as urban misfits and their settlements as fungal or cancerous outgrowths blemishing urban centres. The squatter settlements were often portrayed as powder-kegs for revolution, just waiting for a political demagogue to incite their population.

Such latent threats to state security were the last thing the newly independent state such as Kenya would entertain. These views of the squatter resulted in the indiscriminate demolitions of their settlement, either without providing alternative shelter or coupled with half-hearted attempts to resettle their inhabitants in conventional housing.

At independence, Kenya had a number of squatter settlements in different locations in Nairobi, notably around the present day Kirinyaga Road in Mathare Valley, the Pumwani Road area and the Kaburini area, among others (Obudho and Oduwo, 1989:17 - 30). In an "Operation-Clean-Up" on September 3 1969, 20 NCC policemen burned down 200 shanty buildings in the Kirinyaga Road area in three hours. In November 1970, 49 squatter settlements in Nairobi, containing 700 dwelling units, estimated to be worth over US $300,000.00 and accommodating some 40,000 inhabitants, were razed to the ground and their inhabitants relocated in the Mji wa Huruma area, far from their employment activities.

Other examples of clearance/resettlement are Kariobangi and Pumwani schemes in Nairobi. The Kariobangi scheme, begun in 1963/1964, was intended to provide accommodation for the residents of the Pumwani Road area whose houses were being demolished because they were too close to the Central Business District (CBD). Plots numbering 732 and measuring 40 ft by 45 ft and served with toilets and water lines were laid out. The allottees were allowed to use temporary materials so long as these were phased out within ten years. Furthermore, lodgers could be taken in, but no overcrowding was permitted. Because of the distance from the CBD, many allotees were unwilling to accept this form of resettlement. By 1970, only 15 per cent of the house owners were the original allotees living on their plots, while another 20 per cent were original allotees living elsewhere in Nairobi (some even squatting in irregular settlements) and collecting rents. By any standards, therefore, this scheme was a failure, just like the Pumwani one of the late 1960s.

Some conventional rental houses were also constructed simultaneously with the implementation of the demolition/resettlement policy. However, it was soon realised that even with massive subsidies, majority of the urban poor could still not afford the rents for these flats. They were, therefore, increasingly occupied by middle-income salaried workers. The poor were, implicitly left to fend for themselves in the private housing market. Furthermore, because the rents paid here were sub-economic, the state could not replicate these projects on a massive scale. Therefore, their impact on the urban housing scene was negligible.

By intervening negatively through indiscriminate demolitions, the independence GOK was again attempting to re-assert its authority by preventing the urban poor from taking advantage of the urban development process. President Jomo Kenyatta referred to these settlements as hide-outs for criminals, prostitutes and loafers and threatened to repatriate their inhabitants to the rural areas. Such an attitude was reminiscent of the implementation of the Vagrancy Act by the colonial GOK.

The result of the demolitions era was a drastic reduction in housing stock and the spiralling rents in the squatter settlements that somehow escaped demolition. New policies were needed urgently to deal with this phenomenon as the squatter settlements housed between 30 per cent and 60 per cent of people in most urban centres in the LDCs. Because of the failure of these policies, the new GOK had to intervene once more in the opposite direction. Research findings on squatter settlements in Latin American urban centres concluded that the earlier repressive policies amounted to the art of blaming the victim. The settlement began to be viewed as a solution rather than a problem. These views, despite their temporal and spatial specificity, influenced World Bank thinking on the solution to low-income housing problems. Through its twin package of the- sites- and service and settlement upgrading projects, the World Bank (1975a) has encouraged the LDCs to adopt this *progressive development* policy on low-income housing. This policy was expected to harness the latent resources of the urban poor and pool it with state initiatives in the provision of low-income shelter. The commitment of the World Bank to this policy during the 1970s is evidenced by the huge loans it advanced to LDCs during the period.

Site-and-service and settlement-upgrading schemes, therefore, became the chief policy for providing the urban poor with shelter. This form of intervention resulted from the realisation of the failure of market forces to serve the interests of the urban poor. The state realised that in order to maintain political and social stability, it could not ignore the urban poor who were the majority. These projects have been implemented in the major urban centres in Kenya with funding from donor agencies, such as the World Bank and United States Aid for International Development (USAID).

Although there are various interest groups who have benefited from the implementation of this policy, it cannot be denied that some benefits have acrued to certain sections of the urban poor. Site-and-service schemes, for example, have been lauded for creating additional housing units to the total urban stock. Upgrading, on the other hand, improved the access of the urban poor to basic infrastructural services such as water, sewerage, electricity and other social amenities. However, most evaluation studies of these projects have cast serious doubts as to whether their benefits outweigh their costs and whether the target group actually benefited.

Various interest groups benefit from different aspects of this form of state intervention. The industrial and commercial interests, for example, may benefit from the fact that a labour force which is housed cheaply is less likely to demand higher wages. The building materials sector, on the other hand, benefits by selling materials to self-help builders once this housing sector is incorporated into the mainstream of urban planning. These fractions of capital have interests which often conflict, and when that happens the group that is capable of exerting the greatest force on the state apparatus will claim temporary victory. These contradictions and irreconcilable competition among interested groups may offer a clue to the explanation of the somewhat seemingly contradictory ways in which the state acts in relation to the provision for low-income urban residents. It may explain why sometimes low-income housing is demolished and, at other times, actively supported.

From the middle of the1980s, emphasis once again changed from ensuring the availability of an *enabling environment* back to the market mechanism. Linn (1983) recommended that the governments in the LDCs improve their land and housing markets so as to reduce any massive subsidies that have been responsible for the lack of replication of these projects. He called for the creation of incentives for the private sector to make it view low-income housing production as a viable economic activity in which it can invest. This view has been adopted by the World Bank and other donor agencies and, together with the Structural Adjustment Programmes (SAPs), they now argue that the state should do no more than ensuring that market forces work. In other words, liberalisation policies are being applied even to the low-income housing market. Governments are encouraged to service plots and then sell them off at market prices to developers. It is not clear how this policy is expected to work this time around when its failure to work was one of the reasons for providing an *enabling environment* through site-and-service and settlement-upgrading projects.

Conclusion

Many factors have been cited as being responsible for the poor practical performance of low-income housing projects in Kenya. Such factors are the low-incomes of the target groups, stringent standards of building by-laws, allocation biases and corruption and social and psychological attachment to rural land. However, in Kenya, like in other LDCs, there has been limited work on the structural problems of these forms of state intervention in low-income housing. Since these projects are initiated by the state, the underlying factors behind their initiation would be a pointer towards understanding whom they benefit.

The state in a peripheral capitalist formation such as Kenya must be viewed as an expression of and a tool in the capitalist mode of production that must act in the long-term interests of the most powerful groups or classes. The state, therefore, has a fundamental role to directly support capital accumulation and maintain social and political stability. In other words, it promotes the reproduction of its particular mode of production—in this particular case, peripheral capitalism. In line with such aims, the state may regularise and incorporate low-income settlements into mainstream urban planning, not because it respects their long-term interests but because it would intensify the penetration of the capitalist mode of production.

This is what has happened in most low-income projects in Kenya and has resulted in their failure to achieve their stated objectives. In order to maintain social stability, some piecemeal populist ideology (such as the ethos of private ownership of property) may be propounded to fragment the collective consciousness of the urban poor.

Analysis of the nature of the state in the LDCs has to incorporate certain factors which may be specific to the constitutions of these states; for example, state action in these countries may be significantly influenced by external factors such as the multinational corporations or donor states. For this reason, their interventions in low-income settlements must be interpreted both in the light of internally generated factors as well as exogenous ones.

This chapter has illustrated the changing nature of state intervention in low-income housing in Kenya. Policies, from the colonial period to date have been shown to oscillate from benign neglect and/or demolition to enabling strategies in order to smoothen the operations of the housing market. Some of these projects are not in themselves problematic. The problem (particularly in the post-colonial period) is that they are introduced into political and socio-economic systems whose operations are not conducive to the successful implementation. Consequently, they end up being mere palliatives rather than solutions to urban housing problems. The zig-zag nature of state policy (colonial as well as post-colonial) on low-income housing may be explained in terms of the irreconcilable competition between different interest groups. Given the dialectical relationship between the state and the economy, policy is a tool used by the political elite to serve their interests. It is likely that the state will continue to influence low-income housing production and consumption in Kenya. Therefore, to be able to assess successes and failures of the future policies, a thorough understanding of what informs state action in space and time will continue to be particularly important.

19 Trends in Urban Housing Strategy for Kenya into the Next Century

P.M. Syagga

Introduction

Despite notable progress made by the GOK in most fields of national development programmes since independence, housing seems to remain an elusive subject. From the *First National Development Plan of 1964-1970* to the *Sixth National Development Plan of 1989-1993*, it has been a primary objective of the GOK to provide decent housing for every family. The adequacy of this housing would be determined not only by the shelter and contiguous facilities it provides but also by the entire system of supportive and facilitative infrastructure and services, including accessibility to the work place and social facilities and amenities (Kenya, 1989). As a first step in this direction, the GOK concentrated on the development and strengthening of housing institutions. Thus, in 1964, the support of the United Nations was enlisted in conducting a study of housing needs out of which a national housing policy was issued.

The *Sessional Paper No. 5* of 1966/67 on housing policy, which still remains the blueprint of Kenya's housing policy, declared *inter alia*, that the GOK's objective was to "provide essential housing and a healthy environment to the urban dweller at the lowest possible cost to the occupants" (Kenya, 1966:8). The emphasis has therefore, from the beginning, been on urban housing, leaving rural housing to the initiatives of individual families.

Housing Institutions and Policy

Within the policy framework, the GOK created the National Housing Corporation (NHC) to replace the Central Housing Board which was established in 1953 during the colonial period. The NHC was to be the GOK's chief agency through which public funds intended for low-cost housing would be channelled to local authorities, housing cooperatives and other housing development organisations. Indeed, the NHC was expected to maintain direct communication with local authorities and close links with Non-Governmental Organisations (NGOs), with a view to fostering and encouraging housing development in the private sector of the economy. From its inception, the NHC embarked on building and managing estates and assisting local authorities by providing funds and technical assistance in housing programmes throughout the country. The overall result is that between 1965 and 1987, only a dismal 41,852 housing units in the form of mortgage, tenant purchase, rental and site-and-service schemes had been provided in the urban areas by the NHC, representing 64 per cent of the total formal sector housing. Although the NHC was also expected to finance housing estates in support of industry or agriculture and to encourage housing development in the private sector, there is very little evidence to show any positive actions in this direction. The NHC assumed more the role of developer rather than facilitator of housing development, with the result that many resources remained untapped outside the public sector.

The second development aspect of the housing policy was the establishment of the Rent Tribunal under the *Rent Restriction Act* to deal with complaints by tenants and landlords. Since the inception of the Act in 1960, complaints had been dealt with by courts, but under the policy, only appeals were heard by courts from decisions of the tribunals. The purpose of the Act was and still remains keeping rent levels under review for residential properties so as to prevent exploitation of citizens through unjustified evictions and extortionate charges. However, rent restriction only applied to houses with low rents, revised to cover houses let for Kshs. 2,500.00 per month or less as of January 1, 1981. Any houses let above this figure fall outside the tribunals' control and, in any case, there is no system to ensure that all houses coming into the market are assessed for a standard rent. Housing rents in Kenya are freely determined by market forces of supply and demand both in the formal and informal sectors as well as in the high and low-income groups except for local authority rental estates which charge economic rents only. In this respect, Kenya would appear to differ from countries like Ghana where the effects of rent control have limited rents to a fraction of the market, with the result that housing has been left to deteriorate, landlords demand advances and overcrowding intensifying.

Very few cases ever appear before rent tribunals for assessment and in many cases it is a seller's market on account of the acute housing shortage in the urban area so that landlords use threats to those who are reluctant to pay increased rents. It can, therefore, be safely concluded in the Kenyan context that rent control has not helped the urban poor and that it has also not played any role in holding back housing construction in any given urban centre. Local authority rental estates as well as public service housing (which are heavily subsidised) are in any case exempted from rent regulations and they cater mainly for higher income groups.

The third aspect of the housing policy aimed at institutional development was the establishment of the Housing Research and Development Unit (HRDU) at the University of Nairobi. It was the conviction of the GOK that to get the greatest number of houses built at the cheapest cost possible, research into building techniques and construction costs was of utmost importance. The unit was to examine the comparative cost of materials and skills and how these can be lowered. Since its establishment in 1967, the HRDU has carried out fundamental research work on the use of low-cost technologies and building materials. The HRDU's research on economic, social and technical aspects of housing and human settlements has led to the publication of a wide range of reports from the overall policy issues to specific technical studies. The research findings have, however, not been widely applied and have, therefore, not helped in reducing the cost of housing to the urban dweller largely because of uncompromising building regulations. Since the regulations are based on materials specifications rather than performance specifications, any new materials, however satisfactory, are frowned upon by the urban authorities.

Therefore, although Kenya has had a well set-out housing policy since the mid-1960s, the plight of the urban low-income population has remained acute. While it was estimated that only 7,600 housing units per year were required in the urban areas between 1962 and 1970, but by 1986 a total of 56,000 housing units were needed in the urban areas. Kenya's urban population has been growing at an unprecedented annual rate of 7.3 per cent so that the number of urban centres with more than 2,000 people rose from 17 to 90 between 1948 and 1979. Such centres are now more than to 125. At current rates the urban population estimated at 15 per cent of the total population or 3 million people in 1984 will reach 10 million or 25 per cent of the total national population by the year 2000. While this is one major cause of the inability to provide adequate housing to the urban population, the other causes of the slow rate of housing production have to do with institutional arrangements in the housing delivery mechanisms through successive national development plans.

Institutional Arrangements for Housing Development

When the *First National Development Plan of 1966-70* was prepared, the expanded housing programme was given high priority with special emphasis on production of low-cost housing and home-ownership schemes to reach as many beneficiaries as possible (Kenya, 1966). A separate Ministry of Housing was created in 1966 to determine future housing policy and the NHC was established in 1967 as the executive arm of the government through which all public funds for housing would be channelled. The HFCK was established in 1967 to avail funds for middle and high income mortgage housing whether developed by NHC or other private developers. Housing Research Development Unit (HRDU) was established in 1967 to carry out research on social, economic and technical aspects of housing.

The plan did not make any mention of what role the private sector would play, while the programme gave only an indication of the funds that would come from the GOK. Emphasis was also laid on delivering completed housing units in conventional building materials. The main actor in the process was the NHC which became responsible for carrying out assessments of housing needs and demand in all local authorities (except for Nairobi and Mombasa) and assisting the same in implementing their housing programmes. Housing was at this time seen in Kenya as a welfare good and this could only be provided by the public sector. The capacity of NHC to carry out these tasks was not well-appraised while the Ministry of Housing itself was still at its infancy. While the NHC could help assess the requirements and make the funds available, it was not able to provide technical assistance to all urban centres. The NHC did not have control over urban land nor did the Ministry of Housing. Thus, the process of setting land aside for housing, surveying the land, and preparing the titles belonged to the GOK departments of which had other priorities. Off-site infrastructure was also not readily available, and so land ready for housing development was not always available.

When a review was done at the end of the plan period, it was noted that the public sector produced only 9,500 units against an estimated requirements of 7,600 units per year over the plan period. The majority of the units were, however, rental and tenant purchase housing for the middle and high income groups. The rate of house production was only 25 per cent of the requirements, leaving 75 per cent unsatisfied.

The next plan period, 1970-74, defined the minimum acceptable dwelling unit for low-income groups as having a floor area of 38.5 m^2, consisting of two rooms with separate kitchen, toilet and showers and prescribed that the occupancy rate for such a unit should be 5 occupants. It was also required that the NHC spends 85 per cent of the public funds on developing low-cost housing while HFCK would fund high-cost units. While the low cost housing units were estimated to cost less than Kshs 24,000.00, the actual average cost per unit was Kshs. 36,640.00, which was 53 per cent more than the planned target cost. At the end of the period, only 25,000 units, or 50 per cent of the total planned 50,000 units were built. It was apparent that conventional housing could not be afforded by low-income groups.

While supply was falling behind demand, the GOK continued a policy of slum clearance particularly in Nairobi where, for instance in 1970, 39,000 people were rendered homeless when 6,733 dwellings were demolished. At the same time there was emerging a body of literature on African urban development decrying the failure of African governments to deal adequately with the provision of services, especially housing:

> Accoding to the World Bank, "The malaise of urban centres in the developing countries is only too evident in the squalor of rapidly growing slums and unauthorised settlements, the deterioration in many public services, the extreme shortage of housing". (World Bank, 1972:14)

The World Bank drew attention to the escalating number of urban poor in urban centres of LDCs through studies of proliferating squatter settlements and "informal sector" employment which gradually helped to legitimise their urban poor's attempts to improve their own dwellings through self-help housing and employment initiatives. International aid agencies became more willing to lend funds for low-income housing on a large scale during the 1970s. The governments of LDCs saw it as an effective way of mobilising and managing scarce local resources and large international agencies realised the potential for a rapid increase of housing stock, employment creation and promotion of local building industries.

Thus, in the *Third National Development Plan, 1974-78*, for various reasons, but mainly due to careful analysis, political sensitivity and interventions of international aid agencies such as the World Bank, the unplanned or squatter settlements came to be accepted in Kenya. It was stated that they would not be demolished without providing alternative accommodation. At this time, the urban housing need had been assessed at 160,000 units including 50,000 units due to an accumulated shortfall from previous years. During this plan period, the GOK made a marked shift from providing completed buildings to providing incremental Sites-and-Services (SS) and Settlement-Upgrading (SU) programmes. Out of a total planned 57,000 units, only 12,000 would be conventional middle-and high-income units while the remaining 45,000 (or 79%) would be serviced plots. It was during this plan period that the GOK entered into an agreement with the World Bank to develop 6,000 serviced plots in Dandora in Nairobi to accommodate 36,000 people. From then, SS and SU programmes became accepted modes of housing delivery in Kenya.

The third plan can, therefore, be considered a turning point in the history of housing development in Kenya. First, the GOK recognised the futility of demolishing squatter settlements. In many cases, the settlements came up almost as soon as the old ones were demolished since the displaced persons did not abandon urban life but merely moved to new sites. Squatter settlements grew at an alarming rate in Nairobi — from 500 units in 1952 to 22,000 units in 1972, rising to 110,000 units in 1979, representing approximately 40 per cent of Nairobi's population (Obudho and Oduwo, 1989: 17 - 30, and Kayungo – Male, 1988: 133 - 144). Most of these units were constructed in whatever materials that came in hand while, the areas were devoid of infrastructural services and community facilities. However, their growth was a sign that the housing demand for low-income groups was not being met within the laid down housing policy.

The second aspect of the shift in policy was the acceptance of the development of SS and SU schemes utilising a considerable amount of self-help thereby reducing the financial burden of the GOK. Thirdly there was the allowing of the Nairobi City Council (NCC) and other councils to enter into direct negotiations with the World Bank as the implementing agency with GOK as guarantor to the loans, rather than using the NHC. Within the Council itself, a separate department was established outside existing departments to implement the project. Approval procedures for the allottees were made easier and the project implementation employed community development personnel to work with residents at neighbourhood level giving support and advice on self-help construction and access to bureaucracy.

The fourth shift was that the World Bank-funded projects would be on the basis of full-cost recovery with little or no subsidy. Thus, those who fell below the 35th income percentile were excluded from target beneficiaries because they could not afford to pay back the loan and at the same time construct additional rooms. In effect, the programme completely excluded those most needy at the bottom of the income scale. This trend was continued in other urban areas in the country.

At the end of the plan period, a total of 11,566 units were provided in the public sector including 3,000 serviced plots. The housing situation, however, continued to deteriorate as this number represented only 7.2 per cent of the total requirements. Acute shortage, overcrowding and unauthorised construction devoid of essential services continued to characterise the urban housing scene in Kenya so that by 1983, it was estimated that 35 per cent of all urban households lived in squatter and slum environments. It was recognised that only 30 per cent of urban households had sufficient income to afford minimum cost conventional housing. Besides the lack of purchasing power, it was also recognised that local authorities had a weak financial base and could, therefore, not provide the required infrastructural services and amenities at the expected rate.

By the 1978-83 plan period, urban housing needs had risen from 160,000 to 290,000 units. The shortfall had accumulated from 50,000 units in 1974 to 140,000 units in 1978. It was further emphasised that 90 per cent of all urban development funds allocated for housing through NHC should be used to finance site-and-service projects and other forms of low-income housing. This was because, in the previous periods, NHC spent only 30-40 per cent of its total urban housing resources on low-income groups so that most of the funds went to high and middle income housing. From the survey of the national development programmes, it is evident that the provision of housing in urban areas has lagged behind requirements as shown in Table 19.1.

For the period 1976 to 1982, the number of dwellings produced represented only 20.5 per cent of the new households, not taking into account the accumulated deficit. It is also evident that most of the housing (75 %) was produced by the public sector while the private sector contributed only 25 per cent. The public sector participates in housing development, not only in actual housing production, but more so in the provision of land and related infrastructure and services. The public sector may also participate through regulations such as zoning, building standards and rent control as well as by financial and monetary policies.

In terms of direct housing production, Table 19.1 shows that NHC was responsible for 64 per cent of the total units produced, while housing by the central GOK accounted for 4.2 per cent and other public institutions (such local authorities and parastatal organisations) accounted for another 6.4 per cent. Housing production using public funds has been carried out with subsidies both below market interest rates on loans and with low or no charges for land provision of infrastructure and services have also been heavily subsidised. A number of observations may be further made with respect to the publicly funded housing in Kenya.

Table 19.1: Recorded production of dwellings by public and private sectors compared with the formation of new urban centres

	1976	1977	1978	1979	1980	1981	1982	% Total 1976-1982
Conventional housing by NHC	317	916	1,544	4,085	3,527	2,755	2,938	36.0
Site & service by NHC	1,128	355	1,077	2,389	2,454	2,719	2,550	28.4
Government housing	254	106	359	156	482	471	49	4.2
Other public sector	1,068	193	257	221	481	206	443	6.4
Private sector	791	742	835	2,716	2,065	1,918	2,083	25.0
Total	3,558	2,312	4,072	9,567	9,009	8,069	8,063	100.0
Estimated no. of new households	23,700	25,600	27,600	29,800	32,100	34,700	37,000	
New dwellings units as % of new households	15.0	9.0	14.7	32.1	28.0	23.3	21.5	21.2

Source: Kenya (1989: 16)

Most of the houses developed for low-income groups become private housing after allocation and thereafter operate at the prevailing market prices. It has been shown for instance that 66 per cent of urban housing was on rental market and that only 20 per cent was owner-occupied. This is an indication that housing developed either by private or public funds is put on the rental market, hence that housing is an investment good that may require no subsidy. Even the site-and-service projects and settlement upgrading schemes which are meant for low-income groups and intended for ownership have in effect become rental schemes. A settlement-upgrading project, funded by the World Bank in Migosi, Kisumu (Kenya's third largest town), had 82 per cent of the residents as tenants and only 18 per cent as owner-occupiers. In Dandora, which was the first World Bank project, and in Umoja, which was sponsored by USAID, about 75 per cent of the households are renters. Thus, most of the publicly sponsored housing ends up being rented out for profit.

In the private sector, developers are subjected to a complex web of financial institutions which charge market rates of interest. They are also governed by strict building codes and lengthy bureaucratic planning approvals. Serviced land may not be available to private developers and the process of land acquisition and registration is cumbersome. It is, therefore, not surprising that the formal private sector market contributes so little to housing development. It is possible that with greater encouragement, the private sector can provide more housing.

Whether publicly supported or privately developed, rental housing in the urban areas of Kenya is occupied by the higher-income groups who can afford the going market rents. The majority of the low-income groups cannot afford market rates in Nairobi, for instance, and will, therefore, be adversely affected by any withdrawal of subsidies. They are mainly found in the informal sector housing operating outside building code regulations. In these areas, the infrastructure and the building structures are poor but rents are freely determined and affordable by the majority. Yet these places, where the majority of the population live, are regarded as eyesores by the political elite, professionals and administrators because they do not generally conform to subdivision standards or density zoning; do not receive minimum adequate infrastructure services as stipulated in by-laws and cannot pass conventional building construction tests such as durability, strength, weather soundness and sound insulation, among others.

These settlements need to be recognised and efforts made to provide services that directly relate to the health of the occupants. There is a further need to accept alternative building materials and construction technology that would reduce building costs so as to make proposed units affordable by a larger proportion of urban dwellers. From work carried out at the University of Nairobi (Syagga, 1987 and 1988), it is possible to reduce the cost of a house by 50 per cent through the use of alternative building materials for walls and roofs and use of stabilised rammed murrams for ground floors and compacted hard core in a mesh cage formwork for foundations. Other design aspects could include considering minimum functional room spaces as well as circulation space, reduction or omission of finishes or allowing their future provision and reduction in standard of infrastructure as well as plot sizes.

In later projects, attempts have been made towards settlement upgrading side by side with site-and-service schemes. While in Nairobi most of the projects involved these schemes, other urban centres, notably Mombasa and Kisumu, included settlement upgrading. Like in Nairobi, the urban centres of Mombasa and Kisumu created separate units to deal with this new form of urban development. However, a major drawback in the policy implementation was that the provision of infrastructure and services was based on conventional municipal standards. Other soft costs, including professional fees to consultants, increased the cost further. These only helped to reduce affordability for the target groups. The problem of land acquisition, particularly in Mombasa and Kisumu, was not sorted out and caused delay in implementation of the two projects in late 1970s.

During the *Fifth National Development Plan 1984-1988*, the GOK recognised the adverse effect of the unrealistically high standards advocated in the building code. The plan therefore recommended the adoption of realistic and performance oriented building standards:

> The new low cost building by-law study recommendation shall be implemented by initiating the necessary legislation to permit construction of low-cost housing within urban centres using non-conventional but functional locally produced building materials. (Kenya, 1983: 165)

Although studies have been commissioned to examine the country's by-laws and recommend appropriate changes, no revisions have taken place to date. In the meantime, the housing problem for low-income groups continues to worsen as the population continues to grow. The main concern here is, therefore, the failure of the housing delivery mechanism to provide housing to accommodate required numbers at affordable costs and in appropriate spatial urban locations.

New Directions in Housing and Infrastructure Provision

Given the constraints in the provision of housing and the fact that those intended to benefit from public housing rarely get help, the stage is set for structural adjustment policies or cost recovery principles with the sole purpose of accelerating production of housing stock as well as easing the burden of the public sector. During the 1989-1993 plan period and beyond, the GOK proposes to adopt strategies and seek solutions to problems that have hitherto inhibited housing development. Several documents including *Sessional Paper No. 1 of 1986 on Renewed Economic Growth*, *National Housing Strategy for Kenya 1987-2000* and *Sixth National Development Plan of 1989 -1993* appear unanimous on the proposed new strategies. The following is an outline of the major policy directions.

General policy

The policy of upgrading unplanned settlements rather than demolition will be re-emphasised. Lower income groups will be the focus of GOK action. Additionally, in support of the national economic growth strategy, a greater share of GOK financial and administrative resources will be shifted to small urban centres and rural areas.

Local communities will be increasingly involved in the formulation of housing programmes, preparation of infrastructure and shelter projects, maintenance of facilities and management of the local habitat.

The Cooperative Movement

The cooperative movement has seen the formation of numerous housing cooperatives in recent years and the emergence of an umbrella organisation in the form of National Cooperative Housing Union (NACHU). Cooperatives in general are an extremely important factor in the finance market and thus, have a major role to play in the implementation of national housing strategy.

Land Administration

Local authorities and the NHC will give maximum priority to regularising titles to plots in completed SS and SU projects. Alternative approaches to the public acquisition model will be developed based on the public sector regularising and providing basic services to privately-held land. The private land owners would be required to surrender the land needed to install roads and services and be responsible for house construction. This approach involves the local authorities working in partnership with the private sector.

Financing Housing

The GOK will take steps to ensure that more housing development funds are mobilised. This will be accomplished by restricting GOK borrowing from the financial system, thus releasing an increased pool of funds for housing finance institutions and also by allowing such institutions to gain easier access to pension funds. There will also be a need to develop and expand the capital markets so as to generate a secondary mortgage market through which financial assets can be bought and sold between investors. Appropriate interest rates and lending policies will be evolved to support the market system incentives.

The public sector will be the main provider of funds to finance trunk services, while private developers will finance on site services. In order to stretch available funds to the maximum, resources will be concentrated on those projects which offer lower construction and maintenance costs per beneficiary. This requires that more appropriate engineering standards be adopted that reflect the relative scarcity of capital and the trade-off between lower capital costs and higher maintenance costs later. Whatever the case, the public sector must charge market prices for the services and facilities they provide.

Soft Costs

Since a big proportion of the purchase price for housing comprises soft costs like architectural fees, stamp duty and legal fees, the GOK will investigate the impact of these costs and find ways to reduce them in order to make housing affordable by the low-income groups.

Private Sector

In order to achieve the maximum addition to the stock of housing, an increasing share will have to be developed by the private market, with the GOK acting to facilitate production by specialising in performing a higher volume of those activities which the private sector cannot do or can do only ineffectively such as land transfer and infrastructure provision. The resources of the informal sector will be exploited and encouraged to contribute more to production of acceptable housing by adopting more realistic division and building standards. The public sector will service the land and make it available for housing development to individuals and small investors/developers.

Building and Planning Regulations

The GOK will undertake a general review of these laws to bring them in line with the current demand for housing in urban areas. The GOK has enacted the *Sectional Properties Act* to allow titles to pass to a flat in the block or a room within a condominium development. The purpose is to reduce land and infrastructure costs per occupant through multi-occupation. Since 1995 the building by-laws have been revised although developers have not taken advantage of them.

Rent Regulations

The operation of the *Rent Restriction Act* should be revised not to remove rent control, but to create an ombudsman to make it possible for either landlords or tenants to bring specific grievances and issues like eviction before the *Rent Tribunal* and to allocate maintenance responsibilities more equitably. This is in recognition of the fact that litigation in law courts may be cumbersome and costly. A task force has already been formed to review this registration.

Maintenance

To assist in improving maintenance on the estates, the GOK should undertake a full review of current maintenance deficiencies and develop a strategy for providing local authorities with technical assistance and possible training, to overcome existing backlogs. If new projects are developed, more attention should be given to maintenance and management aspects in the design stage. Policy for the management of maintenance operations for the national housing stock should be formulated. A maintenance fund should be initiated with every housing development in order to cater for future maintenance requirements.

Research and Training

In order to meet its new challenges, the GOK will draw on universities, other research bodies and consultants. The various research bodies will be able to pursue basic as well as applied research in four priority areas, namely, appropriate building materials and technologies; shelter policies and institutions, innovative financing techniques and land tenure and price issues related to housing. In the meantime, research should continue to develop local building materials that are commercially viable and building by-laws should be amended to embody the use of such materials and mortgage lenders should be encouraged to accept units built of these materials in their underwriting. The communication and training functions will be strengthened considerably in order to improve access to research findings by builders, professionals, manufacturers and other interested groups.

Conclusion

Experience gained through the implementation of housing development schemes indicates that the rationale for low-cost housing based on the principle of need rather than ability to pay has resulted in a situation where higher income groups and not target groups have acquired ownership. It is also worth noting that despite the ever-growing population, only 3 per cent of Gross Domestic Product (GDP) is being invested annually in housing. Most of this has been through public sector funding. It is, therefore, understood that shortage of funding is a major constraint to housing development. A number of regulations and building codes exist under the urban planning, land and housing by-law, the *Public Health Act*, and the local government by-laws. To a large extent, these instruments have become major obstacles to rapid housing development. Scarcity of land in urban areas has increasingly become a major constraint to the development of housing and, where available, ever-increasing cost has adversely affected initiatives on the part of individuals and private developers to invest in housing. Besides land, one other constraint concerns the provision of basic infrastructure.

The strategy outlined above is one which changes the role of the GOK from that of direct developer of housing for lower-income households involving subsidies to that of facilitating the development of this housing by private entities charging market prices. It involves increasing housing stock using every available resource. Since the housing must be paid for, it must be affordable. People must, therefore, be enabled to earn income within the projects. An attempt to implement these new policies in housing development has been tried under the Urban III Project in the secondary urban centres of Eldoret, Kitale, Nakuru, Nyeri and Thika, sponsored by the World Bank. The whole project is expected to house 170,000 people. Lower standards have been adopted, particularly in the SU components. Income-generating programmes have been included. It is also evident, particularly in Nairobi, that higher densities are being allowed in the SS programmes.

The SS programmes in Nairobi, sponsored by the World Bank, have encouraged built-storey construction while in Umoja II (sponsored by USAID), condominiums have been developed for multi-occupation. It is assumed in this respect that a sound housing policy should make a substantial contribution to economic development and social welfare for everybody in the economy. Thus, public policy should ensure that private decision-makers behave in a way that maximises social benefits. The likely or actual effects of structural adjustment policies on the housing sector and related infrastructure on various socio-economic groups of the urban population have yet to be determined. However, in order to determine the actual effects of these policies, detailed field surveys will be required when a number of projects have been implemented. What is certain is that with successful implementation of the strategy, the housing stock will increase. The general opinion still holds that these approaches may only marginally help the low-income groups.

20 Public Transport Modes in the City of Nairobi, Kenya

R. A. Obudho and G. O. Aduwo

Introduction

The essence of any urban transportation problem is lack of mobility, severely limited mobility and mobility purchased at a very high social and economic cost. In the City of Nairobi, the current situation of urban transport is alarming. Despite the relatively low levels of private automobile ownership, the city's transportation problems are severe in degree, daily duration and areas affected. These problems are especially felt during the peak demand hours which are often characterised by considerable jostling and stampeding among the travelling public in search of the means of public transport. The chaotic situation is further exacerbated by the carelessness and apparent lack of concern among the Public Service Vehicle (PSV) operators.

For most commuters, the problems faced include increasing walking distance, long waiting hours, severe struggles while getting on and off the available PSVs, insecurity and pick pocketing, accidents, traffic jams, vehicle breakdowns and a general atmosphere of bad temper. Basically, the major problem is that of congestion, mainly witnessed in the form of overloaded buses and *matatus* and traffic jams experienced, especially in the CBD. The centralisation of activities in this zone has ensured that the commuting trend involves a movement to the surrounding expanding residential areas.

Underlying all these problems, however, is the acute shortage of resources in the provision of a long life urban transportation infrastructure sufficient to match the large increase of population and emerging patterns of population distribution and demand for public services. A rapid deterioration in urban transport conditions is in prospect if present trends continue unchecked. The current exceptional rates of urban growth are unlikely to diminish in the near future and may well accelerate. Moreover, as the city expands and the workers live further away from their places of work, a more than proportional expansion of public transport facilities and services is required.

Various methods have been considered and some applied towards achieving this expansion but with minimal success. Among them is the introduction of a railway commuter service, but this can serve only those areas along which the line passes. It has so far proved rather expensive to run and maintain, apart from being unable to meet the demands even of those areas which it serves. Other measures, such as the introduction of the government run Nyayo Bus Service (NBS) — now defunct — to supplement the services offered by the Kenya Bus Services (KBS) had been helpful but insufficient to meet the existing demand, apart from facing such technical problems as lack of parking space in the CBD.

A study conducted by Transurb-Consult (1986) on behalf of the Belgian Government has considered many long-term solutions among which is the introduction of a light rail transport system and the use of busways. However, this study may not be implemented due to lack of funds. A low-cost strategy is important for a city like Nairobi that is growing very fast. Such a strategy involves reshaping the management of the existing infrastructure in order to achieve a more efficient and agreeable equilibrium between demand and supply.

This chapter describes the existing public transportation system in Nairobi and the past and present policy and planning issues. It then suggests alternative policy guidelines towards achieving low-cost solutions to public transport problems.

Policy should not be easily swayed by current popular sentiments. Many of the criticisms levelled at Nairobi's public transport system and the *matatus* in particular are not warranted since they are not based on empirical findings. It is true that there is a need to overhaul the city's public transport system to cope with the ever-increasing demand, but this can only be done if there is a clear understanding of the wide range of problems faced by the system. The value of all policy options needs to be analysed and understood as a basis for suggesting on possible solutions. The degree to which previous attempts at solving the problems have been pursued also needs to be analysed. But, before doing this, a historical background of the city is provided as it relates to the origin and growth of public transport modes and their associated problems.

Origin and Growth of the Transport System

The present role and prospects of Nairobi's public transport system have been shaped by population pressures, urban structures and its general transportation system. The current profile of the city has in turn been shaped by the influences of geography, historical factors and contemporary forces. Many of the current problems that plague public transport in Nairobi are attributed to the high population growth rate, lack of vehicle capacities, energy costs, utilisation of infrastructural facilities, location of high density residential areas, lack of road and vehicle maintenance, road safety, manpower training and development and policy and institutional developments. These problems are not new to Nairobi, nor are they unique from those in other LDCs. In order to understand the problem, one must understand the city's history during both the colonial and post-colonial eras as well as the regional economic, social and geographical interactions between the city and the regional surroundings.

Colonialism and the political legacy of colonial communication patterns have had both negative and positive effects upon the colonial transportation policies and much of the development of post-independence transportation systems. Nairobi's land-use development portrays a classic example of this colonial influence; the street layout, residential locations, CBD location, racial residential separation and architectural peculiarities portray these colonial and alien planning concepts. These concepts have influenced the city's public transport system. As a colonial settlement, the configuration of Nairobi during the colonial period was essentially tripartite in character with the Europeans, Asians and Africans occupying different residential zones and making contacts mainly on official and business matters. Within this configuration, the residential areas of Europeans, sited on the wooded ridges of fertile red soils to the north and west, were well-served by transportation facilities. Typically, the earliest form pattern in Nairobi by the 1920s was dominated by a major trunk road commencing from the CBD to the umland with a spur to the industrial area (Obudho, 1993:91-112 and 1997: 300-343).

Meanwhile the residential areas for Africans were left to develop towards the east away from the major trunk road. This area accommodated the vast majority of Nairobi's population and was characterised by poor transport access both to the city's transport network and within the area itself. The continuous flow of rural migrants into these residential areas accentuated their already high residential density levels, which were incessantly being increased (Table 20.1).

Table 20.1: Population of Nairobi by race for selected years

Year	Africans	Europeans	Asians	Total
1902	6,351	576	3,582	10,509
1926	19,112	1,492	9,260	29,864
1931	26,761	5,195	15,988	47,944
1948	65,939	10,830	41,810	118,579
1962	115,388	21,476	87,454	224,318
1969	421,079	19,185	67,189	507,453
1979	695,353	33,511	108,911	837,775

Sources: Based on population censuses.

Characteristically, however, this growth in the African population did not immediately manifest itself in additional journeys to work from their residential zone to other parts of the city since most migrants confined themselves within their residential locations in the eastern parts of Nairobi. Life revolved around activities within the residential areas, people moving around on foot or on newly-bought bicycles, leaving only the more established members working further away to commute daily.

This state of affairs partly accounted for the relatively trouble-free movement of people during the colonial period and the generally lower demand for public transport services. When the Kenya Bus Service (KBS) started operation in Nairobi in 1934, it could operate only 12 buses. By 1950 these 12 buses were adequate in serving the existing population's public transport demand. Moreover, most residential areas such as Pangani, Landimawe, Muthurwa, Pumwani, Shauri Moyo and Kamukuji were within walking distance of the CBD. Walking was the predominant mode of transport and in 1970, for example, it was estimated that 48 per cent of the commuters in Nairobi walked to their places of work.

The need for public transport services increased as the city expanded and witnessed an increase in the number of well-to-do Africans. Within the African residential zones, mainly in the eastern part of the city, motorisation levels slowly increased as Africans, who were observing the rapid growth in the local population, hired and bought vans mainly used to bring foodstuffs from the rural areas to feed the growing population. This led to the emergence of informal public transport services which became very common as the African residential areas grew and the travel demand increased. It is during this period that the *matatu* emerged, especially in the 1950s when they were used mainly to transport residents of the African neighbourhoods to nearby rural villages. The word *matatu* is derived from the local term *mang'otore matatu* meaning thirty cents which was the standard fare charged then (Aduwo, 1990 and Aduwo and Obudho, 1988 and 1992:120 - 130).

Since the African residential areas were invariably outside the interest and activity spaces of the colonialists, official knowledge about this development was confined to heresay. Africans' transport problems did not figure in any major transportation plans of the time. African areas, high population density and lower level of infrastructure were in stark contrast to the low density developments and good infrastructure provision in the European residential areas. The distribution of access roads within the later areas were both well-planned and well-maintained, providing good access to the city's wider transportation network as a whole and its business and administrative centre in particular. Consequently the predominantly European residents here were very mobile with a high private-vehicle ownership. Indeed, the vehicular growth rate took on dramatic proportions not only in Nairobi, but also in other colonial capital urban centres in Africa (Table 20.2).

By 1928, Nairobi was the most motor-ridden city in the world proportionate to its European population. This high private-vehicular ownership is said to have contributed to the early thinning out of Upper Nairobi. It also presented one of the transportation problems of the day, others being how to improve road access to the industrial area and how to accommodate the increasing motorisation in the CBD (Aduwo and Obudho, 1992: 122).

Table 20.2: Average post-War vehicle growth rates of African colonial urban centres 1960-70

Country	% annual growth
Kenya	8 Nairobi, 6.8% (1960-70)
Tanzania	11
Ghana	12
Nigeria	14 Lagos, 15.5% (196-70)
Uganda	18

Source: Hawkins (1962).

Towards independence and after, economic resources declined and deterioration in infrastructure began. This was partly due to the gradual run-down of the colonial investment in the country, increasing urban population and scarcity of resources in general. It led to a decline in whatever infrastructure existed so that in the face of increased population growth and increase in travel demand, a marked deterioration of Nairobi's transport network and services took place. As independence approached, more and more Africans were assimilated into the roles and functions left vacant by the departing colonialists. There soon emerged elitist groups who developed values and aspirations similar to those of the colonialists. They aspired, for example, to car ownership. But for some civil servants and private sector employees this was further encouraged by the provision of loans to assist in their purchase. Many took residence in the former European areas, thereby accelerating the change from the previous socio-economic and cultural divisions within the city.

This period saw a further increase in car ownership levels with a growth rate of about 6.8 per cent per annum during the period 1960-1970. By 1970 it was estimated that 25 per cent of Nairobi's commuters used private automobiles to reach their places of work. The use of public transport was not predominant with only about 24 per cent user. According to a World Bank estimate (1975b), the number of buses per 1000 population in Nairobi by 1970 stood at only 1.5.

Post-independence movement patterns, with additional travel demands, generated mainly by increased migration from the rural areas, exerted pressure on the city and its infrastructure which were ill-equipped to serve them. A major problem has been the centralisation of activities in the CBD which by 1970 was estimated to employ over 75 per cent of the commuters. This area has for a long time experienced numerous traffic problems, exacerbated by a lack of space within its vicinity. The post-independence period also witnessed a relaxation (not by design) of traffic regulations, parking restrictions and land-use control. Hence, within a few years of independence, much of the previously formalised land-use pattern superimposed on the original settlement structure underwent a process of severe erosion.

Since 1970, the city has expanded tremendously and a new population distribution pattern has emerged. A large percentage of the low-income public transport users today live further away from the CBD, partly due to the introduction of housing schemes such as the site-and-service schemes and the general policy of demolishing squatter settlements sited near the CBD to give way to other developments. Such neighbourhoods as Bahati, Maringo, Ofafa, Ziwani, Landimawe, Kaloleni and Pumwani, which were specifically for African settlement away from the CBD, are now part of the outer core of the CBD. Numerous other neighbourhoods have sprung up in far-off places and developed into congested settlements boasting of hundreds of thousands of residents. Towards the east, low-income settlements have emerged in places like Dandora, Kariobangi, Kayole and Umoja and the process of expansion continues. The city has also expanded to include peri-urban settlements and suburbs such as Kawangware, Ruiru and Kangemi, which are today undergoing the most rapid rates of expansion. Towards the north, new settlements are also rapidly emerging and expanding within the Ruaraka/Kasarani zone, while in the south, the process of residential development has seen the emergence of middle income estates such as Otiende and Ngei, to mention just a few.

This expansion of Nairobi has not been matched by a similar expansion in transport facilities and services. The annual rate of growth of passenger journeys per day is currently increasing by 5.8 per cent per annum, which is high by all standards (Table 20.3).

Table 20.3: Public transport demand in Nairobi: 1985, 1990 and 2000

Year	Passenger journeys	% annual growth per day (000)
1985	676	—
1990	873	5.82
2000	1,393	5.95

Source: Transurb-Consult (1986: 5)

A clear manifestation of the excessively high demand for public transport services is witnessed in the daily stampede and overcrowding which characterises most of the city's transport terminals, especially during the rush hours. For a long time now, *matatus* have operated alongside the KBS as the only major supplier of public transport services. Other services have been introduced to supplement existing services and they include the railway commuter services (which operate during the morning and evening rush hours in areas along the line) and the Nyayo Bus Service (NBS) whose services were offered largely during the rush hours.

The Role of the Public Bus Systems

The KBS has existed as the sole legal supplier of public transport services since it was incorporated as a private company in 1934 with an authorised capital of Kshs. 20,000.00. It was converted into a public company in 1950, jointly owned by the United Transport Overseas Company (UTOC), a British Company managing over 100 other such companies all over the world. Since then, KBS has operated public transport services in Nairobi under various franchises granted by Nairobi City Council (NCC). An agreement was signed in 1966 in which the NCC acquired 21 per cent of the shares of the company, thus formalising the partnership. On the other hand KBS had been keen to develop with local communities' interests. The latest franchise was extended while negotiations took place concerning longer term arrangements.

The buses have been a mixture of the Leyland Guy Victory, the ERF Trailblazers and DAF. These were compatible in terms of service requirements and were specifically adapted for local road conditions. Their engines were designed to minimise fuel consumption, an essential feature in running profitable transport services as well as minimising the overall need for foreign exchange. The assembling and body building has been done locally by the Labh Singh Harnam Singh (LSHS). There previously had been one bus depot situated at Eastleigh, but now a second one has recently been constructed at Riruta.

KBS had a fully staffed engineering department with a staff of over 441. It has been mandatory for each bus to be taken to the workshops for a check-up every ten days, even without having specific mechanical problems. However, a persistent average of 20 buses are down daily due to minor mechanical problems, though it has often taken lengthy periods to repair them. Other problems have included the lack of a cheap source of spare parts and the KBS's inability to maintain crew punctuality and discipline. The result of these has been KBS claims of non-profitability resulting in a tremendous deterioration in service prior to the 1990s.

Despite the operational complexities of running a full-commuter network, KBS deploys over 222 buses daily. This efficiency ratio of 88 per cent is high by any standards. The rate of growth of its fleet has not been consistent with increasing demand as Table 20.4 shows. The decrease in the number of KBS's daily passengers between 1973 and 1990 can be attributed to the tremendous competition offered by the *matatus* whose numbers have consistently increased in the same period. Since 1977, the average daily number of passengers transported by the KBS increased but it continues to face stiff competition from *matatus* whose market share stands at 58 per cent.

Table 20.4: Growth of KBS fleet and passengers, 1962-1990

Year	Fleet size	Growth rate	Average daily passengers (000)	Growth rate (%)
1962	100	3.0	66	2.2
1964	106	9.3	69	12.8
1966	*	*	98	42.2
1967	*	*	105	7.14
1968	146	9.4	116	10.0
1969	*	*	122	5.17
1970	166	9.4	151	23.7
1971	195	17.4	209	38.4
1972	239	22.6	233	11.5
1973	264	5.4	240	3.0
1974	284	7.6	237	-1.2
1975	290	2.1	230	-2.9
1976	288	-0.1	229	-0.4
1977	285	-1.04	229	0.0
1978	291	2.1	250	9.17
1979	317	8.9	270	8.0
1980	310	-2.2	273	1.1
1981	316	1.9	282	3.3
1982	309	-2.2	329	16.7
1985	273	-11.6	382	16.1
1987	300	9.9	410	7.3
1988	313	4.3	387	8.3
1989	274	4.0	387	7.3
1990	272	4.0	364	7.4

* = data not available

Source: Aduwo and Obudho, (1992: 124).

Despite this state of affairs, KBS has increased its services area coverage with operations extending even outside the city's boundaries. Table 20.5 summarises the KBS trading position and shows how the company's operations tend to be inelastic, especially in responding to the ever-increasing demand for public transport in the city.

Matatus

Matatus are small-scale transporters of commuters and goods which are owned and licensed as PSVs. They represent an intermediate form of public transport whose services fall somewhere between the conventional buses and the taxis. They emerged

Table 20.5: KBS's trading position and statistical analysis, 1970-1990

Year	Annual passengers (million)	Total distance covered (km million)	Revenue K.£ (m)	No. of staff (000)
1970	55.2	13.1	1.055	1.112
1971	67.2	15.1	1.295	1.352
1972	85.2	18.2	1.659	1.633
1973	87.6	18.9	1.977	1.762
1974	86.4	22.4	2.492	1.802
1975	84.0	22.7	3.108	1.804
1976	83.6	23.3	3.579	1.837
1977	83.6	23.9	4.312	1.766
1978	91.2	23.6	5.142	1.824
1979	98.4	25.7	5.553	2.126
1980	99.7	24.9	6.912	2.201
1981	103.0	26.3	8.372	2.289
1982	115.6	27.6	9.912	2.295
1983	131.1	28.7	11.990	2.375
1984	125.5	28.2	12.772	2.322
1985	136.0	30.8	14.183	2.300
1986	145.0	32.2	1.638	2.455
1987	142.0	30.9	1.761	2.497
1988	139.3	31.9	1.751	2.481
1989	121.0	26.6	1.806	2.312
1990	118.1	26.6	1.781	2.320

Source: Aduwo and Obudho, (1992: 125).

spontaneously as a result of the inadequacy of buses. Over the last few years, they have assumed an expanding role, especially for those areas which buses do not serve adequately. In the 1960s the total number of *matatus* operating all over Kenya was under 400 and the police perceived them as "private taxis". In 1973, a presidential decree declared that they were a legal form of public transport and could carry fare-paying passengers without having special licences to do so. At the same time, it was emphasised that the existing insurance and traffic regulations were to be complied with. This decree intervened in the anomalous situation of unlicensed *matatus* operating despite the existence of a monopoly franchise for public transport by the KBS. Since then, they have increased in numbers and in the daily number passengers they serve Table 20.6 (Aduwo, 1990; Aduwo and Obudho, 1988 and 1992:126-130 and Obudho, 1993:91-112 and 1997:300-343).

Table 20.6: Growth of matatus in Nairobi and their average daily passengers, 1971-1981

Year	Fleet size	Growth rate (%)	Average daily passengers (000)	Growth rate (%)
1971	217	36.4	38	10.6
1973	375	43.5	47	29.3
1974	538	30.1	63	16.1
1975	700	38.4	74	31.1
1976	696	36.2	101	32.7
1977	1320	8.6	140	15.8
1978	1434	9.3	164	17.3
1979	1567	4.2	195	15.5
1981	1704	5.6	263	54.4

Source: Aduwo and Obudho (1992: 125).

By 1980, *matatus* had captured at least a third of the public transport market in Nairobi. This number has increased so that today they provide a big challenge to the public bus system. Since 1973, their percentage share in the market and the proportionate number of passengers using them has been increasing vis-a-vis that of the KBS (Table 20.7).

Table 20.7: KBS and matatus: Share of the market in Nairobi

Year	Matatus	KBS
1973	16.0	84.0
1974	21.0	79.0
1975	24.0	76.0
1976	31.0	69.0
1977	38.0	62.0
1978	40.0	60.0
1979	41.0	59.0
1980	42.0	58.0
1985	48.0	52.0
1987	49.0	51.0
1988	51.0	46.0
1989	52.0	45.0
1990	52.0	42.0

Sources: Aduwo and Obudho (1992: 125).

Most of those in the *matatu* trade are businessmen who are concerned with the search for profitable returns on their investments. They are expensive vehicles both in their purchase price and their maintenance costs; their increasing number is largely an indication of the lucrativeness of the business. The rapid increase in their numbers is also a strong indication of the high demand for their services due to the inability of other modes to cater for these needs. The sector also plays a significant role in generating employment. In 1985 it was estimated that around 14,000 persons were employed in the *matatu* trade, earning approximately Kshs. 9 million. *Matatus* generate slightly over 1,000 jobs per year and most of those employed earn between Kshs. 900.00 and Kshs. 1,500.00 per month.

Despite the important role which the *matatus* play, they have been an object of persistent public criticism, government restrictions and proposals to phase them out in favour of more modern forms of public transport. They have been viewed as uneconomical, unruly and a hazardous means of transporting the growing urban population. Their operations constitute a gigantic problem in the management of the Nairobi's public transport system. They have been accused of being the cause of the most serious accidents. It is indeed unfortunate that their enormous growth has not been matched by a corresponding growth in their regulation so as to streamline their operations and ensure safety and comfort to commuters and other road users. Numerous complaints have been levelled at the *matatu* operators who have generally shown disregard for traffic regulations. They have often been identified with speeding overloading and being involved in other haphazard operations which make them a major cause of road accidents. Other operational characteristics of *matatus* which people resent include their continuous hooting and touting for passengers, their chaotic parking and abrupt stop their harassment and abuse of commuters and other road users and their general disregard for many other traffic rules which often interfere with smooth flow of traffic in Nairobi.

However, despite these problems, some of which should not be wholly blamed on them. They have often been identified as being beneficial more than most transport "modernisers" assume. It is commonly accepted that they should undergo some improvements and be subjected to the general requirements of good transport system. The following section looks briefly at the past and current planning responses to Nairobi's public transport system especially as relates to *matatus* and the practical measures which have been taken to solve the city's public transport problems.

Past and Current Planning Responses

The internal planning response to the public transport problems in Nairobi since independence has been slow. The GOK has placed greater emphasis on national spatial economic policies rather than on urban development. It has relied on industrialisation and modernisation to achieve national economic goals, which has only enhanced transport problems. Much of the industrial investment in Kenya has taken place in Nairobi, generating additional transport demands on the already inadequate and congested transport system. The technological forces released by industrialisation, modernisation and urbanisation have also generated new travel patterns as well as different travel behaviours and lifestyles all of which have mixed with the old. According to Banjo and Dimitrou (1980) and Obudho (1993: 102-108 and 1997: 330-336), this situation is characteristic of the second generation urban transportation problems faced by all LDCs during their immediate post-independence stage.

Due to the absence of technical and administrative capacity to tackle the problem, countries such as Kenya have often sought assistance from the MDCs. International consultants from these countries have been commissioned to examine the urban transport problems that Nairobi faces and propose means by which they can be tackled. Consultants have been appointed either on the basis of their previous transport planning experience in other LDCs or as a result of the expertise they have acquired in their own countries. In most cases, these consultants have a limited understanding of the local urban movement problems and use standard planning procedures which are based on irrelevant assumptions. They have, for example, failed to concern themselves with the phenomenon of intermediate or informal modes of public transport and lack an understanding of how the system works. As a whole such assumptions have resulted in policies penalising the non-motorised community, the destruction of certain urban forms and structures and the failure to incorporate the informal or intermediate sector into transport plans, among others (Obudho and Otieno, 1995).

While many standardised planning approaches are not completely inappropriate to LDCs, their implications are critical due to limited resources and other crucial problems. Countries like Kenya are, therefore, left with the option of a low-cost strategy, in the use of traffic management schemes. In Kenya ,very little has been attempted along this line despite the tremendously increasing population and demand for public transport facilities.

The first post-independence transportation study of Nairobi was carried out in 1970 by the Nairobi Urban Study Group (NUSG) of the NCC. Research was carried out and a report presented as the *Nairobi Metropolitan Growth Strategy Report*. This report identified the problems of public transport facing Nairobi and made forecasts concerning future demands for public transport and future accomodation of this demand. However, it did not analyse the role of *matatus* but laid emphasis only on the development of buses and a mass transit system. It mistakenly favoured buses due to likely benefits to local employment and the vehicle manufacturing industry forgetting that *matatus* also play a major role in these areas. Other recommendations which were made by this report included a policy of restraint on the ownership and use of private cars in association with measures to encourage public transport usage, progressive reduction in public transport fare, staggering working hours in the CBD in order to spread the effect of traffic peak over a considerable period, and the provision of segregated bus ways and more roads.

In 1978, the Nairobi City Transport Unit was formed and placed within the NCC Engineering Department with a view to controlling both public transport policy and complementary parking policy, among other things. The NCC study further recommended the construction of 27.7 km network of bus priority routes aimed at the regulation and administration of private and public transport, CBD parking supply and traffic circulation. The subsequent *Nairobi Urban Study Report* (1978) comprised a series of recommendations on the creation and improvement of infrastructure, formulation of policy and monitoring procedures. This report pointed out that most of the projects are long overdue and if not implemented as soon as possible, the national energy losses will be colossal.

Since then, the NCC policy documents and studies have recognised the important role which *matatus* play. As a result of a planned World Bank Matatu Assistance Scheme, more studies were carried out under the Kenya Urban Transport Projects (KUTP) in 1979. Nothing was done, however, and none of the proposed assistance to the *matatu* owners and operators ever reached them. The NCC's recognition of *matatus* can be seen in their providing *matatus* with exclusive terminals in the CBD, a process which the *matatu* operators have resented because they see the terminals as an attempt to favour the KBS, which has the exclusive right to operate in some terminals.

In 1980, Leyland (Kenya) Ltd. studied a standard *matatu* prototype in compliance with the specifications laid down by the GOK, but this project was not implemented (though the design was mechanically sound) because it was not financially viable. The study, therefore, recommended the use of different types of *matatus*, though it rejected overloading beyond the manufacturers' specifications. For a long time, the major policy issues in Nairobi's public transport system have been the role of the *matatus* and the KBS franchise agreements. The introduction of NBS and the railway commuter services to supplement the KBS and *matatus* services, especially during the rush hours, were two major practical measures taken to increase the supply of public transport services. The NBS, which was run by the National Youth Service, expanded its operations very fast, operating at one point over 90 buses and serving mainly the densely populated zones. The service collapsed in the mid-1990s.

The Railway Commuter Services (RCS) has operated in the high density residential areas of Kibera, Dandora and Kariobangi as well as suburbs of Kangemi, Limuru and Kamiti. It therefore serves only a few areas which have the line passing through them and does so only by making one morning and evening rush hour trip. The service faces many problems, the most important being high operating costs. Since it started, it has only made losses. Moreover, very few people use it even in the area which it serves because it has few operations and other factors such as the unfavourable location of its terminals and frequent congestion during the peak hours.

A noticeable problem, however, has been the lack of terminal facilities or parking space for buses in the CBD. A study first conducted in April 1984 by

Transurb (for the Belgian Ministry of Foreign Affair's Foreign Trade and Development Cooperation) highlighted three options to deal with the mass transport problems in Nairobi. The first option suggested the use of specially-constructed bus ways while the second recommended introducing a light rail transit system. The third option, which seems to have got GOK approval, introduced the concept of guided or articulated buses. Meanwhile, investigations have continued though it is evident that any modern transportation system in Nairobi will require complementary bus and *matatu* networks.

In its efforts to minimise road accidents caused by such factors as overloading, poor roads, vehicle unroadworthiness, careless driving and other illegal operations often associated with *matatu* operations, the GOK has introduced varied legislation. In 1984, a presidential directive ordered all *matatus* to acquire PSV licences. The Kenya Police Traffic Road Safety Section, which is entrusted with the execution of road safety regulations, released details of the Traffic Amendment Act, 1984, which rules cover driving licences, obstruction, maximum driving hours and other violations. By using such legislation, the police have attempted to clamp down on *matatus* and other PSVs, but with minimal success. Whenever the authorities have become too harsh on the PSV operators, they respond by refusing to offer their services. These strikes have dramatised the plight of the Nairobi's commuters.

A *Daily Nation* editorial states that "no amount of warnings to public transport operators will stop the inexarable growth of recklessness on our road.... something most enduring must be done." A good campaign on road safety would involve the PSVs as much as the travelling public who have seemed to value their lives and comfort less than the immediate need to get to their destinations. The stampede which characterises PSV terminals has caused as much chaos and confusion as the apparent lack of concern among most of PSV operators. Recent measures aimed at reducing chaos such as the use of queues have so far proved successful and popular but more needs to be done for a long-term solution to Nairobi's public transport problem. Because *matatus* increase commuters' accessibility to the city, they should not be considered a problem, but as part of a solution to the current public transportation problems which Nairobi faces.

Conclusion

The inadequacies of Nairobi's urban passenger transport services can be accounted for by a variety of factors. These include the uneconomic spread of the city's morphology great population pressure due to rapid urbanisation the shortage of enough public transport services in the form of buses, *matatus* and the RCS, lack of accessibility to certain parts of the city and inadequate street layout and road space. Apart from these, other problems stem from the inability of the city authorities to implement policy recommendations to solve inherent problems, including the failure to implement efforts geared towards the creation of public transport priority lanes and roads and to change the uniform work schedule which has created the morning and evening rush hours. There are also problems associated with the concentration of activities in the CBD, ensuring that most trips are made there. Nairobi also lacks adequate resources to repair the city's roads and to provide other modern urban transportation facilities. These inadequacies are serious, but not impossible to remedy. The following broad recommendations may help in solving the current urban transport problems which the city faces.

While it is recognised that urban growth is an inevitable process of socio-economic change, the pace of growth of Nairobi should be reduced. The shift of the administrative and other functions from Nairobi to other urban centres could go a long way in reducing population pressure in the city. The DFRDS could ensure the success of this re-allocation process. There is need to strengthen the programme and improve the working and living conditions of people in other urban areas and in the rural areas. This will improve the general standard of living of the people and check the drift of people from rural to urban areas. The current high rate of population growth in Nairobi is the main cause of the city's inability to provide sufficent facilities for its population. The problem of population growth in the city should, therefore, be urgently addressed, bearing in mind that it is largely caused by rural-urban migration.

Within Nairobi, a deliberate de-concentration process should be initiated to remove pressure on the CBD. The relocation of employment and other activities from the CBD to the outskirts needs to be given immediate and practical attention so that some traffic flow can be moved from the CBD to other zones. Such zones have been identified, but they must be firmly established as satellite CBDs. The notable growth of Westlands and Hurlingham centres, for example, is a step in the right direction, but future consideration should be given to establishing such centres closer to the low-income residents who are increasing in numbers and reside mainly in the city's suburbs. Since they are also the major users of public transport services, it is important that their trip patterns be changed to ensure that they do not have to cover long distances and spend a substantial amount of their income commuting to the CBD. The sites which should be given due consideration include Kangemi/Uhuru, Riruta, Kahawa/Kasarani and Embakasi.

The uneconomic use of space in Nairobi should be discouraged. An overview of the built-up sections of Nairobi area shows that the growth of the city's suburbs is amorphous. There exists too much open space between these continuously expanding residential zones and those immediately surrounding the city centre. These suburbs which include Kangemi/Uthiru, Riruta, Roysambu, Kahawa, Embakasi and Njiru are today expanding at a tremendous rate and provide residence to a majority of the city's low-income earners. There is need to intensify and strengthen the city's infilling process so that a more compact city form is achieved in order to enhance accessibility to the road network and public transport services. This would inevitably increase the areal coverage of the city's public transport network by increasing the network's density of access.

The allocation of available financial resources should be sensitive to the needs of the public transport sector because a majority of the city's residents are users of these services. Funds should, therefore, be provided for the procurement of new and more buses, spare parts, fuel and accessories so that there is an adequate number of serviceable buses. Because of high demand, more buses should be introduced to offer an all-round service in the high demand areas. For a long time now, the monopoly enjoyed by the KBS has ensured that it seldoms considers the comfort and safety of its users and until recently has done very little to increase the number of buses and improve on its services. Competition will inevitably lead to a change in this state of affairs. As a whole, the increase in the number of buses will increase their frequency and provide more sitting space. The NCC should also set up a research wing to undertake investigations into traffic patterns in order to provide information on which future bus timetables could be based and upon which bus routes can be prescribed in response to demand.

The present street layout and road space in Nairobi are inadequate. The NCC should embark on a road rehabilitation programme which should not only be concerned with the CBD, but the residential areas as well. In rehabilitating and constructing new roads, the city authorities should ensure that the interests of pedestrians are considered by providing space for pedestrian lanes to reduce the danger of accidents. This concern for the pedestrians is important as the number of private automobiles in the CBD are increasing. There will be a need to restrict private automobiles from the CBD in the near future and to establish Pedestrian Cores. Other important traffic management improvements should include the construction of highways in major traffic flow-zones, the construction of a network of one-way streets, integrated traffic signals, signs and road markings in the CBD, and reserved bus lanes and special roads to service high density residential, industrial and commercial districts.

References

Abu, Lughod, J. (1965). "The Emergence of differential Fertility in Urban Egypt." *The Milbank Memorial Fund Quarterly* vol. 44, no.2 (1965)

Abrams, L. (1964). *Housing in the Modern World: Man's World.* Cambridge, Mass.: MIT Press.

Acland, J.D. (1971). *East African Crops: An Introduction to the production of Field and Plantation Crops in Kenya, Uganda, Tanzania.* London: Longman.

Aduwo, G.O. (1990). *Production, Efficiency and Quality of Service of the Matatu Mode of Public Transport in Nairobi, Kenya: A Geographical Analysis.* Nairobi: Depaetment of Geography, university of Nairobi.

Aduwo, G.O. and R.A. Obudho (1988). "Production, Efficiency and Quality of Urban Transport Systems: A Case of the Matatu Mode of Urban Transport." Paper presented at the First International Conference of Urban Growth and Spatial Planning, Nairobi, Kenya, 13-17 December.

__ (1992). " Urban Transport Systems: A Case of Matatu Mode of Transport in the City of Nairobi, Kenya." *African Urban Quarterly* 7, no.1 and 2 (1992).

Ahawo, D.P.T. (1982). "Age at First Birth and Age at First Marriage: A Study of Adolescent Fertility in Kenya." M.A. Thesis, University of Nairobi.

Aketch S.O.(1990) "Small-holder Food Production: A Geographical Analysis of Maize Production in Rongo Division, South Nyanza District." M. A. Thesis, University of Nairobi.

Allan, A.Y.(1971). "The Influence of Agronomic Factors on Maize Fields in Western Kenya with Special Reference to Time of Planting." Ph.D. Thesis, University of Nairobi.

Anke, R. and K. Knowles. (1980). *An Empirical Analysis of Mortality Differentials in Kenya at Macro and Micro Levels.* Population and Employment Working Paper, No. 60. Geneva: International Labour Organization (ILO).

Ayiemba, E.H.O. (1983) "Nuptial Determinants of Fertility in Western Kenya. Nairobi." Ph.D. Thesis, University of Nairobi.

Badr, G.M. (1966). "The Nile Waters Question: Backgrounds and Recent Developments." *Revue Egyptienne de Doits Internationale* 15, no. 2 (1966).

Banjo G. and H.T. Dimitrou (1980) "Urban Transport Problems in Third World Cities: The Third Generation." Paper presented at the Seminar on Transport Planning in Developing Countries, University of Warwick, 9-16 July 1980.

Behm, H. (1979). "Socio Economic Determinants of Mortality in Latin America." Paper Presented at the Meeting on Mortality, Mexico City, Mexico (1979).

Benard R.H.(1988). *Research Methods in Cultural Anthropology*. London: Sage Publications.

Bongaarts, J. (1982) "The Fertility inhibiting Effects of the Intermediate Fertility Variables." *Studies in Family Planning* vol. 13, no. 6/7 (June-July 1982):17

Brass, W. (1964). "Uses of Census and Survey Data for the Estimation of Vital Rates." Paper Presented at the African Seminar on Vital Statistics, Addis Ababa, Ethiopia.

Buchanan, B. G. et al (1983). "Constructing an Expert System." In *Building Expert Systems*, edited by F. Hayes-Roth et al. Reading, Mass.: Addism-Wesley.

Bumpass, L. et al (1968) "Age and Marital Status to First Birth and the Place of Subsequent Fertility." *Demography* vol. 15, no. 1 (1968).

Bunyasia, C. S. (1985) "The Patterns of Causes of Deaths in Kenya." M.Sc. Thesis, University of Nairobi.

Caldwell, J.C. (1977). "The Economic Rationality of Fertility: An Investigation Illustrated with Nigerian Survey Data." *Population Studies* vol. 31 (1997): 38.

_____ (1979). "Education as a Factor in Mortality Declines: An Examination Illustrated with Nigerian Survey Data." *Population Studies* vo. 33 (1979).

Cohen, D.W. (1985). "Doing Social History From PIM's Doorway." In *Reliving the Past*. Edited by C. Oliver. Zunz Hill. University of North Carolina Press.

Cohen, D. W. and E.S. Atieno Odhiambo (1988). Siaya: The Historical Anthropology of an African Landscape. Nairobi: Heinmann.

Collier A. et al (1967). "Local Variations of Fertility in Taiwan." *Population Studies* vol. 20, no.3 (1967).

David, L. and M. Lakayama (1988). "Human Resources for Environmentally Sound River Basin Management." In *Reflections on the Management of Drainage Basins in Africa*. Edited by C. O. Okidi. Nairobi: Institute for Development Studies.

Day, M.D. (1978). *A Diabetical Approach to State Intervention With Special Reference to Housing Policy*. Department of Geography Occasional Paper No.9. London: London University.

Deeb, M.E. (1987) "Household Structure as Related to Childhood Mortality and Morbidity Among Low Income Areas in Amman." Ph.D. Thesis, The John Hopkins University.

Driver, H.E. (1953). "Statistics in Anthropology." *American Anthropologist*,55 (1953): 42-54.

Dumont, R. (1962). *Tanzania Agriculture After the Arusha Declaration*. Dar es Salaam: Ministry of Economic Affairs and Development Planning.

References

Durkeim, E. (1964). *The Division of Labour in Society*. Glencoe, Ill.: Free Press.

Eckholm, E.P. (1975). "A System Analysis of Rural Areas: The Deterioration of Mountain Environments." *Science* vol. 189 (1975).

Evans-Pritchard, E.E. (1965). *The Position of Women in Primitive Societies and Other Essays*. London: Faber and Faber.

Flegg, A.T. (1982). "Inequality of Income, Literacy and Medical Care as Determinants of Infant Mortality in Regions Under Developed Countries." *Population Studies* vol III (1982).

Fletcher-Cooke, J.(1966). "Parliament, Executive and Civil Service." In *Parliament as an Export*. Edited by Burns. London: George Allen.

Fortes, M. (1949). "Time and Social Structure: An Ashanti Case Study." In *Social Structure: Studies Presented to the A. R. Radcliffe-Brown*, 9th ed. Edited by M. Fortes. Oxford: Clarendon Press.

Gabriel, K.R. (1953). "Fertility of the Jews in Palestine: A Review of Research." *Population Studies*, (1953).

Genesereth, M. and M. Nilson (1987). *Logical Foundations of Artificial Intelligence*. London: Morgan Kaufmann.

Gerhart, J.D. (1974). "The Diffusion of Hybrid Maize in Western Kenya." Ph.D. Thesis, Princeton University.

Gichaga, F.J. (1971). *Practical Training for Engineers in East Africa*. Nairobi: University of Nairobi.

Gluckman, M. (1950). "Kinship and Marriage Among the Lozi of Northern Rhodesia and Zulu of Natal." In *African System of Kinship and Marriage*. Edited by A.R. Radcliffe-Brown and C.D. Forde. London: Oxford University Press.

Goss, K.F. (1979). "Consequences of Diffusion of Innovations." *Rural Sociology* vol 44, no. 4 (1979).

Guariso, G. and D. Whittington (1985). "Implication of Ethiopian Water Development for Egypt and Sudan." Chapel Hill, North Cardina. Mimeo.

Hajnal, J. (1974). "European Marriage Patterns in Perspective." In Population History. Edited by D.V. Glass. London: D.E.C. Eversley.

Harms, H. (1982). "Historical Perspectives on Practice and Purpose of Self-help Housing." In *Self-help Housing: A Critique*. Edited by P. M. Ward. London: Mansell.

Haug, P. (1982). "Applying Living Systems Theory to Ecosystem Simulation." Paper Presented at the Summer Computer Simulation Conference, Denver, Colorado.

Hawkins, E.K. (1962). *Road and Road Transport in an Undeveloped Country: A Case Study*. London: HMSO.

Hayes-Roth F. et al (1983). *Building Expert Systems*. Reading, Mass.: Addison-Wesley.

Heisch, R.B. et al (1958). "The Isolation of the Trypanosoma Rhodensiense From a Bushbuck." *British Journal* vol. 12, no. 1 (1958).

Henin, R. (1979). *Effects of Development on Fertility and Mortality Trends in East Africa: Evidence from Kenya and Tanzania*. Nairobi: Population Studies and Research Institute.

Hewitt, K. (1976). "The Mountain Environment and Geographic Processes." In *Mountain Geomorphology*. Edited by D. Slaymaker and J.H. McPherson. London: Oxford University Press.

Hoselitz, B.F. (1960). *Sociological Factors in Economic Development*. Glencoe, Ill.: Free Press.

Howell, M.L. et al (1988). *Jonglei Canal: Impact and Opportunities*: Cambridge: Cambridge University Press.

Hunter, J.M. (1974). *The Geography of Health and Disease*. Raleigh: University of North Carolina.

Hurst, H.E. (1952). *The Nile: A General Account of the River and the Utilisation of its Waters*. Cairo: Ministry Works.

ILA (1966). *The Helsinki Rules Report by the First-Second Conference of the International Law Association*. Helsinki: International Law Association.

Jarabi, B.O. (1982). "Intra-Urban Mobility and Urban Transportation: A Case Study of Nairobi." M. A. Thesis, University of Nairobi.

Johnson, M. (1981). *The Environmental Law of the Sea*. Gland, Switzerland: International Union for the Conservation of Nature and Natural Resources.

Johnson, S.L. (1970). "Changing Patterns of Maize Utilisation in Western Kenya." *Studies in Third World Societies* vol. 8 (1970).

Jones, M.J. (1974). *Effects of Previous Crops on Maize Yields and Nitrogen Response at Samara, Nigeria*. Samara: Samara Institute of Agricultural Research.

Kanogo, T. (1987). *Squatters and the Roots of Mau Mau*. Nairobi: Heinemann.

Kapila, L. et al (1982). *The Matatu Mode of Public Transport in Metropolitan Nairobi*. Nairobi: Mazingira Institute.

Kasdan, A.R. (1971). "Third World War Environment versus Development." *Record of the Bar Association of the City of New York*, 26 (1971).

Kayongo-Male, D. and P. Onyango (1984). *The Sociology of the Family*. Nairobi: Longman.

Kenya, Government of (1923). *Annual Medical Report for 1922*. Nairobi: Government Press.

___ (1934). *The Kenya (Charter) Land Commission Report*. Nairobi: Government Printer.

___ (1927). *Annual Medical Report for 1926.* Nairobi: Government Printer.

___ (1928). *Annual Medical Report for 1927.* Nairobi: Government Printer.

___ (1929). *Annual Medical Report for 1928.* Nairobi: Government Printer.

___ (1931). *Annual Medical Report for 1930.* Nairobi: Government Printer.

___ (1935). *Annual Medical Report for 1934.* Nairobi: Government Printer.

___ (1966). *Sessional Paper no. 5 of 1966/67 on Housing Policy for Kenya.* Nairobi: Government Printer.

___ (1973). *Kenya Atlas.* Nairobi: Government Printer.

___ (1978). National Water Master Plan of Kenya. Nairobi: Government Printer.

___ (1980a). *Kenya Fertility Survey 1977-78 and World Fertility Survey.* First Report vol.1. Nairobi: Government Printer.

___ (1980b). *Economic Survey.* Nairobi: Government Printer.

___ (1981). *Sessional Paper No. 4 of 1981 on National Food Policy for 1981.* Nairobi: Government Printer.

___ (1982). *Statistical Abstract 1982.* Nairobi: Government Printer.

___ (1983a). *Development Plan 1984-1988.* Nairobi: Government Printer.

___ (1983b). *Economic Survey 1983.* Nairobi: Government Printer.

___ (1986). *Sessional Paper No. 1 of 1986 on Economic Management for Renewed Growth.* Nairobi: Government Printer.

___ (1989). *Development Plan 1989-93.* Nairobi: Government Printer.

___ (1995). *Statistical Abstract of 1995.* Nairobi: Government Printer.

Kibet, M.K.L. (1981). "Mortality Differential in Kenya. Nairobi." M. Sc., Thesis, University of Nairobi.

Kichamu, G.A. (1986). "Mortality Estimation in Kenya With a Special Cause to Vital Registration in Central Province." M.Sc. Thesis, University Of Nairobi.

Kluckhohn, C. (1989). "Theoretical Basis for an Empirical Method of Studying the Acquisition of Culture by Individuals." *Man* 89:1-6.

Krhoda, G.O. (1988) "Water Supply Today and the Year 2000." In *Issues in Population Growth and Resources in Kenya.* Edited by S.H. Ominde. Nairobi: Heinemann.

Krishnamurthy, K.V. (1977). "The Challenge of Africa's Water Development." *Natural Resources Forum* vol. 1 (1977).

Kuyoh, M.A. (1990). "Fertility and Mortality Analysis in Kwale District." M.A. Thesis, University of Nairobi.

Lauer, R.H. (1977). *Persistence in Social Change.* Boston: Allyn and Bacon.

LBDA (1987). *The Study of Integrated Regional Development Master Plan for the Lake Basin Development Area.* Kisumu: Lake Basin Development Authority.

Learmonth, A.T.A. (1978). *Patterns of Disease and Hunger: A Study in Medical Geography.* Newton Abbot, Eng.: David and Cherby.

___ (1988). "Viewpoints on Medical Geography: A Selected Review." *Mapping Disease* vol. 34 (1988).

Lengelle, J.G. (1976). *Anthropogenic Erosion, Swan Hills, Alberta.* Alberta: Environments Conservation Authority.

Levy, M.J. (1966). *Modernization and the Structure of Societies.* Princeton: Princeton University Press

Leys, N. (1973). *Kenya.* London : Frank Cass.

Linn, F. (1983). *Cities in the Developing World: Policies for Their Equitable and Efficient Growth.* London: Oxford University Press.

Lojkine, J. (1976). "Contribution to a Marxist Theory of Capitalist Urbanization." In *Urban Sociology: Critical Essays.* Edited by C. G. Pickvance. London: Methuen.

Lonsdale J.L. (1987). "La Pensee Politique Kikuyu et les Ideologies du Mouvements Mau Mau." *Cahiers d'Etude Africaines* vol. XXVII, no. 3-4 (1987).

Low, D.A. (1971). *Buganda in Modern History.* Berkeley: University of California Press.

Matovu, J.M.K. (1979). "Efficiency of Resource Utilization in Small-scale Farming: A Case Study of Maize and Cotton Production in Machakos and Meru Districts in Kenya." M.Sc. Thesis, University of Nairobi.

McIntyre, L. (1973). "The Lost Empire of the Incas." *National Geographic* vol. 144, no. 6 (1973).

McLelland, D.G. (1961). *The Achieving Society.* New York: The Free Press.

Menken, A.J. (1975). "Biometric Model of Fertility Behaviour." *Social Forces* vol. 54, no. 1 (1975).

Miguda, E. (1987). "The Luo Experience of Mau Mau in Nairobi." M.A. Thesis, University of Nairobi.

Miracle, M.P. (1966). *Maize in Tropical Africa.* London: University of Wisconsin Press.

Moock, P.R. (1973). *Managerial Ability in Small Farm Production: An Analysis of Maize Yields in Vihiga Division of Kenya.* Nairobi: University of Nairobi.

Mordell, D.L. and J. Coales (1983). *A Proposal for the Developing Commonwealth: The Need for Engineers and Technicians and How to Meet it Effectively and Efficiently.* London: Commonwealth.

Mosley, W.H. et al (1981). *Modernization, Child Spacing and Marital Family in Kenya.* Nairobi: Population Studies and Research Institute.

___ (1984). "Analytical Framework for the Study of Child Survival in Developing Countries." *Population and Development Review* Supplement (1984).

Mott, F. (1979). *Infant Mortality in Kenya: Evidence From Kenya Fertility Survey.* Nairobi: Population Studies and Research Institute.

___ (1980). *Infant Mortality in Kenya: Evidence From Kenya Fertility Survey*, Scientific Reports No. 32. London: International Statistical Institute.

Mulligan, H.W. and W.K. Potts (1970). *The African Trypanosomiases.* London: Allen and Unwin.

Muriuki, G. (1974). *A History of the Kikuyu, 1504-1906.* Oxford: Oxford University Press.

Mutai, E.K.I. (1987). "Child Mortality Differentials in Kericho Distict by Location." Diploma Project, Populaion Studies and Research Institute, University of Nairobi.

Mwakwo, A.A. (1982). *After Oil, What Next: Oil Multinationals in Nigeria.* Lagos: Fourth Dimension.

Mwanje, J.I. (1986). "Systems Analysis of Trypanosomiasis Epidemics in Selected Areas of Lake victoria Basin." In *Advances in the Diagnosis, Treatment and Prevention of Immunizable Conference Diseases in Africa.* Edited by S.N. Kinoti et al. Nairobi: University of Nairobi.

Mwanzi, H.A. (1977). *A History of the Kipsigis.* Nairobi: East African Literature.

NCC (1978). "The Nairobi Bus Ways and Feasibility Study." Draft report, Nairobi (1978).

Nerlove, M. and T.P. Schultz (1970). *Love and Life Between the Censuses: A Model of Family Planning in Puerto Rico 1950-1960.* Santa Monica: The Rand Corporation.

Northcott, C.H. (1949). *Africa Labour Efficiency Survey 1947.* London: HMSO.

Nyamwange, F.S. (1982). "Medical Technology, Socio-Economic Status, Demographic Factors and Child Mortality: The Case of Child Mortality Differential in Nairobi." M.Sc. Thesis, University of Nairobi.

Obudho, R.A. (1983). *Urbanisation in Kenya: A Bottoms-Up Approach to Development Planning.* Washington, D.C.: University Press of America.1995).

___ (1993). "Urban Public Transport in Nairobi, Kenya." In *Transports en Afrique Sub-Saharienne.*Lyon: LET.

___ (1997). "Kenya: Urban Communication and Transport in Quest for Research." In *Transport and Communication for Urban Development: Report of the Habitat II Global Workshop.* Nairobi: Habitat.

Obudho, R.A. and M.E. Otieno (1995). "The Urban Non-mortorised Transport in Kenya: The Socio-Cultural, Legal and Policy Constraints." Paper presented at the First World Conference on Transport Research, Sydney, Austrasia, 16-22 July.

Odero-Ogwell, L.A. (1982). "The African Food Problem and the Challenge to International Community." Nairobi, World Food Council. Mimeo.

Ogot, B.A. (1964). "Reintroducing the African Man into the World." *East African Journal* vol. 3 (1964).

___ (1967). *The History of the Southern Luo.* Nairobi: East African Publishing House.

Ojwang, J.B. (1990). *Constitutional Development in Kenya.* Nairobi: ACTS Press.

Ojwang, R. (1992). "The Seasonality of Deaths in Kenya." M.A. Thesis, University of Nairobi.

Okidi, C.O. (1984). "Management of Natural Resources and the Environment for Self Reliance." *Journal of Eastern African Research and Development* vol. 14 (1984).

___ (1986). *Development and the Environment in the Kagera Basin Under the Rusumo Treaty.* Nairobi: Institute for Development Studies.

___ (1988). "Irrigation Activities in Kenya." African Urban Quarterly 3, no. 113 (1988).

Ominde, S.H. (1952). *The Luo Girl from Infancy to Marriage.* London: Macmillan.

___ (1965). "Geography and African Development Nairobi." Inaugural Lecture Delivered at the University of Nairobi, 1965.

___ (1968). *Land and Population Movements in Kenya.* Nairobi: Heinmann.

___ (1968). *Land and Population Movements in Kenya.* Nairobi: Heinmann.

___ (1981). "Population and Resource Crisis: A Kenyan Case Study." *Geojournal* vol. 6, no. 6 (1981).

___ (1988). *Kenya Population Growth and Development for the Year 2000 AD.* Nairobi: Heinemann.

Ominde, S.H. and R.A. Henin (1976). "The Population Studies Research Centre." *Jimla Mutane* vol. 1, no. 2 (1976).

Onaka, A.T. and D. Yaujey (1973). "The Reproductive Time Lost Due to Sexual Union Dissolution in San Jose, Costa Rica." *Population Studies*, 27 (1973).

Ondimu, K.N. (1983). "Socio Economic Determinants of Mortality in Kenya: A Look at KCPS." M.Sc. Thesis, University of Nairobi.

Onyango, R.J. et al (1986). "The Epidemology of T. Rhedesiense Sleeping Sickness in Alego Location Central Nyanza, Kenya: Evidence of Trypanosomes Infection to Man." *Transcript of the Royal Society of Tropical Medicine Hygiene* 60 (1986).

Oucho, J.O. (1974). "Migration Survey in Kisumu Town." M.A. Thesis, University of Nairobi.

Parsons, T. (1951). *The Social System.* New York: The Free Press.

Paul, B.K. (1985). "Approaches to Medical Geography: An Historical Perspective." *Science* 20, no. 4 (1985).

Peatie, L.R. (1979). "Housing Policy in Developing Countries: Two Puzzles." *World Development* vol. 7, no. 11/12 (1979).

Perberdy, J.R. (1972). "Rangeland." In *East Africa: Its Peoples and Resources*. Edited by W.T.W. Morgan. Oxford: Oxford University Press.

Philips, D.R. (1981). "Directions for Medical Geography in the 1980s: Some Observations from the United Kingdom." *Social Science Medicine* vol. 4 (1985).

Poulantzas, N. (1973). *Political Power and Social Classes*. London: New Left Books.

_____ (1975). *Classes in Contemporary Capitalism*. London: New Left Books.

Preston, S.H. (1979). "Causes and Consequences of Mortality Declines in Less Developed Countries During the 20th Century." In *Population and Economic Change in Less Developed Countries*. Edited by . Chicago: University of Chicago Press.

Pyle, G.F. (1977). "International Communication and Medical Geography." *Social Science Medicine* vol. II (1977): 679-682.

Rapemo, J.O. (1991). "Estimation of Fertility Levels for Kenya Using the Reverse Survival Technique." M.Sc. Thesis, University of Nairobi.

Rempel, H. (1978). *The Role of Rural-Urban Migration in the Urbanization and Economic Development Occurring in Kenya*, Research Memorandum RM-78-12. Laxemburg: Austria International Institute for Applied Systems Analysis.

Rogers, A. (1979). "Income Inequality as Determinants of Mortality: An International Cross Section Analysis." *Population Studies* vol.41 (1979).

Rostow, W.W. (1978). *The World Economy: History and Prospect*. Austin: University of Texas Press.

Roxborough, I. (1979). *Theory of Underdevelopment*. London: The Macmillan.

Rummel, R.J. (1967). "Understanding Factor Analysis." *Journal of Conflict Resolution* vol. 2, no. 4 (1967).

Rundquist, F.M. (1984). *Hybrid Maize Diffusion in Kenya*. Lund: CWK Gleeup.

Saunders, P. (1980). *Urban Politics: Sociological Interpretation*. London: Hutchinson

Scargill, D.I. (1975). *Problem Regions of Europe: The Eastern Alps*. London: Oxford University Press.

Selby, M.J. (1976). "Slope Erosion Due to Extreme Rainfall: A Case Study From New Zealand." *Geografiska Annaler* vol. 58A, no. 3 (1976).

Shah, M.M. (1981). "The Kenya Agricultural Model." In *Food for All in a Sustainable World: The IIASA Food and Agricultural Progamme*. Edited by K. Parikh and F. Rabar. Laxemburgh: International Institute for Applied Systems Analysis.

Shah, M.M. et al (1984). *Africa's Growing Dependence on Imported Wheat: Some Implications for Agricultural Policies in Africa*. Luxembourg. International Institute for Applied Systems Analysis.

Shafick, H. (1973). "Childhood Mortality Experience and Fertility Performance in Egypt." In *Population Problems and Prospects*. Edited by A.R. Omran et al. Chapel Hill, N. Car.: Carolina Population Centre.

Shoemaker, C. (1973). "Optimization of Agriculture Pest Management: Biological and Mathematical Background." Mathematical Bioscience vol. 16 (1973).

Simango, V.A. (1976). "Maize Yield in Relation to Rainfall and Solar Energy in Kenya." M.Sc. Thesis, University of Nairobi.

Skalnik, P. (1978). "Uneven and Combined Development in European Mountain Communities." In *Society and Environment: The Crisis in the Mountains*. Edited by D. Pitt. London: Oxford University Press.

Stockwell, E.G. and K.A. Laidlaw (1981). "The Sociology of Development." Interview of Modern Sociology vol. 11, no. 1-2 (1981).

Sullivan, J. (1972). "Models for the Estimation of the Probability of Dying Between Birth and Exact Ages of Early Childhood." *Population Studies* vol. 26, no. 1 (1975).

Swanston D.N. and C.I. Dyrness (1973). "Stability of Steep Land." *Journal of Forestry* vol. 71, no. 5 (1973).

Syagga, P.M. (1987b). "Myths and Reality of Low Cost Housing in Africa." *African Urban Quarterly* vol. 2, no. 3 (1987).

___ (1988). "Social Acceptability of Local Building Materials and Their Application in the Construction of Shelter Built-Form." In Local Building Materials and technology. Edited by W.R. Okot-Uma. London: Commonwealth Science Council.

Tauber, I. (1960). Continuities in the Declining Fertility of Japanese." *The Milbank Memorial Fund Quarterly* vol. 38, no. 3 (1960).

Transurb-Consult (1986). *Study on Urban Transport Needs of Nairobi*, Final Report. Nairobi. Government Printer.

Tridrick, G. (1979). *Kenya: Issues in Agricultural Development.* Nairobi: University of Nairobi.

Trussel, J. and S. Preston (1980). "Estimating the Covariate of Childhood Mortality from Retrospective Reports of Mothers." Paper Presented at the Population Association of America Meeting.

Turner, J.F.C. (1970). "Barrios and Channels for Housing Development in Modernizing Countries." In *Peasants in Cities*. Edited by W.Mangin. Boston: Boston University Press.

Uganda, Government of (1957). *The Equatorial Nile Project and the Nile Waters Agreement of 1929: East Africa's Case.* Entebbe: Government Printer.

UN (1973). *The Determinants and Consequences of Population Trends*, Vol.1. Population Studies. New York: United Nations.

UNDP and WMO (1974). *Report of the Hydrometeorological Survey of the Catchment of Lakes Victoria, Kyoga and Albert: Meteorology and Hydrology of the Basin*, Report No. 1. New York: United Nations Development Programme.

UNECA (1988). Report of the International Meeting on River and Lake Basin Development With Emphasis on the African Region. Addis Ababa: United Nations Economic Commission for Africa.

___ (1979). *Population Dynamics: Fertility and Mortality in Africa*. Proceedings of the Expert Group Meeting on Fertility and Mortality Levels and Trends in Africa and the Policy Implications, Monrovia, Liberia. Addis Ababa: United Nations Economic Commission for Africa.

Vansina, J. (1969). *Oral Traditional: A Study in Historical Methodology*. London: Routledge and Kegan Paul.

Walker, T. (1988). "The Nile Struggles to Keep up the Flow." *Sunday Nutrey Journal* 26 (1988).

Wekwete, T.S. (1988). "Development of Urban Planning in Zimbabwe: An Overview." *Cities* vol. 5, no. 1 (1988).

Were, G.S. (1967). *A History of the Abaluyia of Western Kenya*. Nairobi: East African Publishing House.

Werlin, H.H. (1974). *Government of an African City: A Study of Nairobi*. New York: Africana.

World Bank (1975a). *Housing: A Sector Policy Paper*. Washington, D.C.: World Bank.

___ (1975b). *Urban Transport Sector Policy Paper*. Washington, D.C.: World Bank.

___ (1980). *Kenya: Population and Development*. Washington, D.C.: World Bank.

___ (1986). *Annual Report 1986*. Washington, D. C.: World Bank.

World Commission on Environment and Development (1987). *Our Common Future*. London: Oxford University Press.

Appendix

Simeon Hongo Ominde: A Biographical Note

H.A. Liyai

From Nyahera to Scotland

Professor Simeon Hongo Ominde was born at Nyahera in Kisumu District, Kenya, on June 14 1924. He was the fourth born in a family of seven children. His father, John Owiye was a founder Deacon of the Africa Inland Mission at Ogada in Kisumu Central. His Mother, Lydia Anyim, was one of the first girls to be admitted to the same church for catechism and literacy. This strong religious background was later to influence the young Ominde's ambitions and the pursuit of education. Both parents instilled strict discipline on their children.

At a tender age Ominde was sent to Kima Mission for primary education. He was so small that he was nicknamed "Omutiti"(meaning small in Luhya), but he soon established himself as a serious pupil and was soon borrowing notes form secondary school students at the neighbouring Maseno Secondary School. He completed primary school in 1939 to sit for the national examination which he passed well.

In 1940 he joined Maseno Secondary School, then one of the best schools in Kenya, for pre-O-Level education where his favourite subjects were geography, nature study, Kiswahili, English, biology and physics. He had a blind spot in mathematics which, after developing fresh interest while at Alliance High School, he passed with a distinction in the final examination. Among his teachers at Maseno was Jaramogi Oginga Odinga, Kenyaís first Vice-President, and later doyen of opposition politics.

Ominde joined Alliance High School in 1943 where he distinguished himself in both class work and extra-curricula activities. He sat for Cambridge School Certificate in 1944, which he passed with distinction to gain admission to Makerere University College (Uganda) in 1945.

At Makerere, then the seat of high learning in Eastern Africa, Ominde took interest in research and wide reading, and he subsequently started receiving awards, prizes and honours, a mark that was to continue throughout his life. He recieved the Uganda Bookshop Prize for the essay "The Meaning of Freedom", the Arche and Sturrock Prize for the best students in Arts, and the Arts Research Prize for the manuscript "Luo Girl From Infancy to Childhood". The latter was later published. He was awarded a Diploma in 1947 and later the Teachers Certificate (Division 1) the following year (1948). He returned to Kenya at the end of 1947 and began teaching at his old school (Maseno).

In September 1949, Ominde joined the University of Aberdeen, Scotland, to pursue a Masters degree in geography. He distinguished himself by winning a silver medal of the

Royal Geographical Society Award upon his graduation in 1954, and remained a fellow of the Society. In 1955 he moved to the University of Edinburgh where he studied for a diploma in Education and also obtained a Scottish Teacher's Certificate.

In May 1964, Ominde joined the University College, Nairobi (later the University of Nairobi), and was appointed Professor and Head of Geography Department becoming the first African Professor in East Africa, this put to an end the myth that academic departments could not function unless headed by non-Africans. The following year, he become the first African to be elected Dean of the Faculty of Arts. While at the University, he served as a member of the University Council, the Inter-University Committee for East Africa, and in other innumerable boards and committees.

In the world of academia in one's scholarly career, it is said that one has to profess knowledge and publish or perish. Ominde lived up to his calling, and he delivered a professional inaugural lecture on "Geography and African Development" in 1967. Over the years he has published monographs, articles and reports and has also presented many papers in forums locally and abroad. (The wide range of his scholarly contributions is contained in the Appendix.)

From Scotland, Ominde returned home in 1955 and went to Makerere University College where he had already been appointed an assistant lecturer the previous year. This marked the start of a brilliant career combining teaching, research and publishing. He was one of the pioneer African academics in university education, then dominated by the non-Africans.

Parallel with the struggle for independence, Ominde joined the generation of scholars that sought to establish an indigenous university with full African participation. Consequently, he found himself occupying a unique position of a freedom fighter in the academic battlefield at a time when his contemporaries were battling it out in the political arena.

Research and education

Besides teaching, Ominde started research work as a part of the preparation for his doctorate thesis for the University of London, and was awarded the degree in 1963. His thesis was later published under the title "Land and Population Movements in Kenya" (1968) as an important contribution to the study of economic geography in developing countries. He was promoted to lecturer in 1958 and Senior Lecturer in 1963. In the same year he was elected a representative on the Academic Board of the College. This again marked the beginning of another aspect of academic leadership that was to span his entire career.

In Kenya, the colonial government operated a system of education that favoured racial segregation much to the disadvantage of the African majority. On the eve of Kenya's independence there was an obvious necessity to restructure the entire education system, if only to redress the biases, universalise education and remove all the racial bigotry that colonialism had encouraged.

Based on his experience in Uganda. Ominde had earlier prepared a paper on suggestions for the restructuring of Kenya's education system which he distributed to the African members of the Legislative Council.

Perhaps Ominde's most remarkable contribution to education in Kenya was his steering role in the commission to review the education system. The subsequent Education Commission Report (1964) laid the foundation for a national system of education in Kenya whose philosophy was the creation of a system with equity and national unity, addressing national development needs, thereby scrapping the colonial elitist legacy as well as the racial system of schools.

Professor Ominde also served as a member in another education commision set up in 1976, the National Committee on Educational Objectives and Policies chaired by Peter Gachathi.

Population Research

Ominde's involvement in population research put him at the centre of the "The Great Movement" on population at international level. In 1965, he was one of the people invited by the United Nations to advise on population programmes from which task he got an insight into the direction the United Nations Population Programme was likely to go. He was equally interested in population studies in relation to planning of social economic development, and hence saw need for the establishment and development of an institution to train young scholars in population.

After many years of relentless personal campaigns and lobbying, the Population Studies and Research Institute (PSRI) was established at the University of Nairobi. He was its founder and director, and he steered its research and training programmes for over a decade.

Public Service

Ominde's other public appointments include: Chairman of the Kenya Central Scholarship Board; Member, Commission of Inquiry into the Public Service Structure and Remuneration (1970); Member, National Committee on Human Environment (1971); Member, Government Delegation to the Commonwealth Education Conference on Science and Technology for Development, Vienna (1979); and Chairman, Presidential Committee on Terms and Conditions of Service for the Police Force and Prison Service (1987).

In the private sector, Ominde was the Chairman of a leading book publishing company, Heinemann Kenya Ltd. (now East African Educational Publishers) and a director on the Managing Board of The Nation Group of Newspapers (another leading media company in East Africa.) In 1984, he was appointed Chairman of Kenya Re-insurance Corporation, a position he held until his death.

A Bibliographical Survey

Books and Monographs

The Luo Girl from Infancy to Marriage. London: Macmillan, 1952.

The Ethnic Map of the Republic of Kenya, Occasional Memoir No. 1, Nairobi: University of Nairobi, 1965.

Land and Population Movements in Kenya. London: Heinemann, 1968.

Soil and Land Use Survey of the Kano Plain, Nyanza Province, Kenya, Occasional Memoir Number 2. Nairobi: University of Nairobi, 1973.

The Harambee Movement in Education Development, Occasional Paper 4. Vienna: Vienna Institute for Development, 1974.

Kenya's Population Growth and Development to the Year 2000. Nairobi: Heinemann, 1988.

Population and Development in Kenya. Nairobi: Heinemann, 1984.

Population Growth and Economic Development in Africa. London: Heinemann, 1972 (with C.N. Enjogu).

The Population of Kenya, Tanzania and Uganda. Nairobi: Heinemann, 1975.

Urban Growth in Africa: Rural Urban Drift in Africa, London: Commonwealth Foundation, Occasional Paper No. 29. 1980.

Edited volumes

East Africa UNESCO Source Book in African Geography. Paris: UNESCO, 1965(with R.S. Odingo and F.F. Ojany).

Kenya Education Commission Report, 1964-1965, Volume 1 and 2. Nairobi: Government Printer. (Chairman S.H. Ominde.)

Kenya's Population Growth and Development to the Year 2000. Nairobi: Heinemann, 1988.

Population and Development in Kenya. Co-edited with R.A. Henin and D.F. Sly. Nairobi: Heinemann, 1984.

Nilos Bernados: Colloquium on Regional Inequalities and Development, Brazil. Co-edited with R.S. Odingo. 1971.

Population Growth and Economic Development in Africa. Co-edited with C.N. Ejiogu. London: Heinemann, 1972.

Studies in East African Development: Essays Presented to J.S.K. Baker. London: Heinemann, 1971.

Non-book Material

Kenya: Distribution of Population, 1969 Census. (Map) III x 89 cm on 2 Sheets 64 x 91 cm; Scale 1:1,000,000. Nairobi: Survey of Kenya, 1970. (Compiled by Department of Geography, University of Nairobi, under the Direction of S.H. Ominde.)

Contributions in edited volumes

"Demographic Aspects of Regional Inequalities in Kenya." In *Nilos Bernados: Colloquium on Regional Inequalities and Development, Brazil.* Edited by S.H.Ominde and R.S. Odingo. 1971.

"Demographic, Economic and Social Implications of Rural-Urban Differential in Kenya." In *The Demographical Transition in Tropical Africa: Proceedings of an Expert Group Meeting, Paris, 17-19 November 1970*. Paris: Organization for Economic Co-operation and Development (OECD), 1971.

"The Demographic Transition and Educational Planning in Kenya." In *The Demographic Transition in Tropical Africa: Proceedings of an Expert Group Meeting, Paris, 17-19 November 1970*. Paris: OECD, 1971. "Demography and Ethnic Groups." In *Health and Disease in Kenya*. Edited by L.C. Vogel et al. Nairobi: East African Literature Bureau, 1973.

"Demography and the Law in Africa." In *Law and Population Change in Africa*. Edited by U.U. Uche. Nairobi: East African Literature Bureau, 1976.

"The East African Environment." In *East Africa Past and Present*. Edited by D.J. Stenning. Press: Presence Africaine, 1964.

"Ecology and Man in East Africa." In *Hadith 7: Ecology and History in East Africa*. Edited by B.A. Ogot. Nairobi: Kenya Literature Bureau, 1979.

"Educational Needs and Services in the Lake Victoria Basin." In *Natural Resources and the Development of Lake Victoria Basin in Kenya*. Edited by C.O. Okidi. Nairobi: Institute for Development Studies.

"The Former Scheduled Areas." In *Population and Development in Kenya*. Edited by S.H. Ominde, R.A. Henin and D.F. Sly. Nairobi: Heinemann 1984.

"General Background to the Population Situation in the World and Kenya." In *Population and Development in Kenya*. Edited by S.H. Ominde, R.A. Henin and D.F. Sly. Nairobi: Heinemann, 1984.

"Geography and African Development." In *Studies in East African Development: Essays Presented to J.S.K. Baker*. Edited by S.H. Ominde. London: Heinemann, 1971.

"The Harambee Movement in Educational Development." In *Alternatives in Education*. Edited by H. Steger. Munich: Wilhelm Fiuk Verlag, 1984.

"The Impact of Rapid Population Growth on Socio-Economic Development." In *Centre for African Family Studies: Report on the Sub-Regional Seminars on Population and Development for Parliamentarians from Eastern and Southern Africa, Zimbabwe, 9-15 April 1983*.

"Integration of Environment and Development Planning for Ecological Crisis in Africa." In *Ecological Systems and Ecopolitics: UNESCO Symposium on Social Sciences and Development Planning*. Edited by K.W. Deutch. Paris: UNESCO, 1977.

"Internal Migration of the Economically Active Age Group in Kenya." In *Ostafrikanische (East African Studies): Ferstchrift in Honour of Prof. Dr. Ernst Weight on his 60th Birthday*. Edited by H. Bergert. Nurnberg: Verlag Nurn Berger Press, 1968.

"Introduction: A Bird's Eye View of the Republic in the Modern World." In *Population and Development in Kenya*. Edited by S.H. Ominde, R.A. Henin and D.F. Sly. Nairobi: Heinemann, 1984.

"Kenya." In *Encyclopaedia Britannica*, Volume 10. Chicago: Encyclopedia Britannica, 1971.

"Keynote Address at the First Provincial Planners Seminar." In *Proceedings of the First Provincial Planning Seminar on the Use of Population Data in Economic and Social Planning, Kericho 16-17 August, 1978.* Nairobi: Population Studies and Research Institute, 1978.

"The Land and the People." In *Kenya, An Official Handbook.* Nairobi: East African Publishing House, 1977.

"Migration and Child-Bearing in Kenya." In *Population Growth and Economic Development in Africa.* Edited by S.H. Ominde and C.N. Ejiogu. London: Heinemann, 1972.

"Migration and the Structure of Population in Nairobi." In *Population in African Development*, Volume 1. Edited by P. Cantrelle et al. Dolhain: IUSSP, 1974.

"Migration and Urbanization in the Coastal Region of Kenya." In *Proceedings of the S/21 International Geographical Union Symposium.* New Delhi: International Geographical Union, 1968.

"Migration of 15-44 Year Age Group in Kenya." In *University of East Africa Social Science Council Conference, 1968/69*, Volume 2, Geography Papers. Kampala: Makerere Institute of Social Research, 1969.

"Movements to Towns from Nyanza Province, Kenya." In *Urbanization in Africa Social Change.* Proceedings of the Inaugural Seminar Held in the Centre of African Studies, University of Edinburgh, 5-7, January 1963: 23-33.

"Nairobi City." In *Encyclopaedia Britannica* (Macropaedia—Knowledge in Depth) Volume 12. Chicago: Encyclopaedia, 1971.

"Notes on Importance of Co-ordination of Activities Aimed at Uplifting of Life." In *Family Planning Association of Kenya: Report of the Management Seminar for Senior Volunteers and Staff Held at Safariland Hotel, Nairobi, 1977.* Nairobi: Family Planning Association of Kenya, 1977.

"Number of People and Food Resources." In *Africa Symposium on Food Population and Development, Nairobi, 24-31 October 1974.* Nairobi: International Youth and Student Movement for the United Nations (Africa Secretariat), 1974.

"The Pattern of Land Classification." In *Population and Development in Kenya.* Edited by S.H. Ominde, R.A. Henin and D.F. Sly. Nairobi: Heinemann 1984.

"Population Change and Development Challenge." In *Population Growth and its Relevance to Socio-Economic Development: Proceedings of the First Annual Symposium of the Kenya National Academy for the Advancement of Arts and Sciences.* Edited by J.K.G. Mati and K.A. Buigutt. Nairobi: Kenya National Academy of Advancement of Arts and Sciences, 1980.

"Population Distribution and Urbanization." In *Population and Development in Kenya.* Edited by S.H. Ominde, R.A. Henin and D.F. Sly. Nairobi: Heinemann, 1984.

"Population Geography and the Geographical Study of Migration." In *University of East Africa Social Science Council Conference, 1968/69*, Volume 2, Geography Papers. Kampala: Makerere Institute of Social Research, 1969.

"Population of Kenya." In *Population and Development in Kenya.* Edited by S.H. Ominde, R.A. Henin and D.F. Sly. Nairobi: Heinemann, 1984.

"Population Trends and Health with Special Reference to Kenya." In *Kenya's Population Growth and Development to the Year 2000*. Edited by S.H. Ominde. Nairobi: Heinemann, 1988.

"Problems of Land and Population in the Lake Districts of Western Kenya." In *East African Academy Proceedings of the First Symposium, Makerere University College, June 1963*. Nairobi: Longman, 1964.

"Problems of Land and Population in the Lake Districts of Western Kenya." In *Proceedings of the East African Academy*, Volume 1. Kampala: East African Academy, 1963.

"Regional Distribution of the Employment Problem in Kenya." In *The Spatial Structure of Development in Kenya*. Edited by R.A. Obudho and D.R.F. Taylor. Boulder: Westview Press, 1979.

"Resettlement, Urbanisation, Housing and Community Development." In *Population in African Development*, Volume 1. Edited by P. Canrelle et al. Dolhain: IUSSP, 1974. (With N.O. Addo.)

"Rural Economy in West Kenya." In *Studies in East African Development: Essays Presented to J.S.K. Baker*. Edited by S.H. Ominde. London: Heinemann, 1971.

"Rural Population Patterns and Problems of the Kikuyu, Embu and Meru Districts of Kenya." In *Proceedings of the East African Academy, Second Symposium, University College, Nairobi, June 1964*. Edited by W.B. Banage. Nairobi: East African Academy, 1966.

"Some Aspects of Population Movements in Kenya." In *The Population of Tropical Africa: Proceedings of the First African Population Conference, Ibadan 1966*. Edited by J.C. Caldwell and C. Okonjo. London: Longman, 1968.

"S.J.K. Baker: A Biographical Note." In *Studies in East African Geography and Development: Essays Presented to J.S.K. Baker*. Edited by S.H. Ominde. London: Heinemann, 1971.

"The Semi-Arid and Arid Lands of Kenya." In *Studies in East African Geography and Development: Essays Presented to J.S.K. Baker*. Edited by S.H. Ominde. London: Heinemann, 1971.

"Some Observations on Migration and Regional Variations in Child Bearing Incidence in Kenya." In *Proceedings of the Seminar on Population Growth and Economic Development, Nairobi, 14-22, December 1969*. London: Heinemann, 1972.

"The Structure of Education in Kenya and Some Planning Problems." In *Education, Employment and Rural Development: Proceedings of a Conference Held at Kericho, Kenya, September 1966*. Edited by J.R. Sheffield. Nairobi: East African Publishing House 1967.

"Total Population of Kenya." In *Population and Development in Kenya*. Edited by S.H. Ominde, R.A. Henin and D.F. Sly. Nairobi: Heinemann 1984.

"What Do We Know in the Social Sciences." In *Agricultural Research for Rural Development: Proceedings of the East African Academy*, Volume 6. Edited by S.M. Mbilinyi. Nairobi: EALB, 1973.

Articles in Periodicals

"Advanced Geography of Africa." *Kenyan Geographer* vol. 2, no.2 (1976): 145-147.

"An Appreciation." *Kenyan Geographer* vol. 1, no. 2 (1975): 173-177.

"The City of Nairobi: Population Changes and Patterns." Journal of Eastern African Research and Development vol. 1, no. 1 (1971): 77-87. "Education and Political Development." *East Africa Journal* vol. 2, no. 10 (1966): 38-39.

"Education in Revolutionary Africa." *East Africa Journal* vol. 2, no. 2 (May 1965): 6-14.

"The Kano Plains—A Geographical Challenge." *Africa Scientist*, no. 1 (August 1969): 7-20. "Literacy as a Social Priority in Nation Building." *The Bookshop Bulletin* (December 1960).

"Population Movements to the Main Urban Areas of Kenya." *Cahiers D'Etudes Africaines* vol. 5, no. 20 (1965): 593-617.

"The Population Studies Research Centre." *Jimla Mutane* vol. 1, no. 2 (1976): 304-312. (With R.A. Henin.)"Population Trends and Demographic Factors in Africa with Special References to Unemployment." *Jimla Mutane* vol. 1, no. 1 (1979): 175-190.

"Population and Resource Crisis: A Kenya 'Case Study'." *Geojournal* vol. 6, no. 6 (1981): 537-556.

"The Population Factor in Kenya's Economic Development." *Bulletin of the International Institute for Labour Studies*, No. 39 (1967): 14-28.

"The Population Studies Research Centre." *Jimla Mutane* vol. 1, no. 2 (1976): 304-312. (With R.A. Henin.)

"Regional Patterns of Fertility in Kenya." *Kenya Geographer* vol. 1, no. 1 (1975): 13-29.

"Spatial Population Change: Kenya Case Study." Geojournal vol. 8, no. 5 and 6 (1977): 231-241.

Unpublished Papers and Reports

"African Relevance of Environmental Education Training and Research." Paper presented at UN Workshop on Environmental Education and Training in African Universities, organised by Association of African Universities, 1978.

"Case Study: Urbanisation and Environment in Kenya." Paper presented at the United Nations Conference on Human Environment, Nairobi, 1971 (with A.N. Ligale and A.B. Cahusac).

"Central and Eastern Provinces." Keynote Address to Provincial Planning Officers, Central and Eastern Province, University Of Nairobi May, 1979.

"Central Province: Population Pattern, 1962-1979." Nairobi: Population Studies and Research Institute, 1981.Mimeo.

"The Changing Population Pattern of the Central Province, Kenya." Paper presented at the West African Regional Conference, Commonwealth Association of Geographers, Accra, 1970. "Coast Province: Population Profile." Nairobi: Population Studies and Research Institute, 1981.Mimeo.

"The Dangers of Population Explosion to the Development in Kenya." Paper presented at the National Rotary Club, 31 August, 1983.

"The Demographic Factor and Issues in Human Resource Planning and Use." Keynote Address, National Study Seminar on Population, Human Resources and Development, held at Tom Mboya Labour College, Kisumu, 10-12 February, 1985.

"Demographic Overview." Central Bureau of Statistics/PSRI, Nairobi, November, 1982.Mimeo.

"Demographic Situation in Kenya." Paper presented to the National Council for Science and Technology, Nairobi, March, 1980.

"Dynamics in Kenya." Paper presented at the Kenya Medical Association Annual Scientific Conference, Mombasa 24-26, April, 1986.

"Economic Development and Population Aspects With Special Reference to Kenya." Paper Presented at the Kenya Institute of Administration, 1 September, 1978.

"Environment Development and Peace: A Kenyan Study." Paper presented at the Vasterhaninge Seminar, Sweden, 1973.

"Environmental Issues in Population." Paper presented at the African Regional Workshop on Environmental Education for Adult Education, Nairobi, 4 December, 1978.

"Environmental Problems Found Mainly in the Developing Countries." Paper presented at the United Nations Symposium on Population Resources and Environment, Stockholm, 26 September-5 October 1973.

"Family Life Education." Keynote Address to the 24th Annual Delegates Conference, Kenya National Union of Teachers, Nairobi, 16 December, 1981.

"Geography and African Development." Inaugural Lecture Delivered at the Nairobi University College 1965.

"Implications of Urban Growth of Kenya." Paper presented to the Kenya Economic Association, Nairobi 3 April, 1980.

"Kenya Population Profile." Nairobi, Population Studies and Research Institute, 1981.

"Kenya's Population Trends." Paper presented at the National Seminar on Human Settlement, Nairobi, 7-11 March 1988.

"Kisumu Multi-Purpose Survey, Preliminary Report." Paper presented to the Population Policy Research Programme Conference, Bellagio, Italy, 1973.

"The Lake Basin Demographic Profile and Agricultural Land Availability." Nairobi, Population Studies and Research Institute, 1981.

"Land and Population." Paper Presented at the ILO Seminar on Economic and Social Development in Southern Africa, 1977.

"Migration and Urbanisation in the Coastal Region of Kenya." Paper presented to the Department of Geography, University College, Nairobi. "Notes: Scope and Limitation of Population Census Data in Kenya." Paper presented at the Department of Geography's, University of Nairobi, Seminar, 1973.

"Nyanza and Western Provinces." Keynote Address to Provincial Planning Officers from Nyanza and Western Provinces, Population Studies and Research Institute, August, 1978.

"Population and Development." Paper presented to the Seminar on Moral and Ethical Issues in Population Dynamics and Development, Accra, 31 March-4 April, 1974.

"Population and Family Planning: The African Dimension." Paper presented at the Senior Staff Course in Integrated Family Welfare, Nairobi, 2 March-24 April, 1981.

"Population and the State of the Environment." Nairobi, Population Studies and Research Institute, 1979.Mimeo.

"Population Change and Socio-Economic Development in East Africa." Nairobi, Population Studies and Research Institute, 1980.

"Population Change in African Countries." Paper presented at International Geographical Congress, Tokyo, 29 August-6 September, 1980.

"Population Distribution in Kenya." Nairobi, Population Studies and Research Institute, 1982.

"Population Distribution in the Coast Province, 1962-1979." Nairobi, Population Studies and Research Institute, 1981.

"Population Distribution in the Eastern Province, 1962-1979." Nairobi, Population Studies and Research Institute, 1981.

"Population Distribution in the Lake Basin Authority Area." Nairobi, Population Studies and Research Institute, 1981.

"Population Explosion in Kenya: A Reality or A Myth." Paper presented at the Seminar on Population and Development, Mwea Lodge, Uganda, 24-30 June, 1984.

"Population Factor in Economic Development in Kenya." Lecture at the Seminar on Labour Problems in Economic Development, International Institute for Labour Studies, East Africa, Nairobi, 18 April, 1967. "Population Growth and Resource Development in Africa." A Kenyan Case Study Paper presented at IGU Regional Conference, Lagos, 1-8 August, 1978.

"Population Growth, Human Settlement and Development Strategy in Kenya." Paper presented at the National Seminar on Human Settlements, Nairobi, 21-23 May, 1975.

"The Population Issue in Kenya." Paper presented at the First Annual Medical Research Institute, (KEMRI) and Kenya Trypanosomiasis Research Institute (KETRI), Nairobi, January-February, 1980.

"Population Problems and Urbanisation in Kenya." Paper presented to the Postgraduate Medical Seminar on Population Studies, December, 1973. "Population Problems of Nyanza and Western Regions of Kenya, With Special Reference to Uganda." Lecture to the Members of the Uganda Society, 1963.

"Population Research in Institutions in Kenya." Paper Presented at the United Nations, 1979.

"Population Trends and Demographic Factors in East Africa With Special Reference to Unemployment." Paper presented to the ILO/IDS Seminar, Limuru, 1973.

"Problems of Population in Africa: Philosophy, Man and the Environment." Paper presented at the World Conference of Philosophy, Nairobi, 21-25 July, 1991.

"Report on Kisumu Population, Health Nutrition and Family Planning." Nairobi, Population Studies and Research Institute, April 1983. (With A.B.C. Ocholla-Ayayo and J.C. Oyieng.)

"The Rift Valley Province: Population Pattern, 1962-1979." Nairobi, Population Studies and Research Institute, 1981.

"The Role of Medical Profession in Family Health and Population Dynamics in Kenya." Paper presented at the Kenya Medical Association Annual Scientific Conference, Mombasa, 24-26 April, 1986.

"The Role of Social Studies in African Education with Special Reference to Kenya." Paper presented to the Conference of African Educators EDC and CREDU, Mombasa, 1986.

"Science and Technology for Development in Kenya." Paper presented at the National Council for Science and Technology, Nairobi, 1980.

"Some Aspects of Urban Growth in Africa." Population Studies and Research Institute, 1970.Mimeo.

"Some Population Characteristics of the Main Urban Centres in Kenya." Paper prepared for the International Population Conference, IUSSP, Liege, Belgium, August, 1973.

"Spatial Population Change as a Framework for Rural-Urban Interaction." Keynote Address, Seminar on Migration Remittances and Rural Development in Kenya, Sponsored by IDRC and Ford Foundation, Nairobi, 1984.

"Spatial Population Change in Kenya." Nairobi, Population Studies and Research Institute, 1980.

"The Teacher in the Community." Report of the Conference of Teachers Education for East Africa, University College, Nairobi, 1965.

"Urban Growth in Africa." Paper present at the Commonwealth Association of Planners, Africa Region, Conference, Nairobi, 10-16 February, 1974. "Urban Rural Problems of the African Environment." Paper presented to the Economic Commission for Africa Seminar on Human Environment, Addis Ababa, August, 1971.

"Urbanisation in Africa: A Case Study." Nairobi, Population Studies and Research Institute, 1980.

"Urbanisation in Africa: A Kenyan Case Study." ILO Study Course on Social and Economic Development in Southern Africa, 1980.

Index

Abagusii, 101
Aberdeen University, 5
Aberta, Swan Hills of, 36
Absentee landowners, problems of 230
Acid rain, 60
Aduwo, G.O., 271
Aetiological aspects, 94
African demography, 6
African economics, 204
African geography, 3
African history, 20
African labour efficiency survey, 181
African land development (ALDEV), 74
African migrant workers 177
African phenomena, 21
  study of, 18
African:
  professors, 18
  scholars, 25
  scholarship, 20
Africa's drainage basins, 41, 43
  agricultural density, 191
  agricultural engineer, 185
Africanisation, 21
Ainsworth, John, 178
Akech, S.O., 214
Alps, 29, 32
Anthropodogical:
  data, 160
  techniques, 162
Anthropologists, 163
Anthropology, goals of, 162
Aquatic weeds, 73
Aquifers, 68
Arable land, 195
Arid and semi-arid lands, 195, 202, 207
Arithemetic density, 197
Artificial intelligence (AI), 98
ASAL *see* Arid and semi-arid lands.
Aswan High Dam, 47, 49, 56
Atlas mountains, 31
Axany Samuel, 10
Ayiemba, Elias H.O., 21, 99

Baker, Samuel J.K., 3
Bilhazia, 73
Blue Nile, 39

Bowers, L.B., 9
Box-plot, 143
Brass coefficients, 137
Brass-type models, 132
British Columbia, 35
Brundtland Report, 41
Bufumbino Mountain, 36, 37
Bura irrigation schemes, 54

Cambridge School Certificate Examination, 20
Canadian Rockies, 32
Cattle, 242
Caucasus mountains, 31
Census data, 158
Central Business District (CBO), 245
Central Province:
  life expectancy, 139
Child mortality:
  to calculate child mortality, 132
Civil engineer, 185, 191
Civil Registration Demonstration Project, 147
Cohen David William, 11
Colonial Kenya:
  health of migrant
  labour recruitment, 176
  labourers, 175
  medical examination, 176
  state intervention, 253
Commission of Environment Law (CEL), 40
Confirmatory Data Analysis (CDA), 142
Contraceptive method, 137
Contraceptive pills, 163
Co-operative movement, 267
Crop production, 239
Curative services, 132

Data:
  analysis, 158
  collection, 158
Deforestation, 35
Democratic Republic of Congo, 41
Demographic techniques, 152
Developmental problems, 187
Disease Systems Analyst (DSA), 83, 85-93
Disease theory, 93
Dispute testimony, 163
District Land Registry, 221

Index

Districts Focus for Rural Development
  (DFRDS), 114, 117, 199, 126, 127, 137
Ditdessa River, 51
Drainage basins, 67
Drought-resistant crops, 223
Dual economy, 199
Dumont, Rene, 54

East Africa, study of population in, 7
East African highlands, 30
East African Railways and Harbours, 59
East African Royal Report, 252
Ecological degradation, 34, 35
Economy: market, 30
  subsistence, 30
Edinburgh University, 20
Education:
  mortality differential, 118
Educational differential, 117
Egerton University, 184
Egypt, 39, 44
  irrigation in, 49
8-4-4 education system, 189
Elders, Council of, 163
Electrical engineers, 185, 191
Elgon, 37
Eliot, Charles, 175
Engineering education, 183
Engineering Registration Board (ERB), 186
Entrophication, 73
Environmental:
  conservation, 41
  factors, 131
  problems, 73
  stress, 41, 48
Epidemic diseases, 119
Epidemic logical models, 87
Ethiopia, 39, 41
Europe, 31
European Settlement, 178
European Union, 56
Evans-Pritchard E.C., 11, 12, 13
Exploratory Data Analysis (EDA), 142, 143

Family:
  immortalisation, 100
  planning, 111, 112
Fertility:
  Malthusian analytical of, 99
  nuptial determinants of, 99, 161
Fifth National Development Plan 1984-1988, 266
First National Development Plan 1964-70; 258, 260Finke, 81

Flash floods, 77
Fletcher-Cooke, John, 19
Flow regime, 43
Focus Group Panel (FGP), 161, 162
Food production
  political system of, 201
  problems of, 203
  role of state, 208
  state intervention, 208
Foreign exchange, 205
Forest:
  destruction, 35
  removal, 34
Francis, Carey, 217

G. austeni, 62
G. brevipal pis, 62
G. longi pennis, 62
Geography and African Development, 3, 6
Ghali Boutros, 50
Gichaga, Francis J., 183
Glossina G. pallialipesi, 62
Glossina palpalis, 89
Gross Domestic Product (GDP), 203
Group testimony, 163

Hancok Memorial Prize, 8
Health Systems Analyst (HAS), 83, 85-93
Hickes, E.W., 178
Hippocrates, 81
History of the Southern Luo, 20
Housing Research Development Unit (HRDU), 259, 260
Human African Trypanosomiasis (HAT), 90, 91
Human ecology, 94
Human geography, 3
Hurst, H.E., 54
Hybrid maize, 226, 229
Hydroment Survey, 59
Hydrometeorological Survey Team, 58
Hydro-power production, 43

Impact zone, 90
Inca civilisation, 29
Indian Ocean, 64, 134
Infant Mortality Rate (IMR), 137, 141
Infective cell, 90, 91
Institute of Engineers of Kenya (IEK), 186
Instrumentalism, 248
International rivers
  management of, 59
Ionomorphosis, 40
Irrigation schemes:
  health hazard, 73

310

Index

Jimla Mutane, 5, 6
Jomo Kenyatta University of Agriculture and Technology, 184
Jonglei Canal, 47, 56

Kabaka Mutesa, 8
Kabaka Crisis, 8
Kagera Basin Organisation (KBO), 55
Kagera Basin, 56
Kenya Bus Services (KBS), 274
Kenya Contraceptive Prevalence Survey (KCPS), 101
Kenya Fertility Survey (KFS), 101, 102, 116, 117, 120
Kenya highlands, 32
Kenya Police Traffic Road Safety, 284
Kenya Soil Survey, 72
Kenya Urban Transport Project, 283
Kenya
   population distribution, 209
Kenyatta, Jomo, 254
Kenya-Uganda Railway (KUR), 175, 176
Kiano, Julius, 9
Krhoda, George, O., 62

Labh Singh Harnam Singh (LSHS), 277
Labour commission (1912-13), 179
Labour:
   casual, 232
   intensive use of, 239
   international division of, 237
Lake Alberta, 44
Lake Basin Development Authority (LBDA), 53
Lake Eyassi, 54
Lake Kyoga, 44
Lake Superior, 44
Lake Victorian plateau, 219
Lambwe Valley ecosystem, 90, 98
Land adjudication, 111
Land administration, 267
Land and Population Movements in Kenya, 3, 6
Landslide, 34
Leyland (Kenya), 283
Life Table:
   construction of, 141
Limnologists, 61
Linear regression model, 153
Livestock development, 74
London, University of, 20

Mackay Report, 16
Macoloo, G.C., 247

Magadi, 176
Makerere University, 8
Malaria, 73
Malo Shadrack, 10
Marital status differentials, 120
Market mechanism, 256
Marriage:
   institution of, 99
Marxist theory, 249
Maseno school, 3, 9
Master and Servant Ordinances, 176
Matatus, 271, 272, 278
Maternal education, 130, 131
Mau Mau, 9
Medical geography, 81-83, 88, 89
Mediterranean sea, 48
Migrant labour, 175
Milk production, 206
Mji Kumi, 164, 172
Mobile clinic, 163
Modernisation:
   analysis, 237
   theory, 235, 239
Moi University, 184
Monogamous unions, 137
Monsoonal system, 64
Morgan, W.T.W, 3
Mortality differentials, history of, 115
Mortality levels:
   calculation of, 141
Mount Elgon, 36
Mountainous environments, 29, 32
Muganzi, Zibeon, 2, 21, 113
Multi-Phase Focus Panel methods, 164, 172
Multiple regession:
   analysis, 154
   model, 147
Muriuki, Godfrey, 20
Mwanje Justus, I, 81
Mwanzi, H., 20
Mwea Irrigation Scheme, 73

Nairobi City Council, 251, 262
Nairobi Urban Study Group (NUSG), 282, 283
Nairobi's child mortality, 134
Nairobi's public transport system, 272
National demographic surveys, 158
National development
   scholarship in, 17
Native labour commission, 178, 179
Natural disasters, 35
New Zealand Highlands, 32
NGO, 258
Nile Basin, 46

Index

Nilotic Kavirondo, 11
Njonjo, Charles, 9
North America, 31
Northcott Commission, 181
Nyayo Bus Services (NBS), 272

Obara D.A., 203
Observation, techniques of, 161
Obudho, R.A., 21, 271
Ochieng, William, 20
Ocholla-Ayayo A.B.C., 160
Ochuka, Festo, 15
Odaga, Asenath, 14
Odero-Ogwell, L.A., 42
Odhiambo E.S. Atieno, 8, 129
Odingo, R.S., 3, 21
Ofafa, Ambrose, 8
Ogendo, R.B., 21
Ogot, B.A., 10
Ojany, F.F., 21
Ojiji, deconstruction, 14
    the world of, 11
Ojwang J.B., 17
Okidi, C.O., 39
Okoth-Ogendo, 13
Ominde, Mary, 10
Ominde, Simeon Hongo, 3, 5, 16, 183
    and the Luo girl, 8, 9, 10, 11, 13, 14, 15, 16
    contribution, 6
    philosophy, 5
    secondary education, 9
Onegos, 9
Oneko, Achieng, 8
Open system thermodynamics, 85
Optimal Control Theory, 92
Ordinary Least Squared (OLS), 148, 153
    basic problems in, 150
Oucho, J.O., 21, 195
Ouma, F.O., 129
Owara-Ojungu, P.H., 29
Owen Falls Dam, 44, 48

Palanieri, A., 51
Pastoralists, 74
Perverra, 54
Philosophy, Western 18
Physiological density, 197
Pim, 11, 12
Pluralism, 248
Polygamous unions, 137
Population movement, role of, 175
Population pressure, 201
Population Studies and Research Institute,
    (PSRI), 5, 103, 158

Pre-cambrian rock system, 66, 68
Pressure point, 48
Primary acceleration, 34
Private sector, 268
Public bus system, role of, 277
Public Health Act, 269
Public Service Vehicle (PSV), 271
Pull factors, 32
Pull-push concept, 30, 33
Push factors, 32

Qualitative information, 160

Railway Commuter Service, 271, 283
Razorback, 35
Regional differentials, 120
Regression analysis, 147
Relative autonomy, 249
Rent Restriction Act, 259
Rent Tribunal, 259, 268
Resource Adoption (RA) 33
River Nzoia, 53
River Semiliki, 44
Royal Scottish Geographical Society, 20
Ruma National Park, 91, 92
Rural exodus, 231
Rural-urban migration, 203
Ruwenzori [Mountains], 37

Sadat, Anwar, 50
Sahel zone, 41
Salinisation, 74
Scholarship, concept of, 17
Sclossina pallidipes, 89
Sectional Properties Act, 268
Sediments, 66
Sessional Paper No. 1 of 1986, 266
Sessional Paper No. 5 of 1966-67, 258
Simba, 12
Sindiga Isaac, 175
Single mothers, childs mortality for, 131
Site and service [housing] schemes, 251
Siwindhe, 11, 12
Sixth National Development Plan, 1989-93;
    258, 266
Sleeping sickness, 64, 89
Small pox, 81
Smith sound, 54, 55
Sorghum, 239
South Nyanza District, 217
Sudd, 44
Spatial demography, 197
Spatial heterogeneity, 87

312

Sponge effect, 199
Squatters, study of, 175
State intervention, nature of, 247
Steady and state thermodynamics, 85
Stem-leagrolot, 146
Stratified Focus Group Panels, 164
Structural Adjustment Programmes (SAPs), 256
Structuralism, 248
Suda, Colletta A., 234
Sudan Peoples Liberation Army (SPLA), 56
Syagga, P.M., 258
Systems analysis:
    foundation of, 93
    steps, 88

T. brucei, 90
Irrigation initiatives, Tanzania, 54
Third National Development Plan 1974-78, 262
Thucydides, 81
Tilapia esculenta technology, transfer of, 192
Total Fertility Rates, (TFR), 103, 105, 109
Trevor-Roper, Hugh, 20
Trussell's coefficients, 137
    model, 140
    technique, 134
Tryponosoma brucei, 89
Trypanosoma gambleinse, 89
Tryponosome, 64
Trypanosomiasis, management of, 93
Trypanosomiasis Information System (TIS), 93, 95, 96, 97
Tsetse fly, 62
    control of, 91
    distribution of, 72
Tuberculosis:
    immunisation against, 113
Typhoid, 73

Ulluguru mountains, 31
UNESCO, 61
United States Bureau of Reclamation, 51, 52
United Transport Overseas Company (UTOC), 277
University of Aberdeen (Scotland), 20
University of Nairobi, 5, 24, 184, 265
Urban centres, 269
Urban housing, 250
Urban transportation problem, 271
Urban wages, 250
Urbanisation, 112
USAID, 255, 264

Vectors, 89
Vembere Plateau, 54
Venereal disease, 181

Water resource development, 62
Were, Gideon, 20
Western education, 9
Western Kenya:
    fertility, 99
    recorded environmental hazards, 104
Western scholarship, 21, 22
Widowed mothers:
    child mortality for, 137
World Bank, 199, 255, 275
World Food Council, 42

Zebu, 89
Zero Wildlife Strategy, 92
Zoonosis, 8